Mobile Robotics

Series Editor
Hisham Abou Kandil

Mobile Robotics

Luc Jaulin

ISTE PRESS

ELSEVIER

First published 2015 in Great Britain and the United States by ISTE Press Ltd and Elsevier Ltd

ISTE Press Ltd
27-37 St George's Road
London SW19 4EU
UK

www.iste.co.uk

Elsevier Ltd
The Boulevard, Langford Lane
Kidlington, Oxford, OX5 1GB
UK

www.elsevier.com

British Library Cataloguing-in-Publication Data
A CIP record for this book is available from the British Library
Library of Congress Cataloging in Publication Data
A catalog record for this book is available from the Library of Congress
ISBN 978-1-78548-048-5

Printed and bound in the UK and US

Contents

Introduction

A *mobile robot* can be defined as a mechanical system capable of moving in its environment in an autonomous manner. For that purpose, it must be equipped with:

– *sensors* that will help in gaining knowledge of its surroundings (which it is more or less aware of) and determine its location;

– *actuators* which will allow it to move;

– an *intelligence* (or algorithm, regulator), which will allow it to compute, based on the data gathered by the sensors, the commands to send to the actuators in order to perform a given task.

Finally, to this we must add the *surroundings* of the robot which correspond to the world in which it evolves and its *mission* which is the task it has to accomplish. Mobile robots have been constantly evolving, mainly from the beginning of the 2000s, in military domains (airborne drones [BEA 12], underwater robots [CRE 14], etc.), and even in medical and agricultural fields. They are in particularly high demand for performing tasks considered to be painful or dangerous to humans. This is the case, for instance, in mine-clearing operations, the search for black boxes of damaged aircraft on the ocean bed and planetary exploration. Artificial satellites, launchers (such as Ariane V), driverless subways and elevators are examples of mobile robots. Airliners, trains and cars evolve in a continuous fashion toward increasingly autonomous systems and will very probably become mobile robots in the decades to follow.

Mobile robotics is the discipline which looks at the design of mobile robots [LAU 01]. It is based on other disciplines such as automatic control, signal processing, mechanics, computing and electronics. The aim of this book is to give an overview of the tools and methods of robotics which will aid in the design of mobile robots. The robots will be modeled by *state equations*, i.e., a set of first order (mostly nonlinear) differential equations. These state equations can be obtained by using the laws of mechanics. It is not in our objectives to teach, in detail, the methods of robot modeling (refer to [JAU 05] and [JAU 15] for more information on the subject), merely to recall its principles. By *modeling*, we mean obtaining the state equations. This step is essential for simulating robots as well as designing controllers. We will however illustrate the principle of modeling in Chapter 1 on deliberately three-dimensional (3D) examples. This choice was made in order to introduce important concepts in robotics such as Euler angles and rotation matrices. For instance, we will be looking at the dynamics of a wheel and the kinematics of an underwater robot. Mobile robots are strongly nonlinear systems and only a nonlinear approach allows the construction of efficient controllers. This construction is the subject of Chapters 2 and 3. Chapter 2 is mainly based on control methods that rely on the utilization of the robot model. This approach will make use of the concept of *feedback linearization* which will be introduced and illustrated through numerous examples. Chapter 3 presents more pragmatic methods which do not use the state model of the robot and which will be referred to as *without model* or *mimetic*. The approach uses a more intuitive representation of the robot and is adapted to situations in which the robots are relatively simple to remotely control, such as in the case of cars, sailing boats or airplanes. Chapter 4 looks at *guidance*, which is placed at a higher level than control. In other words, it focuses on guiding and supervising the system which is already under control by the tools presented in Chapters 2 and 3. Therefore there will be an emphasis on finding the instruction to give to the controller in order for the robot to accomplish its given task. The guidance will then have to take into account the knowledge of the surroundings, the presence of obstacles and the roundness of the Earth. The nonlinear control and guidance methods require good knowledge of the state variables of the system, such as those which define the position of the robot. These position variables are the most difficult to find and Chapter 5 focuses on the problem of *positioning*. It introduces the classical nonlinear approaches that have been used for a very long time by humans for positioning, such as observing beacons, stars, using a compass or counting steps. Although positioning can be viewed as a particular case of state observation, the specific methods derived from it warrant a separate chapter. Chapter 6 on *identification* focuses on finding, with a certain precision, non-measured quantities (parameters and position) from other, measured quantities. In order to perform this identification, we will mainly be looking at the so-called *least squares* approach which consists of finding the vector of variables that minimizes the sum of the squares of the errors. Chapter 7 presents the *Kalman filter*. This filter can be seen as a state observer for dynamic linear systems with coefficients that vary in time.

The MATLAB code related to the exercises of this book together with explanatory videos can be found at the following address:

`www.ensta-bretagne.fr/jaulin/isterob.html`

1

Three-dimensional Modeling

This chapter presents the three-dimensional (3D) modeling of a solid (non-articulated) robot. Such a modeling is used to represent an airplane, a quadcopter, a submarine and so forth. Through this modeling, we will introduce a number of fundamental concepts in robotics such as state representation, rotation matrices and Euler angles. The robots, whether mobile, manipulator or articulated, can generally be put into a state representation form:

$$\begin{cases} \dot{\mathbf{x}}(t) = \mathbf{f}(\mathbf{x}(t), \mathbf{u}(t)) \\ \mathbf{y}(t) = \mathbf{g}(\mathbf{x}(t), \mathbf{u}(t)) \end{cases}$$

where $\mathbf{x}$ is the state vector, $\mathbf{u}$ is the input vector and $\mathbf{y}$ is the vector of measurements [JAU 05]. We will call *modeling* the step which consists of finding a more or less accurate state representation of the robot in question. In general, constant parameters may appear in the state equations (such as the mass and moment of inertia of a body, viscosity, etc.). In such cases, an identification step might prove to be necessary. We will assume that all of the parameters are known. Of course, there is no systematic methodology that can be applied for modeling a mobile robot. The aim of this chapter is to present the tools which allow us to reach a state representation of 3D solid robots in order for the readers to acquire a certain experience which will be helpful when modeling his/her own robots. This modeling will also allow us to recall a number of important concepts in Euclidean geometry, which are fundamental in mobile robotics. This chapter begins by recalling a number of important concepts in kinematics which will be useful for the modeling.

1.1. Rotation matrices

For 3D modeling, it is essential to have a good understanding of the concepts related to rotation matrices, which are recalled in this section. It is by using this tool that we will perform our coordinate system transformations and position our objects in space.

1.1.1. *Definition*

Let us recall that the jth column of the matrix of a linear application of $\mathbb{R}^n \to \mathbb{R}^n$ represents the image of the jth vector $\mathbf{e}_j$ of the standard basis (see Figure 1.1). Thus, the expression of a rotation matrix of angle θ in the plane $\mathbb{R}^2$ is given by:

$$\mathbf{R} = \begin{pmatrix} \cos\theta & -\sin\theta \\ \sin\theta & \cos\theta \end{pmatrix}$$

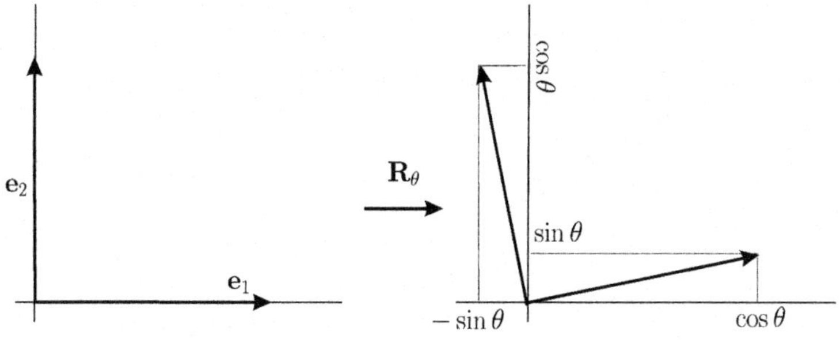

Figure 1.1. *Rotation of angle θ in a plane*

Concerning rotations in the space $\mathbb{R}^3$ (see Figure 1.2), it is important to specify the axis of rotation. We distinguish three main rotations: the rotation around the Ox axis, the rotation around the Oy axis and the rotation around the Oz axis.

The associated matrices are, respectively, given by:

$$\mathbf{R}_x = \begin{pmatrix} 1 & 0 & 0 \\ 0 & \cos\theta_x & -\sin\theta_x \\ 0 & \sin\theta_x & \cos\theta_x \end{pmatrix}, \quad \mathbf{R}_y = \begin{pmatrix} \cos\theta_y & 0 & \sin\theta_y \\ 0 & 1 & 0 \\ -\sin\theta_y & 0 & \cos\theta_y \end{pmatrix},$$

$$\mathbf{R}_z = \begin{pmatrix} \cos\theta_z & -\sin\theta_z & 0 \\ \sin\theta_z & \cos\theta_z & 0 \\ 0 & 0 & 1 \end{pmatrix}$$

Let us recall the formal definition of a rotation. A rotation is a linear application which is an isometry (in other words, it preserves the scalar product) and a movement (it does not change the orientation in space).

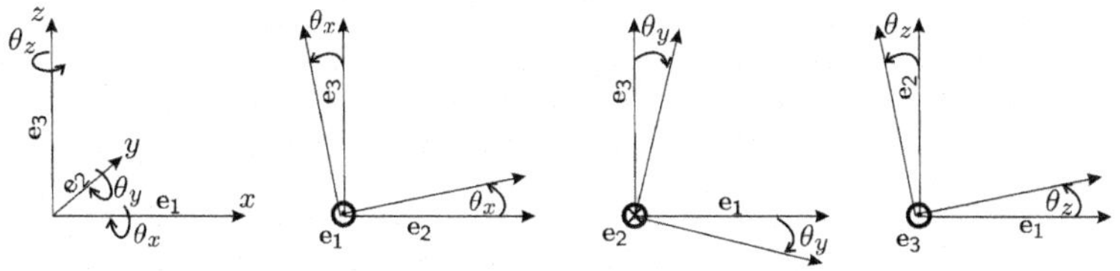

Figure 1.2. *Rotations in $\mathbb{R}^3$ following various viewing angles*

THEOREM 1.1.– A matrix $\mathbf{R}$ is a rotation matrix if and only if:

$$\mathbf{R}^{\mathrm{T}} \cdot \mathbf{R} = \mathbf{I} \text{ and } \det \mathbf{R} = 1$$

PROOF.– The scalar product is preserved by $\mathbf{R}$ if, for any $\mathbf{u}$ and $\mathbf{v}$ in $\mathbb{R}^n$, we have:

$$(\mathbf{Ru})^{\mathrm{T}} \cdot (\mathbf{Rv}) = \mathbf{u}^{\mathrm{T}}\mathbf{R}^{\mathrm{T}}\mathbf{Rv} = \mathbf{u}^{\mathrm{T}}\mathbf{v}$$

in other words, if $\mathbf{R}^{\mathrm{T}}\mathbf{R} = \mathbf{I}$. The symmetries relative to a plane, as well as all the other improper movements (isometries that change the orientation of space, such as a mirror), also verify the property $\mathbf{R}^{\mathrm{T}}\cdot\mathbf{R} = \mathbf{I}$. The condition $\det \mathbf{R} = 1$ allows us to be limited to the isometries which are movements . The set of rotation matrices of $\mathbb{R}^n$ forms a group referred to as a *special orthogonal group* (special because $\det \mathbf{R} = 1$, orthogonal because $\mathbf{R}^{\mathrm{T}}\cdot\mathbf{R} = \mathbf{I}$). ■

1.1.2. *Rotation vector*

If $\mathbf{R}$ is a rotation matrix depending on time t, by differentiating the relation $\mathbf{RR}^{\mathrm{T}} = \mathbf{I}$, we obtain:

$$\dot{\mathbf{R}} \cdot \mathbf{R}^{\mathrm{T}} + \mathbf{R} \cdot \dot{\mathbf{R}}^{\mathrm{T}} = \mathbf{0}$$

Thus, the matrix $\dot{\mathbf{R}} \cdot \mathbf{R}^{\mathrm{T}}$ is a skew-symmetric matrix (in other words, it satisfies $\mathbf{A}^{\mathrm{T}} = -\mathbf{A}$ and therefore its diagonal contains only zeroes, and for each element of $\mathbf{A}$, we have $a_{ij} = -a_{ji}$). Therefore, we may write, in the case where $\mathbf{R}$ is of dimension 3×3:

$$\dot{\mathbf{R}} \cdot \mathbf{R}^{\mathrm{T}} = \begin{pmatrix} 0 & -\omega_z & \omega_y \\ \omega_z & 0 & -\omega_x \\ -\omega_y & \omega_x & 0 \end{pmatrix} \qquad [1.1]$$

The vector $\omega = (\omega_x, \omega_y, \omega_z)$ is called the *rotation vector* associated with the pair $\left(\mathbf{R}, \dot{\mathbf{R}}\right)$. It must be noted that $\dot{\mathbf{R}}$ is not a matrix with good properties (for instance the fact of being skew-symmetric). However, the matrix $\dot{\mathbf{R}} \cdot \mathbf{R}^{\mathrm{T}}$ has the [1.1] structure since it allows us to be positioned within the coordinate system in which the rotation is performed and this, due to the change of basis performed by $\mathbf{R}^{\mathrm{T}}$. We will define the *vector product* between two vectors ω and $\mathbf{x} \in \mathbb{R}^3$ as follows:

$$\omega \wedge \mathbf{x} = \begin{pmatrix} \omega_x \\ \omega_y \\ \omega_z \end{pmatrix} \wedge \begin{pmatrix} x_1 \\ x_2 \\ x_3 \end{pmatrix} = \begin{pmatrix} x_3\omega_y - x_2\omega_z \\ x_1\omega_z - x_3\omega_x \\ x_2\omega_x - x_1\omega_y \end{pmatrix}$$

$$= \begin{pmatrix} 0 & -\omega_z & \omega_y \\ \omega_z & 0 & -\omega_x \\ -\omega_y & \omega_x & 0 \end{pmatrix} \begin{pmatrix} x_1 \\ x_2 \\ x_3 \end{pmatrix}$$

1.1.3. *Adjoint*

For each vector $\omega = (\omega_x, \omega_y, \omega_z)$, we may adjoin the skew-symmetric matrix:

$$\mathbf{Ad}(\omega) \stackrel{\text{def}}{=} \begin{pmatrix} 0 & -\omega_z & \omega_y \\ \omega_z & 0 & -\omega_x \\ -\omega_y & \omega_x & 0 \end{pmatrix}$$

which can be interpreted as the matrix associated with a vector product by the vector ω.

PROPOSITION 1.1.– If $\mathbf{R}(t)$ is a rotation matrix that depends on time, its rotation vector is given by:

$$\omega = \mathbf{Ad}^{-1}\left(\dot{\mathbf{R}} \cdot \mathbf{R}^{\mathrm{T}}\right) \quad [1.2]$$

PROOF.– This relation is a direct consequence of equation [1.1]. ■

PROPOSITION 1.2.– If $\mathbf{R}$ is a rotation matrix in $\mathbb{R}^3$ and if $\mathbf{a}$ is a vector of $\mathbb{R}^3$, we have:

$$\mathbf{Ad}\left(\mathbf{R} \cdot \mathbf{a}\right) = \mathbf{R} \cdot \mathbf{Ad}\left(\mathbf{a}\right) \cdot \mathbf{R}^{\mathrm{T}} \quad [1.3]$$

which can also be written as:

$$\left(\mathbf{R} \cdot \mathbf{a}\right) \wedge = \mathbf{R} \cdot \left(\mathbf{a} \wedge\right) \cdot \mathbf{R}^{\mathrm{T}}$$

PROOF.– Let $\mathbf{x}$ be a vector of $\mathbb{R}^3$. We have:

$$\begin{aligned} \mathbf{Ad}\left(\mathbf{R} \cdot \mathbf{a}\right) \cdot \mathbf{x} &= \left(\mathbf{R} \cdot \mathbf{a}\right) \wedge \mathbf{x} = \left(\mathbf{R} \cdot \mathbf{a}\right) \wedge \left(\mathbf{R} \cdot \mathbf{R}^{\mathrm{T}}\mathbf{x}\right) \\ &= \mathbf{R} \cdot \left(\mathbf{a} \wedge \mathbf{R}^{\mathrm{T}} \cdot \mathbf{x}\right) = \mathbf{R} \cdot \mathbf{Ad}\left(\mathbf{a}\right) \cdot \mathbf{R}^{\mathrm{T}} \cdot \mathbf{x}. \end{aligned}$$ ■

PROPOSITION 1.3.– (duality). We have:

$$\mathbf{R}^{\mathrm{T}}\dot{\mathbf{R}} = \mathbf{Ad}\left(\mathbf{R}^{\mathrm{T}}\omega\right) \quad [1.4]$$

This relation expresses the fact that the matrix $\mathbf{R}^{\mathrm{T}}\dot{\mathbf{R}}$ is associated with the rotation matrix ω, associated with $\mathbf{R}\left(t\right)$ but expressed in the coordinate system associated with $\mathbf{R}$ whereas $\dot{\mathbf{R}} \cdot \mathbf{R}^{\mathrm{T}}$ is associated with the same vector, but this time expressed in the coordinate system of the standard basis.

PROOF.– We have:

$$\mathbf{R}^{\mathrm{T}}\dot{\mathbf{R}} = \mathbf{R}^{\mathrm{T}}\left(\dot{\mathbf{R}} \cdot \mathbf{R}^{\mathrm{T}}\right)\mathbf{R} \overset{[1.2]}{=} \mathbf{R}^{\mathrm{T}} \cdot \mathbf{Ad}\left(\omega\right) \cdot \mathbf{R} \overset{[1.3]}{=} \mathbf{Ad}\left(\mathbf{R}^{\mathrm{T}}\omega\right)$$ ■

1.1.4. *Coordinate system change*

Let $\mathcal{R}_0 : (\mathbf{o}_0, \mathbf{i}_0, \mathbf{j}_0, \mathbf{k}_0)$ and $\mathcal{R}_1 : (\mathbf{o}_1, \mathbf{i}_1, \mathbf{j}_1, \mathbf{k}_1)$ be two coordinate systems and let the point $\mathbf{u}$ be a vector of $\mathbb{R}^3$ (see Figure 1.3). We have the following relation:

$$\begin{aligned}\mathbf{u} &= x_0\mathbf{i}_0 + y_0\mathbf{j}_0 + z_0\mathbf{k}_0\\ &= x_1\mathbf{i}_1 + y_1\mathbf{j}_1 + z_1\mathbf{k}_1\end{aligned}$$

where (x_0, y_0, z_0) and (x_1, y_1, z_1) are, respectively, the coordinates of $\mathbf{u}$ in $\mathcal{R}_0$ and $\mathcal{R}_1$.

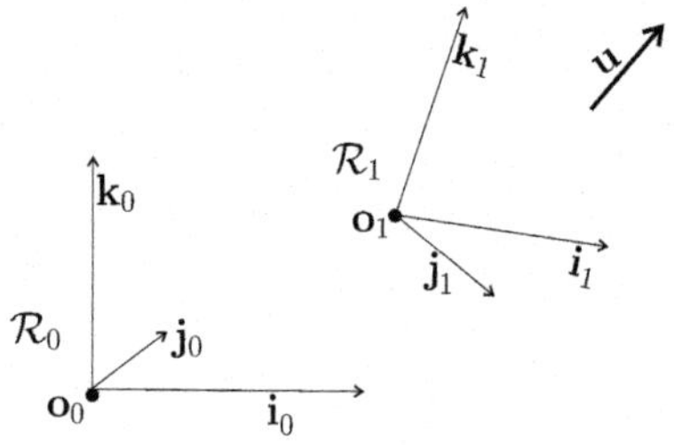

Figure 1.3. *Changing the coordinate system $\mathcal{R}_0$ to the system $\mathcal{R}_1$*

Thus, for any vector $\mathbf{v}$, we have:

$$\langle x_0\mathbf{i}_0 + y_0\mathbf{j}_0 + z_0\mathbf{k}_0, \mathbf{v}\rangle = \langle x_1\mathbf{i}_1 + y_1\mathbf{j}_1 + z_1\mathbf{k}_1, \mathbf{v}\rangle$$

By taking, respectively, $\mathbf{v} = \mathbf{i}_0, \mathbf{j}_0, \mathbf{k}_0$, we obtain the following three relations:

$$\begin{cases}\langle x_0\mathbf{i}_0 + y_0\mathbf{j}_0 + z_0\mathbf{k}_0, \mathbf{i}_0\rangle &= \langle x_1\mathbf{i}_1 + y_1\mathbf{j}_1 + z_1\mathbf{k}_1, \mathbf{i}_0\rangle\\ \langle x_0\mathbf{i}_0 + y_0\mathbf{j}_0 + z_0\mathbf{k}_0, \mathbf{j}_0\rangle &= \langle x_1\mathbf{i}_1 + y_1\mathbf{j}_1 + z_1\mathbf{k}_1, \mathbf{j}_0\rangle\\ \langle x_0\mathbf{i}_0 + y_0\mathbf{j}_0 + z_0\mathbf{k}_0, \mathbf{k}_0\rangle &= \langle x_1\mathbf{i}_1 + y_1\mathbf{j}_1 + z_1\mathbf{k}_1, \mathbf{k}_0\rangle\end{cases}$$

However, since the basis $(\mathbf{i}_0, \mathbf{j}_0, \mathbf{k}_0)$ of $\mathcal{R}_0$ is orthonormal, $\langle \mathbf{i}_0, \mathbf{i}_0\rangle = \langle \mathbf{j}_0, \mathbf{j}_0\rangle = \langle \mathbf{k}_0, \mathbf{k}_0\rangle = 1$ and $\langle \mathbf{i}_0, \mathbf{j}_0\rangle = \langle \mathbf{j}_0, \mathbf{k}_0\rangle = \langle \mathbf{i}_0, \mathbf{k}_0\rangle = 0$. Thus, these three relations become:

$$\begin{cases}x_0 = x_1\langle \mathbf{i}_1, \mathbf{i}_0\rangle + y_1\langle \mathbf{j}_1, \mathbf{i}_0\rangle + z_1\langle \mathbf{k}_1, \mathbf{i}_0\rangle\\ y_0 = x_1\langle \mathbf{i}_1, \mathbf{j}_0\rangle + y_1\langle \mathbf{j}_1, \mathbf{j}_0\rangle + z_1\langle \mathbf{k}_1, \mathbf{j}_0\rangle\\ z_0 = x_1\langle \mathbf{i}_1, \mathbf{k}_0\rangle + y_1\langle \mathbf{j}_1, \mathbf{k}_0\rangle + z_1\langle \mathbf{k}_1, \mathbf{k}_0\rangle\end{cases}$$

Or in matrix form:

$$\underbrace{\begin{pmatrix} x_0 \\ y_0 \\ z_0 \end{pmatrix}}_{=\mathbf{u}|_{\mathcal{R}_0}} = \underbrace{\begin{pmatrix} \langle \mathbf{i}_1, \mathbf{i}_0 \rangle & \langle \mathbf{j}_1, \mathbf{i}_0 \rangle & \langle \mathbf{k}_1, \mathbf{i}_0 \rangle \\ \langle \mathbf{i}_1, \mathbf{j}_0 \rangle & \langle \mathbf{j}_1, \mathbf{j}_0 \rangle & \langle \mathbf{k}_1, \mathbf{j}_0 \rangle \\ \langle \mathbf{i}_1, \mathbf{k}_0 \rangle & \langle \mathbf{j}_1, \mathbf{k}_0 \rangle & \langle \mathbf{k}_1, \mathbf{k}_0 \rangle \end{pmatrix}}_{=\mathbf{R}_{\mathcal{R}_0}^{\mathcal{R}_1}} \cdot \underbrace{\begin{pmatrix} x_1 \\ y_1 \\ z_1 \end{pmatrix}}_{\mathbf{u}|_{\mathcal{R}_1}} \qquad [1.5]$$

We can see a rotation matrix $\mathbf{R}_{\mathcal{R}_0}^{\mathcal{R}_1}$ appear whose columns are the coordinates of $\mathbf{i}_1, \mathbf{j}_1, \mathbf{k}_1$ expressed in the absolute system $\mathcal{R}_0$. In other words:

$$\mathbf{R}_{\mathcal{R}_0}^{\mathcal{R}_1} = \left(\begin{array}{c|c|c} \mathbf{i}_1|_{\mathcal{R}_0} & \mathbf{j}_1|_{\mathcal{R}_0} & \mathbf{k}_1|_{\mathcal{R}_0} \end{array} \right)$$

This matrix depends on time and links the system $\mathcal{R}_1$ to $\mathcal{R}_0$. The matrix $\mathbf{R}_{\mathcal{R}_0}^{\mathcal{R}_1}$ is often referred to as a *direction cosine matrix* since its components involve the direction cosines of the basis vectors of the two coordinate systems. Likewise, if we had several systems $\mathcal{R}_0, \ldots, \mathcal{R}_n$ (see Figure 1.4), we would have:

$$\mathbf{u}|_{\mathcal{R}_0} = \mathbf{R}_{\mathcal{R}_0}^{\mathcal{R}_1} \cdot \mathbf{R}_{\mathcal{R}_1}^{\mathcal{R}_2} \cdot \ \ldots \ \cdot \mathbf{R}_{\mathcal{R}_{n-1}}^{\mathcal{R}_n} \cdot \mathbf{u}|_{\mathcal{R}_n}$$

Let us consider, for instance, the situation of a robot moving in a 3D environment. Let us call $\mathcal{R}_0 : (\mathbf{o}_0, \mathbf{i}_0, \mathbf{j}_0, \mathbf{k}_0)$ its reference frame (for example, the frame of the robot at an initial time). The position of the robot is represented by the vector $\mathbf{p}(t)$ expressed in $\mathcal{R}_0$ and its attitude (in other words, its orientation) by the rotation matrix $\mathbf{R}(t)$ which represents the coordinates of the vectors $\mathbf{i}_1, \mathbf{j}_1, \mathbf{k}_1$ of the coordinate system $\mathcal{R}_1$ of the robot expressed in the coordinate system $\mathcal{R}_0$, at time t. It follows that:

$$\mathbf{R}(t) = \left(\begin{array}{c|c|c} \mathbf{i}_1|_{\mathcal{R}_0} & \mathbf{j}_1|_{\mathcal{R}_0} & \mathbf{k}_1|_{\mathcal{R}_0} \end{array} \right) = \mathbf{R}_{\mathcal{R}_0}^{\mathcal{R}_1}(t)$$

This matrix can be returned by a precise attitude unit positioned on the robot. If the robot is also equipped with a *Doppler velocity log* (or DVL) which provides it with its speed vector $\mathbf{v}_r$ relative to the ground or seabed, expressed in the coordinate system $\mathcal{R}_1$ of the robot, then the speed vector $\mathbf{v}$ of the robot satisfies:

$$\underbrace{\mathbf{v}|_{\mathcal{R}_0}}_{\dot{\mathbf{p}}(t)} \overset{[1.5]}{=} \underbrace{\mathbf{R}_{\mathcal{R}_0}^{\mathcal{R}_1}}_{\mathbf{R}(t)} \cdot \underbrace{\mathbf{v}|_{\mathcal{R}_1}}_{\mathbf{v}_r(t)}$$

in other words:

$$\dot{\mathbf{p}}(t) = \mathbf{R}(t) \cdot \mathbf{v}_r(t) \qquad [1.6]$$

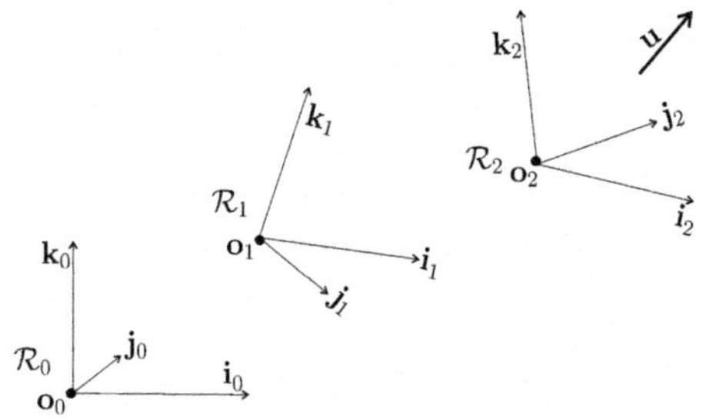

Figure 1.4. *Composition in the coordinate system changes*

Dead reckoning consists of integrating this state equation from the knowledge of $\mathbf{R}(t)$ and $\mathbf{v}_r(t)$.

1.2. Euler angles

1.2.1. *Definition*

In the related literature, the angles proposed by Euler in 1770 to represent the orientation of solid bodies in space are not uniquely defined. We mainly distinguish between the roll-yaw-roll, roll-pitch-roll and roll-pitch-yaw formulations. It is the latter that we will choose since it is imposed in the mobile robotics language. Within this roll-pitch-yaw formulation, the Euler angles are sometimes referred to as *Cardan angles*. Any rotation matrix of $\mathbb{R}^3$ can be expressed in the form of the product of three matrices as follows:

$$\mathbf{R}(\psi,\theta,\varphi) = \underbrace{\begin{pmatrix} \cos\psi & -\sin\psi & 0 \\ \sin\psi & \cos\psi & 0 \\ 0 & 0 & 1 \end{pmatrix}}_{\mathbf{R}_\psi} \underbrace{\begin{pmatrix} \cos\theta & 0 & \sin\theta \\ 0 & 1 & 0 \\ -\sin\theta & 0 & \cos\theta \end{pmatrix}}_{\mathbf{R}_\theta} \underbrace{\begin{pmatrix} 1 & 0 & 0 \\ 0 & \cos\varphi & -\sin\varphi \\ 0 & \sin\varphi & \cos\varphi \end{pmatrix}}_{\mathbf{R}_\varphi},$$

in developed form:

$$\begin{pmatrix} \cos\theta\cos\psi & -\cos\varphi\sin\psi+\sin\theta\cos\psi\sin\varphi & \sin\psi\sin\varphi+\sin\theta\cos\psi\cos\varphi \\ \cos\theta\sin\psi & \cos\psi\cos\varphi+\sin\theta\sin\psi\sin\varphi & -\cos\psi\sin\varphi+\sin\theta\cos\varphi\sin\psi \\ -\sin\theta & \cos\theta\sin\varphi & \cos\theta\cos\varphi \end{pmatrix} \quad [1.7]$$

$$\underbrace{\qquad}_{\mathbf{i}_1|_{\mathcal{R}_0}} \qquad \underbrace{\qquad\qquad}_{\mathbf{j}_1|_{\mathcal{R}_0}} \qquad \underbrace{\qquad\qquad}_{\mathbf{k}_1|_{\mathcal{R}_0}}$$

The angles ψ, θ, φ are the *Euler angles* and are, respectively, called the *heading*, *elevation* and *bank*. The terms *yaw, pitch and roll* are often employed, although they correspond, respectively, to variations of heading, elevation and bank.

NOTE 1.1.– Given a rotation matrix $\mathbf{R}$, we can easily find the three Euler angles by solving, following equation [1.7], the equations:

$$\begin{cases} -\sin\theta = r_{31} \\ \cos\theta\sin\varphi = r_{32} \quad \cos\theta\cos\varphi = r_{33} \\ \cos\theta\cos\psi = r_{11} \quad \cos\theta\sin\psi = r_{21} \end{cases}$$

By imposing $\theta \in [-\frac{\pi}{2}, \frac{\pi}{2}]$, $\varphi \in [-\pi, \pi]$, $\psi \in [-\pi, \pi]$, we find:

$$\theta = -\arcsin r_{31}, \ \varphi = \text{atan2}(r_{32}, r_{33}) \text{ and } \psi = \text{atan2}(r_{21}, r_{11})$$

Here, atan2 is the two-argument arctangent function defined by:

$$\theta = \text{atan2}\,(y, x) \Leftrightarrow \theta \in]-\pi, \pi] \text{ and } \exists r > 0 \mid \begin{cases} x = r\cos\theta \\ y = r\cos\theta \end{cases} \quad [1.8]$$

1.2.2. *Derivative of an Euler matrix*

Let us consider a rotation matrix expressed using its time-dependent Euler angles:

$$\mathbf{R}(t) = \mathbf{R}(\psi(t), \theta(t), \varphi(t))$$

Let us try to express $\dot{\mathbf{R}}(t)$, or equivalently $\dot{\mathbf{R}}(t)\mathbf{R}^{\mathrm{T}}(t)$. We would indeed prefer to express this derivative in the coordinate system associated with $\mathbf{R}\,(t)$. For this, we could of course differentiate expression [1.7] term-by-term but the calculations are rather cumbersome and, moreover, we would risk not being able to simplify the obtained expression. We have:

$$\begin{aligned}\dot{\mathbf{R}}\mathbf{R}^{\mathrm{T}} &= \tfrac{d}{dt}\left(\mathbf{R}_\psi \cdot \mathbf{R}_\theta \cdot \mathbf{R}_\varphi\right) \cdot \mathbf{R}_\varphi^{\mathrm{T}} \cdot \mathbf{R}_\theta^{\mathrm{T}} \cdot \mathbf{R}_\psi^{\mathrm{T}} \\ &= \left(\dot{\mathbf{R}}_\psi \cdot \mathbf{R}_\theta \cdot \mathbf{R}_\varphi + \mathbf{R}_\psi \cdot \dot{\mathbf{R}}_\theta \cdot \mathbf{R}_\varphi + \mathbf{R}_\psi \cdot \mathbf{R}_\theta \cdot \dot{\mathbf{R}}_\varphi\right) \cdot \mathbf{R}_\varphi^{\mathrm{T}} \cdot \mathbf{R}_\theta^{\mathrm{T}} \cdot \mathbf{R}_\psi^{\mathrm{T}} \\ &= \dot{\mathbf{R}}_\psi \cdot \mathbf{R}_\psi^{\mathrm{T}} + \mathbf{R}_\psi \cdot \dot{\mathbf{R}}_\theta \cdot \mathbf{R}_\theta^{\mathrm{T}} \cdot \mathbf{R}_\psi^{\mathrm{T}} + \mathbf{R}_\psi \cdot \mathbf{R}_\theta \cdot \dot{\mathbf{R}}_\varphi \cdot \mathbf{R}_\varphi^{\mathrm{T}} \cdot \mathbf{R}_\theta^{\mathrm{T}} \cdot \mathbf{R}_\psi^{\mathrm{T}}\end{aligned}$$

However, following equation [1.2], we have:

$$\begin{cases} \dot{\mathbf{R}}_\psi \mathbf{R}_\psi^{\mathrm{T}} = \mathbf{Ad}(\dot{\psi}\mathbf{k}) = \dot{\psi}\mathbf{Ad}(\mathbf{k}) \\ \dot{\mathbf{R}}_\theta \mathbf{R}_\theta^{\mathrm{T}} = \mathbf{Ad}(\dot{\theta}\mathbf{j}) = \dot{\theta}\mathbf{Ad}(\mathbf{j}) \\ \dot{\mathbf{R}}_\varphi \mathbf{R}_\varphi^{\mathrm{T}} = \mathbf{Ad}(\dot{\varphi}\mathbf{i}) = \dot{\varphi}\mathbf{Ad}(\mathbf{i}) \end{cases}$$

Therefore, we have:

$$\begin{aligned}\dot{\mathbf{R}}\mathbf{R}^{\mathrm{T}} &= \dot{\psi} \cdot \mathbf{Ad}(\mathbf{k}) + \dot{\theta} \cdot \mathbf{R}_\psi \cdot \mathbf{Ad}(\mathbf{j}) \cdot \mathbf{R}_\psi^{\mathrm{T}} + \dot{\varphi} \cdot \mathbf{R}_\psi \cdot \mathbf{R}_\theta \cdot \mathbf{Ad}(\mathbf{i}) \cdot \mathbf{R}_\theta^{\mathrm{T}} \cdot \mathbf{R}_\psi^{\mathrm{T}} \\ &\overset{[1.3]}{=} \dot{\psi} \cdot \mathbf{Ad}(\mathbf{k}) + \dot{\theta} \cdot \mathbf{Ad}(\mathbf{R}_\psi \cdot \mathbf{j}) + \dot{\varphi} \cdot \mathbf{Ad}(\mathbf{R}_\psi \cdot \mathbf{R}_\theta \cdot \mathbf{i})\end{aligned} \quad [1.9]$$

Note the linear dependence on $(\dot{\psi}, \dot{\theta}, \dot{\varphi})$.

1.2.3. *Rotation vector of an Euler matrix*

Let us consider a solid body moving in a coordinate system $\mathcal{R}_0$ and a coordinate system $\mathcal{R}_1$ attached to this body (see Figure 1.5). The conventions chosen here are those of the *Society of Naval and Marine Engineers* (SNAME). The two coordinate systems are assumed to be orthonormal. Let $\mathbf{R}(t) = \mathbf{R}(\psi(t), \theta(t), \varphi(t))$ be the rotation matrix that links the two systems. We need to find the instantaneous rotation vector ω of the solid body relative to $\mathcal{R}_0$ as a function of $\psi, \theta, \varphi, \dot{\psi}, \dot{\theta}, \dot{\varphi}$. We have:

$$\begin{aligned}\omega_{|\mathcal{R}_0} &\overset{[1.2]}{=} \mathbf{Ad}^{-1}\left(\dot{\mathbf{R}} \cdot \mathbf{R}^{\mathrm{T}}\right) \\ &\overset{[1.9]}{=} \mathbf{Ad}^{-1}\left(\dot{\psi} \cdot \mathbf{Ad}(\mathbf{k}) + \dot{\theta} \cdot \mathbf{Ad}(\mathbf{R}_\psi \cdot \mathbf{j}) + \dot{\varphi} \cdot \mathbf{Ad}(\mathbf{R}_\psi \cdot \mathbf{R}_\theta \cdot \mathbf{i})\right) \\ &= \dot{\psi} \cdot \mathbf{k} + \dot{\theta} \cdot \mathbf{R}_\psi \cdot \mathbf{j} + \dot{\varphi} \cdot \mathbf{R}_\psi \cdot \mathbf{R}_\theta \cdot \mathbf{i}\end{aligned}$$

Thus, after having calculated the quantities $\mathbf{k}$, $\mathbf{R}_\psi \mathbf{j}$ and $\mathbf{R}_\psi \cdot \mathbf{R}_\theta \cdot \mathbf{i}$ in the coordinate system $\mathcal{R}_0$, we have:

$$\omega|_{\mathcal{R}_0} = \dot{\psi} \cdot \begin{pmatrix} 0 \\ 0 \\ 1 \end{pmatrix} + \dot{\theta} \cdot \begin{pmatrix} -\sin\psi \\ \cos\psi \\ 0 \end{pmatrix} + \dot{\varphi} \cdot \begin{pmatrix} \cos\theta\cos\psi \\ \cos\theta\sin\psi \\ -\sin\theta \end{pmatrix}$$

And from this, we get the result:

$$\omega|_{\mathcal{R}_0} = \begin{pmatrix} 0 & -\sin\psi & \cos\theta\cos\psi \\ 0 & \cos\psi & \cos\theta\sin\psi \\ 1 & 0 & -\sin\theta \end{pmatrix} \begin{pmatrix} \dot{\psi} \\ \dot{\theta} \\ \dot{\varphi} \end{pmatrix} \quad [1.10]$$

Note that this matrix is singular when $\cos\theta = 0$. Therefore, we will make sure to never have an elevation θ equal to $\pm\frac{\pi}{2}$.

1.3. Kinematic model of a solid robot

A robot (airplane, submarine and boat) can often be considered a solid whose inputs are the (tangential and angular) accelerations. Indeed, these are analytic functions of the forces that are at the origin of the robot's movement. Here, we will consider that the inputs of the kinematic model are the tangential accelerations and the angular speeds. The reason for this is that these are directly measurable (if expressed in the robot's coordinate system) and that we may consider them to be directly controllable (even if a rotation can take a substantial amount of time for larger structures). The state vector for a kinematic model is composed of the vector $\mathbf{p} = (p_x, p_y, p_z)$ that gives the coordinates of the center of the robot expressed in the absolute inertial coordinate system $\mathcal{R}_0$, the three Euler angles (ψ, θ, φ) and the speed vector $\mathbf{v}_r$ of the robot expressed in its own coordinate system. The inputs of the system are for one the acceleration $\mathbf{a}_r = \mathbf{a}_{\mathcal{R}_1}$ of the center of the robot expressed in its own coordinate system and second, the vector $\omega_r = \omega_{\mathcal{R}_1/\mathcal{R}_0|\mathcal{R}_1} = (\omega_x, \omega_y, \omega_z)$ corresponding to the rotation vector of the robot relative to $\mathcal{R}_0$ expressed in the coordinate system $\mathcal{R}_1$ of the robot. It is indeed conventional to express $\mathbf{a}$, ω in the coordinate system of the robot since these quantities are generally measured by the robot itself via the sensors attached on it. They are, therefore, naturally expressed in the frame of the robot. The first state equation is:

$$\dot{\mathbf{p}} \overset{[1.6]}{=} \mathbf{R}(\psi, \theta, \varphi) \cdot \mathbf{v}_r$$

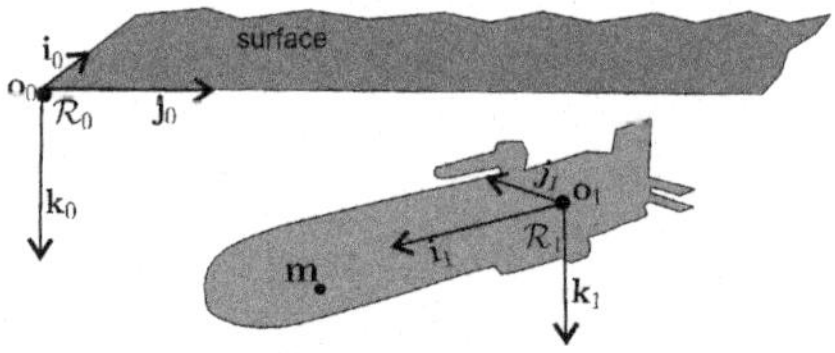

Figure 1.5. *The coordinate system $\mathcal{R}_1 : (\mathbf{o}_1, \mathbf{i}_1, \mathbf{j}_1, \mathbf{k}_1)$ attached to the robot*

In order to express $\mathbf{v}_r$, let us differentiate this equation. We obtain:

$$\ddot{\mathbf{p}} = \dot{\mathbf{R}} \cdot \mathbf{v}_r + \mathbf{R} \cdot \dot{\mathbf{v}}_r$$

with $\mathbf{R} = \mathbf{R}(\psi, \theta, \varphi)$, in other words:

$$\dot{\mathbf{v}}_r = \underbrace{\mathbf{R}^{\mathrm{T}} \cdot \ddot{\mathbf{p}}}_{\mathbf{a}_r} - \mathbf{R}^{\mathrm{T}} \dot{\mathbf{R}} \cdot \mathbf{v}_r \overset{[1.4]}{=} \mathbf{a}_r - \mathbf{Ad}\left(\omega_{\mathcal{R}_1/\mathcal{R}_0|\mathcal{R}_1}\right) \cdot \mathbf{v}_r$$

Thus:

$$\dot{\mathbf{v}}_r = \mathbf{a}_r - \omega_r \wedge \mathbf{v}_r$$

constitutes the second state equation. Finally, we also need to express $\dot{\psi}, \dot{\theta}, \dot{\varphi}$ as a function of the state variables. The relation:

$$\omega_{|\mathcal{R}_0} = \mathbf{R}(\psi, \theta, \varphi) \cdot \omega_{|\mathcal{R}_1}$$

becomes, following equation [1.10]:

$$\begin{pmatrix} 0 & -\sin\psi & \cos\theta\cos\psi \\ 0 & \cos\psi & \cos\theta\sin\psi \\ 1 & 0 & -\sin\theta \end{pmatrix} \begin{pmatrix} \dot{\psi} \\ \dot{\theta} \\ \dot{\varphi} \end{pmatrix} = \mathbf{R}(\psi, \theta, \varphi) \cdot \omega_r$$

By isolating in this expression the vector $\left(\dot{\psi}, \dot{\theta}, \dot{\varphi}\right)$, we obtain the third state equation. By bringing together the three state equations, we obtain the following kinematic model for the robot:

$$\begin{cases} \dot{\mathbf{p}} = \mathbf{R}(\psi, \theta, \varphi) \cdot \mathbf{v}_r \\ \dot{\mathbf{v}}_r = \mathbf{a}_r - \omega_r \wedge \mathbf{v}_r \\ \begin{pmatrix} \dot{\psi} \\ \dot{\theta} \\ \dot{\varphi} \end{pmatrix} = \begin{pmatrix} 0 & \frac{\sin\varphi}{\cos\theta} & \frac{\cos\varphi}{\cos\theta} \\ 0 & \cos\varphi & -\sin\varphi \\ 1 & \tan\theta\sin\varphi & \tan\theta\cos\varphi \end{pmatrix} \cdot \omega_r \end{cases} \qquad [1.11]$$

On a horizontal plane: for a robot moving on a horizontal plane, we have $\varphi = \theta = 0$. Equation [1.11] gives us $\dot{\psi} = \omega_{r3}$, $\dot{\theta} = \omega_{r2}$ and $\dot{\varphi} = \omega_{r1}$. In such a case, there is a perfect correspondence between the components of ω_r and the differentials of the Euler angles. There are singular cases, for instance when $\theta = \frac{\pi}{2}$ (this is the case when the robot points upward), in which the differentials of the Euler angles cannot be defined. Using the rotation vector is often preferred since it does not have such singularities.

Dead reckoning: for dead reckoning (in other words, without external sensors), there are generally extremely precise laser gyrometers (around $0.001\ deg/s.$). These make use of the Sagnac effect (in a circular optical fiber turning around itself, the time taken by light to travel an entire round-trip depends on the path direction). Using three fibers, these gyrometers generate the vector $\omega_r = (\omega_x, \omega_y, \omega_z)$. There are also accelerometers capable of measuring the acceleration $\mathbf{a}_r$ with a very high degree of precision. In pure inertial mode, we determine our position by differentiating equations [1.11] only using the acceleration $\mathbf{a}_r$ and the rotation speed ω_r, both expressed in the coordinate system of the robot. In the case where we are measuring the quantity $\mathbf{v}_r$ (also expressed in the frame of the robot) with a *DVL*, we only need to integrate the first and the last of these three equations. Finally, when the robot is a correctly ballasted submarine or a terrestrial robot moving on a relatively plane ground, we know *a priori* that on average the bank and elevation are equal to zero. Therefore, we may incorporate this information through a Kalman filter in order to limit the drift in positioning. An efficient inertial unit integrates an amalgamation of all the available information.

Inertial unit: a pure inertial unit (without hybridization and without taking into account Earth's gravity) represents the robot by the kinematic model of Figure 1.6, which itself uses the state equations given in [1.11]. This system is written in the form $\dot{\mathbf{x}} = \mathbf{f}(\mathbf{x}, \mathbf{u})$, where $\mathbf{u} = (\mathbf{a}_r, \omega_r)$ is the vector of the measured inertial inputs (accelerations and rotation speeds viewed by an observer on the ground, but expressed in the frame of the robot) and $\mathbf{x} = (\mathbf{p}, \mathbf{v}_r, \psi, \theta, \varphi)$ is the state vector. For the moment, we use a numerical integration method such as the Euler method. This leads to replacing the differential equation $\dot{\mathbf{x}} = \mathbf{f}(\mathbf{x}, \mathbf{u})$ with the recurrence:

$$\mathbf{x}(t + dt) = \mathbf{x}(t) + dt \cdot \mathbf{f}(\mathbf{x}(t), \mathbf{u}(t))$$

Dynamic modeling: for the dynamic modeling of a submarine, the reference work is the book of Fossen [FOS 02]. In order to obtain a dynamic model, it is sufficient to take the kinematic equations and to consider that the angular and tangential accelerations caused by forces and dynamic performance. These quantities become the new inputs of our system. The link between the accelerations and forces is made by Newton's second law (or the fundamental principle of dynamics). Thus, for instance if $\mathbf{f}$ is the net force resulting from the external forces expressed in the inertial frame and m is the mass of the robot, we have:

$$m\ddot{\mathbf{p}} = \mathbf{f}$$

$$
\begin{aligned}
\dot{\mathbf{p}} &= \mathbf{R}(\psi, \theta, \varphi) \cdot \mathbf{v}_r \\
\dot{\mathbf{v}}_r &= \mathbf{a}_r - \boldsymbol{\omega}_r \wedge \mathbf{v}_r \\
\begin{pmatrix} \dot{\psi} \\ \dot{\theta} \\ \dot{\varphi} \end{pmatrix} &= \begin{pmatrix} 0 & \dfrac{\sin\varphi}{\cos\theta} & \dfrac{\cos\varphi}{\cos\theta} \\ 0 & \cos\varphi & -\sin\varphi \\ 1 & \tan\theta\sin\varphi & \tan\theta\cos\varphi \end{pmatrix} \cdot \boldsymbol{\omega}_r
\end{aligned}
$$

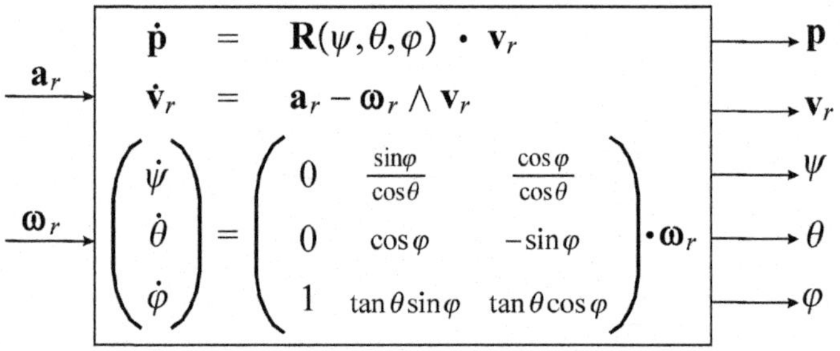

Figure 1.6. *Kinematic model used by an inertial unit*

This relation viewed from the inertial frame but expressed in the coordinate system of the robot gives $m\mathbf{a}_r = \mathbf{R}^{\mathrm{T}}\mathbf{f}$, in other words:

$$\mathbf{a}_r = \frac{1}{m}\mathbf{R}^{\mathrm{T}} \cdot \mathbf{f}.$$

Therefore, we have that the tangential acceleration (which appears as an input of the kinematic model) is an algebraic function of the forces acting on the robot.

1.4. Exercises

EXERCISE 1.1.– Properties of the adjoint matrix

Let us consider the vector $\omega = (\omega_x, \omega_y, \omega_z)$ and its adjoint matrix $\mathbf{Ad}(\omega)$.

1) Show that the eigenvalues of $\mathbf{Ad}(\omega)$ are $\{0, \|\omega\|i, -\|\omega\|i\}$. Give an eigenvector associated with 0. Discuss.

2) Show that the vector $\mathbf{Ad}(\omega)\,\mathbf{x} = \omega \wedge \mathbf{x}$ is a vector perpendicular to ω and $\mathbf{x}$, such that the trihedron $(\omega, \mathbf{x}, \omega \wedge \mathbf{x})$ is direct.

3) Show that the norm of $\omega \wedge \mathbf{x}$ is surface of the parallelogram $\mathcal{A}$ mediated by ω and $\mathbf{x}$.

EXERCISE 1.2.– Jacobi identity

The Jacobi identity is written as:

$$\mathbf{a} \wedge (\mathbf{b} \wedge \mathbf{c}) + \mathbf{c} \wedge (\mathbf{a} \wedge \mathbf{b}) + \mathbf{b} \wedge (\mathbf{c} \wedge \mathbf{a}) = \mathbf{0}$$

1) Show that this identity is equivalent to:

$$\mathbf{Ad}\,(\mathbf{a} \wedge \mathbf{b}) = \mathbf{Ad}\,(\mathbf{a})\,\mathbf{Ad}\,(\mathbf{b}) - \mathbf{Ad}\,(\mathbf{b})\,\mathbf{Ad}\,(\mathbf{a})$$

where $\mathbf{Ad}\,(\omega)$ is the adjoint matrix of the vector $\omega \in \mathbb{R}^3$.

2) In the space of skew-symmetric matrices, the Lie bracket is defined as follows:

$$[\mathbf{A}, \mathbf{B}] = \mathbf{A} \cdot \mathbf{B} - \mathbf{B} \cdot \mathbf{A}$$

Show that:

$$\mathbf{Ad}\,(\mathbf{a} \wedge \mathbf{b}) = [\mathbf{Ad}\,(\mathbf{a})\,, \mathbf{Ad}\,(\mathbf{b})]$$

3) An *algebra* is an algebraic structure $(\mathcal{A}, +, \times, \cdot)$ over a body $\mathbb{K}$, if (1) $(\mathcal{A}, +, \cdot)$ is a vector space over $\mathbb{K}$; (2) the multiplication rule $\times$ of $\mathcal{A} \times \mathcal{A} \to \mathcal{A}$ is left-distributive and right-distributive with respect to $+$ and (3) for all $\alpha, \beta \in \mathbb{K}$, and for all $x, y \in \mathcal{A}$, $\alpha \cdot x \times \beta \cdot y$, $(\alpha\beta) \cdot (x \times y)$. Note that in general, an algebra is non-commutative ($x \times y \neq y \times x$) and non-associative ($(x \times y) \times z \neq x \times (y \times z)$). A *Lie algebra* $(\mathcal{G}, +, [\,]\,, \cdot)$ is a non-commutative and non-associative algebra in which multiplication, denoted by a so-called Lie bracket, verifies (1) $[,]$ that is bilinear, in other words linear with respect to each variable; (2) $[x, y] = -[y, x]$ (antisymmetry) and (3) $[x, [y, z]] + [y, [z, x]] + [z, [x, y]] = 0$ (Jacobi relation). Verify that the set $\left(\mathbb{R}^3, +, \wedge, \cdot\right)$ forms a *Lie algebra*.

EXERCISE 1.3.– Varignon's formula

Let us consider a solid body whose center of gravity remains at the origin of a Galilean coordinate system and is rotating around an axis Δ with a rotation vector of ω. Give the equation of the trajectory of a point $\mathbf{x}$ of the body.

EXERCISE 1.4.– Rodrigues' formula

Let us consider a solid body whose center of gravity remains at the origin of a Galilean coordinate system and is rotating around an axis Δ. The position of a point $\mathbf{x}$ of the body satisfies the state equation (Varignon's formula):

$$\dot{\mathbf{x}} = \omega \wedge \mathbf{x}$$

where ω is parallel to the axis of rotation Δ and $||\omega||$ is the rotation speed of the body (in rad.s^{-1}).

1) Show that this state equation can be written in the form:

$$\dot{\mathbf{x}} = \mathbf{A}\mathbf{x}$$

Explain why the matrix $\mathbf{A}$ is often denoted by $\omega\wedge$.

2) Give the expression of the solution of the state equation.

3) Deduce from this that the expression of the rotation matrix $\mathbf{R}$ with angle $||\omega||$ around ω is given by the following formula, referred to as *Rodrigues' formula*:

$$\mathbf{R} = e^{\omega\wedge}$$

4) Calculate the eigenvalues of $\mathbf{A}$ and show that ω is the eigenvector associated with the zero eigenvalue. Discuss.

5) What are the eigenvalues of $\mathbf{R}$?

6) Using the previous questions, give the expression of a rotation around the vector $\omega = (1,\ 0,\ 0)$ of angle α.

7) Write a program in MATLAB, `eulermat(phi,theta,psi)` that uses Rodrigues' formula to return the Euler matrix.

EXERCISE 1.5.– Geometric approach to Rodrigues' formula

Let us consider the rotation $\mathcal{R}_{\mathbf{n},\varphi}$ of angle φ around the unit vector $\mathbf{n}$. Let $\mathbf{u}$ be a vector that we will subject to this rotation. The vector $\mathbf{u}$ can be decomposed as follows:

$$\mathbf{u} = \underbrace{< \mathbf{u}, \mathbf{n} > \cdot \mathbf{n}}_{\mathbf{u}_{||}} + \underbrace{\mathbf{u} - < \mathbf{u}, \mathbf{n} > \cdot \mathbf{n}}_{\mathbf{u}_{\perp}}$$

where $\mathbf{u}_{||}$ is collinear to $\mathbf{n}$ and $\mathbf{u}_{\perp}$ is in the plane $P_{\perp}$ orthogonal to $\mathbf{n}$ (see Figure 1.7).

1) Prove Rodrigues' formula given by:

$$\mathcal{R}_{\mathbf{n},\varphi}(\mathbf{u}) = < \mathbf{u}, \mathbf{n} > \cdot \mathbf{n} + (\cos\varphi)(\mathbf{u} - < \mathbf{u}, \mathbf{n} > \cdot \mathbf{n}) + (\sin\varphi)(\mathbf{n} \wedge \mathbf{u})$$

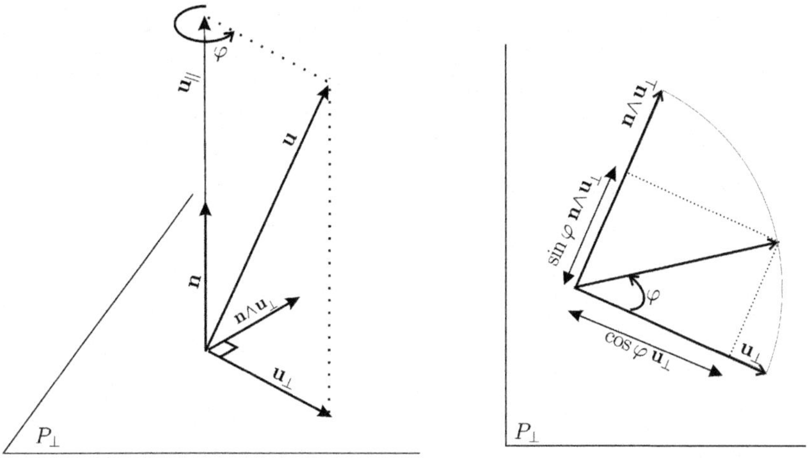

Figure 1.7. *Rotation of the vector* $\mathbf{u}$ *around the vector* $\mathbf{n}$*; a): perspective view; b): view from above*

2) Using the double vector product formula $\mathbf{a}\wedge(\mathbf{b}\wedge\mathbf{c}) = (\mathbf{a}^{\mathrm{T}}\mathbf{c})\cdot\mathbf{b} - (\mathbf{a}^{\mathrm{T}}\mathbf{b})\cdot\mathbf{c}$, on the element $\mathbf{n}\wedge(\mathbf{n}\wedge\mathbf{u})$, show that Rodrigues' formula can also be written as:

$$\mathcal{R}_{\mathbf{n},\varphi}(\mathbf{u}) = \mathbf{u} + (1-\cos\varphi)(\mathbf{n}\wedge(\mathbf{n}\wedge\mathbf{u})) + (\sin\varphi)(\mathbf{n}\wedge\mathbf{u})$$

Deduce from this that the matrix associated with the linear operator $\mathcal{R}_{\mathbf{n},\varphi}$ is written as:

$$\mathbf{R}_{\mathbf{n},\varphi} = \begin{pmatrix} 1\,0\,0 \\ 0\,1\,0 \\ 0\,0\,1 \end{pmatrix} + (1-\cos\varphi)\begin{pmatrix} -n_y^2-n_z^2 & n_xn_y & n_xn_z \\ n_xn_y & -n_x^2-n_z^2 & n_yn_z \\ n_xn_z & n_yn_z & -n_x^2-n_y^2 \end{pmatrix} + (\sin\varphi)\begin{pmatrix} 0 & -n_z & n_y \\ n_z & 0 & -n_x \\ -n_y & n_x & 0 \end{pmatrix}$$

3) Conversely, we are given a rotation matrix $\mathbf{R}_{\mathbf{n},\varphi}$ for which we wish to find the axis of rotation $\mathbf{n}$ and the angle of rotation φ. Give an expression for $\mathbf{R}_{\mathbf{n},\varphi} - \mathbf{R}_{\mathbf{n},\varphi}^{\mathrm{T}}$ and use it to obtain $\mathbf{n}$ and φ as a function of $\mathbf{R}_{\mathbf{n},\varphi}$. For a geometric illustration, Figure 1.8 might prove to be useful.

4) Using a Maclaurin series development of sinφ and cosφ, show that:

$$\mathbf{R}_{\mathbf{n},\varphi} = \exp\left(\varphi \cdot \mathbf{Ad}\left(\mathbf{n}\right)\right)$$

which sometimes written as:

$$\mathbf{R}_{\mathbf{n},\varphi} = \exp\left(\varphi \cdot \mathbf{n}\wedge\right)$$

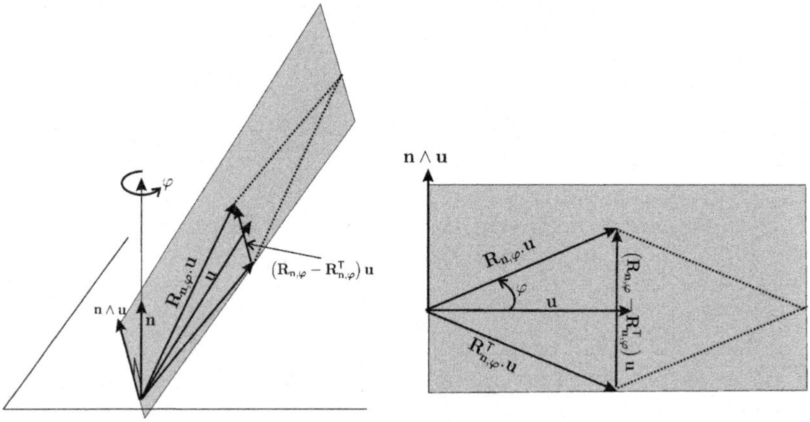

Figure 1.8. *Left we have a view of the rotation of angle φ around $\mathbf{n}$; right we have a visualization of the section corresponding to the Rodrigues rhombus*

EXERCISE 1.6.– Schuler oscillations

One of the fundamental components of an inertial unit is the inclinometer. This sensor gives the vertical direction. Traditionally, we use a pendulum (or a plumb line) for this. However, when we are moving, due to the accelerations the pendulum starts to oscillate and it can no longer be used to measure the vertical direction. Here, we are interested in designing a pendulum for which any horizontal acceleration does not lead to oscillations. Let us consider a pendulum with two masses m at each end, situated at a distance ℓ_1 and ℓ_2 from the axis of rotation of the rod (see Figure 1.9). The axis moves over the surface of the Earth. We assume that ℓ_1 and ℓ_2 are small in comparison to the Earth's radius r.

1) Find the state equations of the system.

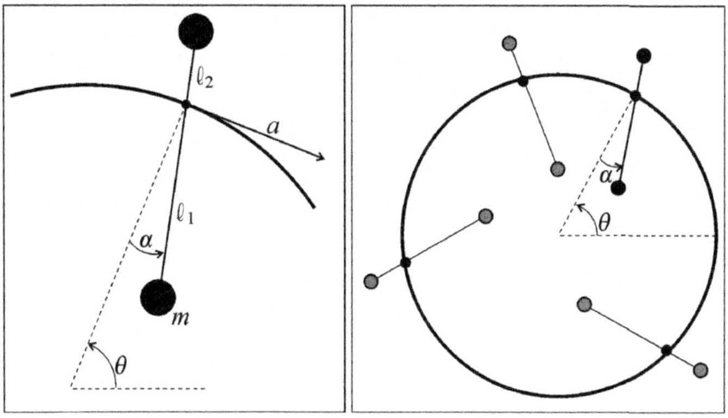

Figure 1.9. *Inclinometer pendulum moving over the surface of the Earth*

2) Let us assume that $\alpha = \dot{\alpha} = 0$. For which values of ℓ_1 and ℓ_2 does the pendulum remain vertical, for any horizontal movement of the pendulum? What values does ℓ_2 have to take if we let $\ell_1 = 1$ m and if we take $r =$ 6 400 km for the Earth's radius?

3) Let us assume that, as a result of disturbances, the pendulum starts to oscillate. The period of these oscillations is called the *Schuler period*. Calculate this period.

4) Simulate the system graphically by taking a Gaussian white noise as the acceleration input. Since the system is conservative, an Euler integration method will not perform well (the pendulum would gain energy). A higher order integration scheme, such as that of *Runge Kutta*, should be used. This is given by:

$$\mathbf{x}(t+dt) \simeq \mathbf{x}(t) + dt. \\ \times \left(\frac{\mathbf{f}\left(\mathbf{x}(t), \mathbf{u}(t)\right)}{4} + \tfrac{3}{4}\mathbf{f}(\mathbf{x}(t) + \frac{2dt}{3}\,\mathbf{f}(\mathbf{x}(t), \mathbf{u}(t)), \mathbf{u}(t + \tfrac{2}{3}dt)) \right)$$

For an easier graphical representation, we will take $r = 10$ m, $\ell_1 = 1$ m and $g = 10$ ms^{-2}. Discuss the results.

EXERCISE 1.7.– Brake detector

We will now look at a problem that involves basis changes and rotation matrices. A car is preceded by another car $\mathbf{m}$ (which we will assume to be a point). We attach to this car the coordinate system $\mathcal{R}_1 : (\mathbf{o}_1, \mathbf{i}_1, \mathbf{j}_1)$ as represented in Figure 1.10. The coordinate system $\mathcal{R}_0 : (\mathbf{o}_0, \mathbf{i}_0, \mathbf{j}_0)$ is a ground frame assumed to be fixed.

This car is equipped with the following sensors:

– several odometers placed on the rear wheels allowing us to measure the speed v of the center of the rear axle;

– a gyro giving the angular speed of the car $\dot{\theta}$, as well as the angular acceleration $\ddot{\theta}$;

– an accelerometer placed at $\mathbf{o}_1$ allowing us to measure the acceleration vector (α, β) of $\mathbf{o}_1$ expressed in the coordinate system $\mathcal{R}_1$;

– using two radars placed at the front, our car is capable of (indirectly) measuring the coordinates (a_1, b_1) of the point $\mathbf{m}$ in the coordinate system $\mathcal{R}_1$ as well as the first two derivatives $\left(\dot{a}_1, \dot{b}_1\right)$ and $\left(\ddot{a}_1, \ddot{b}_1\right)$.

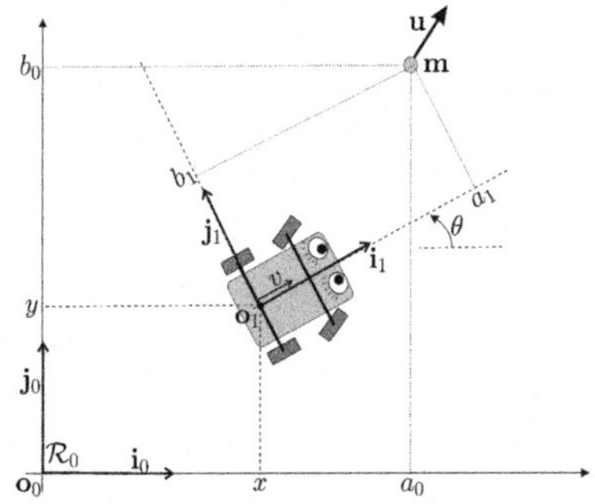

Figure 1.10. *Car trying to detect whether the point* $\mathbf{m}$ *is braking or not*

However, the car is not equipped with a positioning system (such as a Global Positioning System (GPS)) that would allow it to gain knowledge of $x, y, \dot{x}, \dot{y}$. It does not have a compass for measuring the angle θ. The quantities in play are the following:

Measured $v, \dot{\theta}, a_1, b_1, \alpha, \beta$.

Unknown $x, y, \dot{x}, \dot{y}, \theta, a_0, b_0$

When a quantity is measured, we assume that its differentials are also measured, but not its primitives. For example, $\dot{a}_1, \dot{b}_1, \ddot{a}_1, \ddot{b}_1$ are considered to be measured since a_1, b_1 are measured. However, $\dot{\theta}$ is measured but θ is not. We do not know the state equations of our car. The goal of this problem is to find a condition on the measured variables (and their derivatives) that will allow us to tell whether the point $\mathbf{m}$ is braking or not. We understand that such a condition would allow us to build a warner informing us that the preceding vehicle is braking, even when its rear brake lights are not visible (fog, trailer without brake lights) or defective.

1) By expressing the Chasle relation ($\mathbf{o}_0\mathbf{m} = \mathbf{o}_0\mathbf{o}_1 + \mathbf{o}_1\mathbf{m}$) in the coordinate system $\mathcal{R}_0$, show the basis change formula:

$$\begin{pmatrix} a_0 \\ b_0 \end{pmatrix} = \begin{pmatrix} x \\ y \end{pmatrix} + \mathbf{R}_\theta \begin{pmatrix} a_1 \\ b_1 \end{pmatrix}$$

where $\mathbf{R}_\theta$ is a rotation matrix.

2) Show that $\mathbf{R}_\theta^{\mathrm{T}}\,\dot{\mathbf{R}}_\theta$ is a skew-symmetric matrix and find its expression.

3) Let $\mathbf{u}$ be the speed vector of the point $\mathbf{m}$ viewed by a fixed observer. Find an expression $\mathbf{u}_{|\mathcal{R}_1}$ of the speed vector $\mathbf{u}$ expressed in the coordinate system $\mathcal{R}_1$. Express $\mathbf{u}_{|\mathcal{R}_1}$ as a function of the measured variables $v, \dot{\theta}, a_1, \dot{a}_1, b_1, \dot{b}_1$.

4) We will now use the accelerometer of our vehicle, which gives us the acceleration vector (α, β) of $\mathbf{o}_1$, expressed in $\mathcal{R}_1$. By differentiating two times $\mathbf{m}_{|\mathcal{R}_0}$, give the expression of the acceleration $\mathbf{a}_{|\mathcal{R}_0}$ of $\mathbf{m}$ in the coordinate system $\mathcal{R}_0$. Deduce from this its expression $\mathbf{a}_{|\mathcal{R}_1}$ in the coordinate system $\mathcal{R}_1$. Give the expression of $\mathbf{a}_{|\mathcal{R}_1}$ only as a function of the measured variables.

5) Find a condition on the measurements $\left(v, \dot{\theta}, a_1, b_1, \dot{a}_1, \dot{b}_1, \ddot{a}_1, \ddot{b}_1, \alpha, \beta\right)$ which allows us to detect whether the vehicle in front is braking or not.

EXERCISE 1.8.– Modeling an underwater robot

The robot we will be modeling is the Redermor (*greyhound of the sea* in the Breton language). It is represented in Figure 1.11. It is an entirely autonomous underwater robot. This robot, developed by GESMA (*Groupe d'Etude Sous-Marine de l'Atlantique* – Atlantic underwater research group), has a length of 6 m, a diameter of 1 m and a weight of 3 800 kg. It has a very efficient propulsion and control system with the aim of finding mines on the seabed.

Let us build a local coordinate system $\mathcal{R}_0 : (\mathbf{o}_0, \mathbf{i}_0, \mathbf{j}_0, \mathbf{k}_0)$ over the area traveled by the robot. The point $\mathbf{o}_0$ is placed on the surface of the ocean. The vector $\mathbf{i}_0$ indicates north, $\mathbf{j}_0$ indicates east and $\mathbf{k}_0$ is oriented toward the center of the Earth. Let $\mathbf{p} = (p_x, p_y, p_z)$ be the coordinates of the center of the robot expressed in the coordinate system $\mathcal{R}_0$. The state variables of the underwater robot are its position $\mathbf{p}$ in the coordinate system $\mathcal{R}_0$, its tangential v and its three Euler angles ψ, θ, φ. Its inputs are the tangential acceleration $\dot{v}$ as well as three control surfaces which act, respectively, on $\omega_x, \omega_y, \omega_z$. More formally, we have:

$$\begin{cases} u_1 &= \dot{v} \\ vu_2 &= \omega_y \\ vu_3 &= \omega_z \\ vu_4 &= \omega_x \end{cases}$$

where the factor v preceding u_1, u_2, u_4 indicates that the robot is only able to turn when it is advancing. Give the kinematic state model for this system.

Figure 1.11. Redermor *built by GESMA (Groupe d'Etude Sous-Marine de l'Atlantique – Atlantic underwater research group), on the water surface, still close to the boat it was launched from*

EXERCISE 1.9.– 3D robot graphics

Drawing two-dimensional (2D) or 3D robots or objects on the screen is widely used for simulation in robotics. The classic method (used by OPENGL) relies on modeling the posture of objects using a series of affine transformations (rotations, translations and homotheties) of the form:

$$\mathbf{f}_i : \begin{array}{l} \mathbb{R}^n \to \mathbb{R}^n \\ \mathbf{x} \;\; \mapsto \mathbf{A}_i\mathbf{x} + \mathbf{b}_i \end{array}$$

with $n = 2$ or 3. However, the manipulation of compositions of affine functions is less simple than that of linear applications. The idea of the transformation in *homogeneous coordinates* is to transform a system of affine equations into a system of linear equations. Note first that an affine equation of the type $\mathbf{y} = \mathbf{A}\mathbf{x} + \mathbf{b}$ can be written as:

$$\begin{pmatrix} \mathbf{y} \\ 1 \end{pmatrix} = \begin{pmatrix} \mathbf{A} & \mathbf{b} \\ 0 & 1 \end{pmatrix} \begin{pmatrix} \mathbf{x} \\ 1 \end{pmatrix}$$

Thus, we will define the *homogeneous transformation* of a vector as follows:

$$\mathbf{x} \mapsto \mathbf{x}_\mathrm{h} = \begin{pmatrix} \mathbf{x} \\ 1 \end{pmatrix}$$

Thus, an equation of the type:

$$\mathbf{y} = \mathbf{A}_3 \left(\mathbf{A}_2 \left(\mathbf{A}_1 \mathbf{x} + \mathbf{b}_1 \right) + \mathbf{b}_2 \right) + \mathbf{b}_3$$

involving the composition of three affine transformations, can be written as:

$$\mathbf{y}_\mathrm{h} = \begin{pmatrix} \mathbf{A}_3 & \mathbf{b}_3 \\ \mathbf{0} & 1 \end{pmatrix} \begin{pmatrix} \mathbf{A}_2 & \mathbf{b}_2 \\ \mathbf{0} & 1 \end{pmatrix} \begin{pmatrix} \mathbf{A}_1 & \mathbf{b}_1 \\ \mathbf{0} & 1 \end{pmatrix} \mathbf{x}_\mathrm{h}$$

A *pattern* is a matrix with two or three rows (depending on the object being in the plane or space) and n columns representing the n vertices of a rigid polygon embodying the object. It is important that the union of all the segments formed by two consecutive points of the pattern forms all the vertices of the polygon that we wish to represent.

1) Let us consider the underwater robot (or *autonomous underwater vehicle* for (AUV)) whose pattern in homogeneous coordinates is the following:

$$\begin{pmatrix} 0 & 0 & 10 & 0 & 0 & 10 & 0 & 0 \\ -1 & 1 & 0 & -1 & -0.2 & 0 & 0.2 & 1 \\ 0 & 0 & 0 & 0 & 1 & 0 & 1 & 0 \\ 1 & 1 & 1 & 1 & 1 & 1 & 1 & 1 \end{pmatrix}$$

Draw this pattern in perspective view on a piece of paper.

2) The state equations of the robot are the following:

$$\begin{cases} \dot{p}_x &= v\cos\theta\cos\psi \\ \dot{p}_y &= v\cos\theta\sin\psi \\ \dot{p}_z &= -v\sin\theta \\ \dot{v} &= u_1 \\ \dot{\psi} &= \frac{\sin\varphi}{\cos\theta}\cdot v\cdot u_2 + \frac{\cos\varphi}{\cos\theta}\cdot v\cdot u_3 \\ \dot{\theta} &= \cos\varphi\cdot v\cdot u_2 - \sin\varphi\cdot v\cdot u_3 \\ \dot{\varphi} &= -0\cdot 1\ \sin\varphi + \tan\theta\cdot v\cdot(\sin\varphi\cdot u_2 + \cos\varphi\cdot u_3) \end{cases}$$

where (φ, θ, ψ) are the three Euler angles. The inputs of the system are the tangential acceleration u_1, the pitch u_2 and the yaw u_3. The state vector is, therefore, equal to $\mathbf{x} = (p_x, p_y, p_z, v, \psi, \theta, \varphi)$. Give a MATLAB function capable of drawing the robot in 3D together with its shadow, in the plane x-y. Verify that the drawing is correct by moving the six degrees of freedom of the robot one-by-one. Use the `plot3` function of MATLAB to obtain a 3D representation such as the one shown in Figure 1.12.

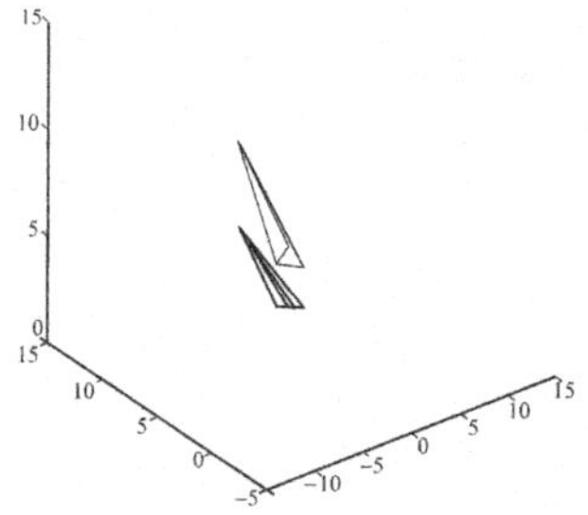

Figure 1.12. *3D representation of the robot together with its shadow in the horizontal plane*

3) By using the relation:

$$\omega|_{\mathcal{R}_0} = \begin{pmatrix} 0 & -\sin\psi & \cos\theta\cos\psi \\ 0 & \cos\psi & \cos\theta\sin\psi \\ 1 & 0 & -\sin\theta \end{pmatrix} \begin{pmatrix} \dot{\psi} \\ \dot{\theta} \\ \dot{\varphi} \end{pmatrix}$$

draw the instantaneous rotation vector of the robot.

4) Simulate the robot in MATLAB in various conditions using a Euler method.

EXERCISE 1.10.– Manipulator robot

A manipulator robot, such as *Staubli* represented in Figure 1.13, is composed of several rigid arms. We retrieve the coordinates of the end effector, at the extremity of the robot, using a series of geometric transformations. We can show that a parametrization with four degrees of freedom allows us to represent these transformations. There are several possible parametrizations, each with its own advantages and disadvantages. The most widely used one is probably the *Denavit–Hartenberg* parametrization. In the case where the articulations are rotational joints (as is the case of the Staubli robot where the joins can turn), the parametrization represented by the figure might prove to be practical since it makes drawing the robot easier. This transformation is the composition of four elementary transformations: (1) a translation of length r following z; (2) a translation of length d following x; (3) a rotation of α around y and (4) a rotation of θ (the variable activated around z.) Using the figure for drawing the arms and the photo for the robot, perform a realistic simulation of the robot's movement in MATLAB.

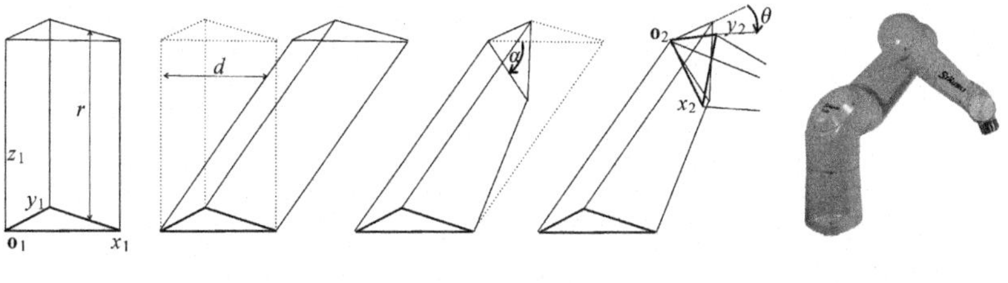

Figure 1.13. *Parametrization for the direct geometric model of a manipulator robot*

EXERCISE 1.11.– 3D modeling of a wheel

1) Let us consider the wheel rolling in a plane as shown in Figure 1.14 and for which we seek to obtain the state equations. In this figure, $\mathbf{u}$ is the unit vector indicating the movement direction of the point of contact $\mathbf{p}$. The wheel is assumed to roll without friction, and therefore the ground reaction force $\mathbf{r}$ is orthogonal to $\mathbf{n}$.

First, we need to define the Euler angles in the context of the wheel, where the concepts of elevation and bank are meaningless. Let us choose for ψ the angle of the horizontal projection of the wheel axis (indicating the horizontal direction to the left of $\mathbf{n}$). For θ, we will take the wheel dishing and for φ, the angle of the wheel made on itself. The reason for this choice is that the angle θ will be within the interval $\left[-\frac{\pi}{2}, \frac{\pi}{2}\right]$ in accordance with what happens with Euler angles. Thus, the singularities of the equations will correspond to physical singularities. Indeed, the matrix involved in [1.10] is singular if $\cos\theta = 0$ and in such a case, the wheel can no longer roll. We will assume that the wheel is solid, similar to a homogeneous disk of mass m and radius ρ. Its inertia matrix is given by:

$$\mathbf{I} = \begin{pmatrix} \frac{m\rho^2}{2} & 0 & 0 \\ 0 & \frac{m\rho^2}{4} & 0 \\ 0 & 0 & \frac{m\rho^2}{4} \end{pmatrix}$$

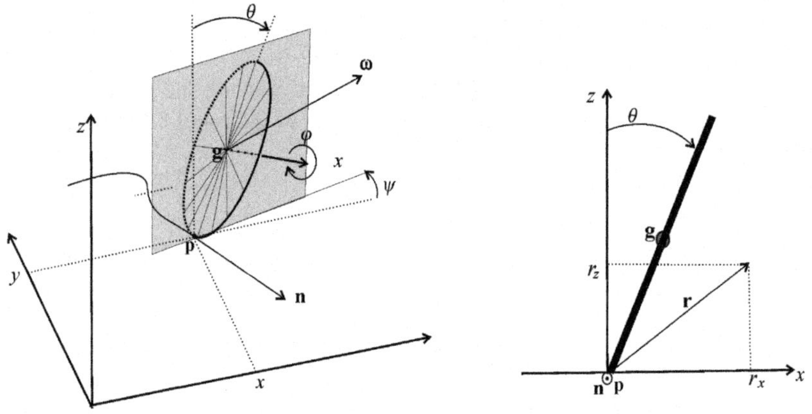

Figure 1.14. *Wheel rolling in a plane. a) Three-dimensional view; b) section in a plane perpendicular to* $\mathbf{u}$ *following plane in gray*

Give the state equations of this wheel.

2) Let us assume that we are only able to move masses in the plane of the wheel, while conserving the cylindrical symmetry of the wheel. The center of gravity is, therefore, always in the center of the wheel and we are only able to influence the rotation vector ω following the wheel axis. Is it possible to control the trajectory of the wheel as well as its speed?

EXERCISE 1.12.– Mechanical stability of an underwater robot

Let us consider a homogeneous body in water with the same density as water. Its center of gravity is denoted by $\mathbf{g}$. We place on this body, at a given point denoted by $\mathbf{a}$, a point mass (and therefore of infinite density) exerting a force $\mathbf{f}$, as shown in Figure 1.15(a).

1) What is the condition for rotational equilibrium (in other words, one which does not lead to a rotation of the body)?

2) Let us assume that we have rotational equilibrium. Under what condition do we have a stable equilibrium?

3) Consider an underwater robot that we wish to evolve in a horizontal plane. At equilibrium under water, we notice that the robot is slightly leaning forward as shown in Figure 1.15(b). What needs to be done in order to counter this?

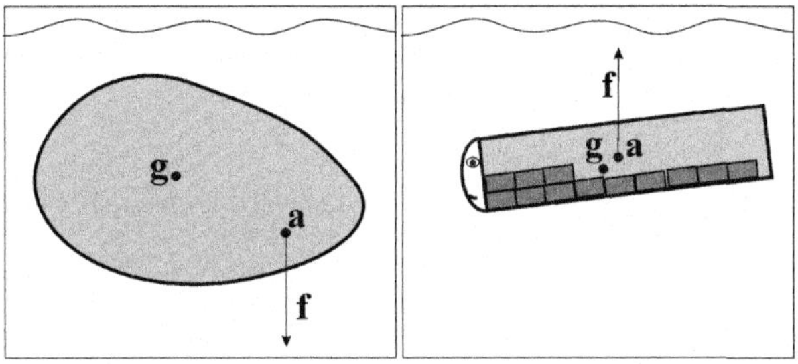

Figure 1.15. *a) Submerged homogeneous body; b) unbalanced robot*

4) The robot is now in horizontal equilibrium, which is what we wanted. However, this equilibrium is unstable, in other words the robot tends to turn around as soon as it departs from its position of equilibrium. What needs to be done in order to have stability?

5) The robot is now correctly in equilibrium with respect to Archimedes' force. We give it an initial horizontal speed and we let go of it. The robot submerges. Why? What needs to be done to keep it from submerging?

6) The robot is now in stable equilibrium relative to Archimedes' principle and the drag. We now act on the propeller, but the robot submerges once again. Why? What needs to be done to avoid this?

7) The heading of the robot is now controlled using its compass, and it is trying to move closer to a concrete wall parallel to the compass. It then starts to strongly oscillate while remaining in a horizontal plane. Why? How do we avoid this?

8) Figure 1.16 represents a robot entirely in equilibrium with various configurations for the position of its propellers. Is configuration (a) more stable or more maneuverable than configuration (b)? Is configuration (c) more stable or more maneuverable than configuration (d)?

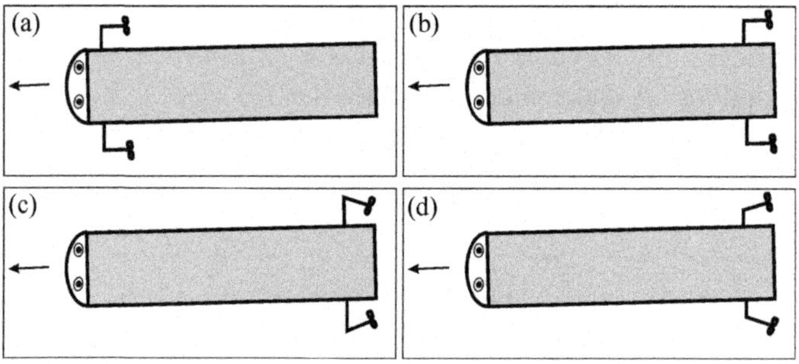

Figure 1.16. *An underwater robot viewed from above with various configurations for its propellers*

1.5. Corrections

CORRECTION FOR EXERCISE 1.1.– (Properties of the adjoint matrix)

1) The characteristic polynomial of the matrix $\mathbf{Ad}(\omega)$ is calculated relatively easily. It is given by:

$$s^3 + \left(\omega_x^2 + \omega_y^2 + \omega_z^2\right) s = s\left(s^2 + \left(\omega_x^2 + \omega_y^2 + \omega_z^2\right)\right)$$

From this, we obtain the eigenvalues $\{0, ||\omega||i, -||\omega||i\}$. Finally, we have:

$$\begin{pmatrix} 0 & -\omega_z & \omega_y \\ \omega_z & 0 & -\omega_x \\ -\omega_y & \omega_x & 0 \end{pmatrix} \begin{pmatrix} \omega_x \\ \omega_y \\ \omega_z \end{pmatrix} = \begin{pmatrix} 0 \\ 0 \\ 0 \end{pmatrix}$$

and therefore the eigenvector associated with 0 is ω. The matrix $\mathbf{Ad}(\omega)$ is associated with a speed vector field of a turning coordinate system with ω. Since the axis ω does not move, we will have $\mathbf{Ad}(\omega) \cdot \omega = \mathbf{0}$.

i) We will show that $\mathbf{x} \perp (\omega \wedge \mathbf{x})$. For this, it is sufficient to prove that $\mathbf{x}^{\mathrm{T}}\mathbf{Ad}(\omega)\,\mathbf{x} = 0$. Thus, we will have $\mathbf{x} \perp \mathbf{Ad}(\omega)\,\mathbf{x}$. We have:

$$
\begin{aligned}
2\mathbf{x}^{\mathrm{T}}\mathbf{Ad}(\omega)\,\mathbf{x} &= \mathbf{x}^{\mathrm{T}}\mathbf{Ad}(\omega)\,\mathbf{x} + \mathbf{x}^{\mathrm{T}}\mathbf{Ad}^{\mathrm{T}}(\omega)\,\mathbf{x} \text{ (since they are scalar)} \\
&= \mathbf{x}^{\mathrm{T}}\left(\mathbf{Ad}(\omega) + \mathbf{Ad}^{\mathrm{T}}(\omega)\right)\mathbf{x} \\
&= 0 \text{(since } \mathbf{Ad}(\omega) \text{ is antisymmetric)}
\end{aligned}
$$

ii) We will show that $\omega \perp (\omega \wedge \mathbf{x})$. We have:

$$
\omega^{\mathrm{T}}\mathbf{Ad}(\omega)\,\mathbf{x} = \mathbf{x}^{\mathrm{T}}\mathbf{Ad}^{\mathrm{T}}(\omega)\,\omega
$$

However, ω is the eigenvector associated with 0 of the matrix $\mathbf{Ad}(\omega)$ and therefore also for its transpose. Therefore, $\mathbf{Ad}^{\mathrm{T}}(\omega) \cdot \omega = \mathbf{0}$, which gives us $\omega \perp \left(\mathbf{Ad}^{\mathrm{T}}(\omega) \cdot \omega\right) = 0$.

iii) It is easily shown that:

$$
\det\left(\begin{array}{c|c|c} \omega & \mathbf{x} & \omega \wedge \mathbf{x} \end{array}\right) = \|\omega \wedge \mathbf{x}\|^2
$$

For this, we need to develop the two expressions above and verify their equality. The positivity of the determinant implies that the trihedron $(\omega, \mathbf{x}, \omega \wedge \mathbf{x})$ is direct.

2) The parallelepiped carried by $\omega, \mathbf{x}$ and $\omega \wedge \mathbf{x}$ has a volume of:

$$
v = \det\left(\begin{array}{c|c|c} \omega & \mathbf{x} & \omega \wedge \mathbf{x} \end{array}\right) = \|\omega \wedge \mathbf{x}\|^2
$$

However, since $\omega \wedge \mathbf{x}$ is orthogonal to $\mathbf{x}$ and ω, the volume of the parallelepiped is equal to the surface of its basis $\mathcal{A}$ multiplied by its height $h = \|\omega \wedge \mathbf{x}\|$, in other words:

$$
v = \mathcal{A} \cdot \|\omega \wedge \mathbf{x}\|
$$

By equating the two expressions for v, we obtain $\mathcal{A} = \|\omega \wedge \mathbf{x}\|$.

CORRECTION FOR EXERCISE 1.2.– (Jacobi identity)

1) We have:

$$\underbrace{\mathbf{a} \wedge (\mathbf{b} \wedge \mathbf{c})}_{=\mathbf{Ad}(\mathbf{a})\mathbf{Ad}(\mathbf{b})\cdot\mathbf{c}} + \underbrace{\mathbf{c} \wedge (\mathbf{a} \wedge \mathbf{b})}_{=-(\mathbf{a}\wedge\mathbf{b})\wedge\mathbf{c}=-\mathbf{Ad}(\mathbf{a}\wedge\mathbf{b})\cdot\mathbf{c}} + \underbrace{\mathbf{b} \wedge (\mathbf{c} \wedge \mathbf{a})}_{=-\mathbf{b}\wedge(\mathbf{a}\wedge\mathbf{c})=-\mathbf{Ad}(\mathbf{b})\mathbf{Ad}(\mathbf{a})\cdot\mathbf{c}} = \mathbf{0}$$

Therefore, we have, for all $\mathbf{c}$:

$$\mathbf{Ad}(\mathbf{a})\,\mathbf{Ad}(\mathbf{b}) \cdot \mathbf{c} - \mathbf{Ad}(\mathbf{a} \wedge \mathbf{b}) \cdot \mathbf{c} - \mathbf{Ad}(\mathbf{b})\,\mathbf{Ad}(\mathbf{a}) \cdot \mathbf{c} = \mathbf{0}$$

in other words:

$$\mathbf{Ad}(\mathbf{a} \wedge \mathbf{b}) = \mathbf{Ad}(\mathbf{a})\,\mathbf{Ad}(\mathbf{b}) - \mathbf{Ad}(\mathbf{b})\,\mathbf{Ad}(\mathbf{a})$$

2) We simply have:

$$\mathbf{Ad}(\mathbf{a} \wedge \mathbf{b}) = \mathbf{Ad}(\mathbf{a})\,\mathbf{Ad}(\mathbf{b}) - \mathbf{Ad}(\mathbf{b})\,\mathbf{Ad}(\mathbf{a}) = [\mathbf{Ad}(\mathbf{a}), \mathbf{Ad}(\mathbf{b})]$$

3) The verification is trivial and we will not give it here. Let us note that this result enables us to deduce that the set of skew-symmetric matrices equipped with addition, bracket and standard outer product is also a Lie algebra.

CORRECTION FOR EXERCISE 1.3.– (Varignon's formula)

The position of a point $\mathbf{x}$ of the solid body satisfies the state equation:

$$\dot{\mathbf{x}} = \mathbf{Ad}(\omega) \cdot \mathbf{x}$$

where ω is parallel to the axis of rotation Δ and $||\omega||$ is the rotation speed of the solid (in rad.s^{-1}). After integrating this linear state equation, we find:

$$\mathbf{x}(t) = e^{\mathbf{Ad}(\omega)t} \cdot \mathbf{x}(0)$$

We could also have obtained this formula by applying Rodrigues' formula studied in exercise 1.4. This property can be interpreted by the fact that $\mathbf{Ad}(\omega)$ represents a rotation movement, whereas its exponential represents the result of this movement (in other words, a rotation).

CORRECTION FOR EXERCISE 1.4.– (Rodrigues' formula)

1) It is sufficient to verify that:

$$\omega \wedge \mathbf{x} = \begin{pmatrix} 0 & -\omega_z & \omega_y \\ \omega_z & 0 & -\omega_x \\ -\omega_y & \omega_x & 0 \end{pmatrix} \mathbf{x}$$

2) The solution of the state equation is:

$$\mathbf{x}(t) = e^{\mathbf{A}t}\mathbf{x}(0)$$

3) At time t, the solid body has turned by an angle of $||\omega|| \cdot t$ and therefore for $t = 1$, it has turned by $||\omega||$. Therefore, the rotation $\mathbf{R}$ of angle $||\omega||$ around ω is given by:

$$\mathbf{R} = e^{\mathbf{A}} = e^{\omega \wedge}$$

4) The characteristic polynomial of $\mathbf{A}$ is $s^3 + \left(\omega_x^2 + \omega_y^2 + \omega_z^2\right) s$. The eigenvalues of $\mathbf{A}$ are $0, i||\omega||, -i||\omega||$. The eigenvectors associated with the eigenvalue 0 are collinear to ω. This is logical since the points of the axis of rotation have zero speed.

5) The eigenvalues of $\mathbf{R}$ are obtained using the eigenvalue correspondence theorem and are therefore equal to $e^0, e^{i||\omega||}, e^{-i||\omega||}$.

6) The expression of a rotation around the vector $\omega = (1,\ 0,\ 0)$ of angle α is:

$$\mathbf{R} = \exp \begin{pmatrix} 0 & 0 & 0 \\ 0 & 0 & -\alpha \\ 0 & \alpha & 0 \end{pmatrix} = \begin{pmatrix} 1 & 0 & 0 \\ 0 & \cos\alpha & -\sin\alpha \\ 0 & \sin\alpha & \cos\alpha \end{pmatrix}$$

7) Rodrigues' formula tells us that the rotation matrix around the vector ω of angle $\varphi = ||\omega||$ is given by:

$$\mathbf{R}_\omega = \exp \begin{pmatrix} 0 & -\omega_z & \omega_y \\ \omega_z & 0 & -\omega_x \\ -\omega_y & \omega_x & 0 \end{pmatrix}$$

The MATLAB script returning the Euler matrix is, therefore, the following:

```
eulermat(phi,theta,psi)
Apsi=psi*[0 -1 0;1 0 0;0 0 0];
Atheta=theta*[0 0 1;0 0 0;-1 0 0];
Aphi=phi*[0 0 0;0 0 -1;0 1 0];
R=expm(Apsi)*expm(Atheta)*expm(Aphi);
```

CORRECTION FOR EXERCISE 1.5.– (Geometric approach to Rodrigues' formula)

1) We have:

$$\begin{aligned}\mathcal{R}_{\mathbf{n},\varphi}(\mathbf{u}) &= \mathcal{R}_{\mathbf{n},\varphi}\left(\mathbf{u}_{||}+\mathbf{u}_{\perp}\right)\\ &= \mathcal{R}_{\mathbf{n},\varphi}\left(\mathbf{u}_{||}\right)+\mathcal{R}_{\mathbf{n},\varphi}\left(\mathbf{u}_{\perp}\right) \text{ (by linearity of the rotation operator)}\\ &= \mathbf{u}_{||}+(\cos\varphi)\,\mathbf{u}_{\perp}+(\sin\varphi)\,(\mathbf{n}\wedge\mathbf{u}_{\perp}) \text{ (see figure, on the right)}\\ &= <\mathbf{u},\mathbf{n}>\cdot\mathbf{n}+(\cos\varphi)\,(\mathbf{u}-<\mathbf{u},\mathbf{n}>\cdot\mathbf{n})\\ &\quad +(\sin\varphi)\,(\mathbf{n}\wedge(\mathbf{u}-<\mathbf{u},\mathbf{n}>\cdot\mathbf{n}))\\ &= <\mathbf{u},\mathbf{n}>\cdot\mathbf{n}+(\cos\varphi)\,(\mathbf{u}-<\mathbf{u},\mathbf{n}>\cdot\mathbf{n})+(\sin\varphi)\,(\mathbf{n}\wedge\mathbf{u})\end{aligned}$$

Hence Rodrigues' formula:

$$\mathcal{R}_{\mathbf{n},\varphi}(\mathbf{u}) = (\cos\varphi)\cdot\mathbf{u}+(1-\cos\varphi)\,(<\mathbf{u},\mathbf{n}>\cdot\mathbf{n})+(\sin\varphi)\,(\mathbf{n}\wedge\mathbf{u})$$

2) We have:

$$\begin{aligned}\mathbf{n}\wedge(\mathbf{n}\wedge\mathbf{u}) &= \left(\mathbf{n}^{\mathrm{T}}\mathbf{u}\right)\cdot\mathbf{n}-\left(\mathbf{n}^{\mathrm{T}}\mathbf{n}\right)\cdot\mathbf{u}\\ &= \left(\mathbf{n}\cdot\mathbf{u}^{\mathrm{T}}\right)\mathbf{n}-||\mathbf{n}||^{2}\mathbf{u} = <\mathbf{u},\mathbf{n}>\mathbf{n}-\mathbf{u}\end{aligned}$$

Therefore:

$$<\mathbf{u},\mathbf{n}>\cdot\mathbf{n} = \mathbf{n}\wedge(\mathbf{n}\wedge\mathbf{u})+\mathbf{u}$$

Therefore, Rodrigues' formula can also be written as:

$$\begin{aligned}\mathcal{R}_{\mathbf{n},\varphi}(\mathbf{u}) &= \mathbf{n}\wedge(\mathbf{n}\wedge\mathbf{u})+\mathbf{u}+(\cos\varphi)\,(\mathbf{u}-\mathbf{n}\wedge(\mathbf{n}\wedge\mathbf{u})+\mathbf{u})\\ &\quad +(\sin\varphi)\,(\mathbf{n}\wedge\mathbf{u})\\ &= \mathbf{u}+(1-\cos\varphi)\,(\mathbf{n}\wedge(\mathbf{n}\wedge\mathbf{u}))+(\sin\varphi)\,(\mathbf{n}\wedge\mathbf{u})\end{aligned}$$

The operator $\mathcal{R}_{\mathbf{n},\varphi}$ can be represented by the following rotation matrix:

$$\mathbf{R}_{\mathbf{n},\varphi} = \mathbf{I} + (1 - \cos\varphi)\left(\mathbf{Ad}^2(\mathbf{n})\right) + (\sin\varphi)(\mathbf{Ad}(\mathbf{n}))$$

or, in a developed form:

$$\mathbf{R}_{\mathbf{n},\varphi} = \begin{pmatrix} 1 & 0 & 0 \\ 0 & 1 & 0 \\ 0 & 0 & 1 \end{pmatrix} + (1 - \cos\varphi)\begin{pmatrix} -n_y^2 - n_z^2 & n_x n_y & n_x n_z \\ n_x n_y & -n_x^2 - n_z^2 & n_y n_z \\ n_x n_z & n_y n_z & -n_x^2 - n_y^2 \end{pmatrix} + (\sin\varphi)\begin{pmatrix} 0 & -n_z & n_y \\ n_z & 0 & -n_x \\ -n_y & n_x & 0 \end{pmatrix}$$

3) We have:

$$\mathbf{R}_{\mathbf{n},\varphi} - \mathbf{R}_{\mathbf{n},\varphi}^{\mathrm{T}} = 2(\sin\varphi)\begin{pmatrix} 0 & -n_z & n_y \\ n_z & 0 & -n_x \\ -n_y & n_x & 0 \end{pmatrix}$$

The vectors $\mathbf{R}_{\mathbf{n},\varphi} \cdot \mathbf{u}$ and $\mathbf{R}_{\mathbf{n},\varphi}^{\mathrm{T}} \cdot \mathbf{u}$ form the two sides of a rhombus (Rodrigues' rhombus) whose vector:

$$\left(\mathbf{R}_{\mathbf{n},\varphi} - \mathbf{R}_{\mathbf{n},\varphi}^{\mathrm{T}}\right)\mathbf{u} = 2(\sin\varphi) \cdot \mathbf{n} \wedge \mathbf{u}$$

corresponds to the diagonal of this rhombus.

4) Let us recall that the Maclaurin series development of the sine and cosine functions is written as:

$$\begin{aligned} \sin\varphi &= \varphi - \tfrac{\varphi^3}{3!} + \tfrac{\varphi^5}{5!} - \tfrac{\varphi^7}{7!} + \ldots \\ \cos\varphi &= 1 - \tfrac{\varphi^2}{2!} + \tfrac{\varphi^4}{4!} - \tfrac{\varphi^6}{6!} + \ldots \end{aligned}$$

Let $\mathbf{H} = \mathbf{Ad}(\mathbf{n})$. Since $\mathbf{n}$ is an eigenvector of $\mathbf{H}$ associated with the zero eigenvalue, we have $\mathbf{H}\left(\mathbf{n} \cdot \mathbf{n}^{\mathrm{T}}\right) = \mathbf{0}$. Moreover:

$$\mathbf{H}^2 = \left(\mathbf{n} \cdot \mathbf{n}^{\mathrm{T}} - \mathbf{I}\right)$$

Therefore:

$$\begin{aligned}
&\mathbf{H}^3 = \mathbf{H} \cdot \left(\mathbf{n} \cdot \mathbf{n}^{\mathrm{T}} - \mathbf{I}\right) = -\mathbf{H} \\
&\mathbf{H}^4 = \mathbf{H} \cdot \mathbf{H}^3 = -\mathbf{H}^2 \\
&\mathbf{H}^5 = \mathbf{H} \cdot \mathbf{H}^4 = \mathbf{H}\left(-\mathbf{H}^2\right) = -\mathbf{H}^3 = \mathbf{H} \\
&\mathbf{H}^6 = \mathbf{H} \cdot \mathbf{H}^5 = \mathbf{H}^2 \\
&\mathbf{H}^7 = \mathbf{H} \cdot \mathbf{H}^6 = \mathbf{H}^3 = -\mathbf{H} \quad \ldots
\end{aligned}$$

Thus, Rodrigues' formula is written as:

$$\begin{aligned}
\mathbf{R}_{\mathbf{n},\varphi} &= \mathbf{I} + (\sin\varphi) \cdot \mathbf{H} + (1 - \cos\varphi) \cdot \mathbf{H}^2 \\
&= \mathbf{I} + \left(\varphi - \frac{\varphi^3}{3!} + \frac{\varphi^5}{5!} - \frac{\varphi^7}{7!} + \ldots\right) \cdot \mathbf{H} \\
&\quad + \left(\frac{\varphi^2}{2!} - \frac{\varphi^4}{4!} + \frac{\varphi^6}{6!} + \ldots\right) \cdot \mathbf{H}^2 \\
&= \mathbf{I} + \varphi \cdot \mathbf{H} + \frac{\varphi^2}{2!} \cdot \mathbf{H}^2 - \frac{\varphi^3}{3!} \cdot \mathbf{H} - \frac{\varphi^4}{4!} \cdot \mathbf{H}^2 + \frac{\varphi^5}{5!} \cdot \mathbf{H} \\
&\quad + \frac{\varphi^6}{6!} \cdot \mathbf{H}^2 - \frac{\varphi^7}{7!} \cdot \mathbf{H} + \ldots \\
&= \mathbf{I} + \varphi \cdot \mathbf{H} + \frac{\varphi^2}{2!} \cdot \mathbf{H}^2 + \frac{\varphi^3}{3!} \cdot \mathbf{H}^3 + \frac{\varphi^4}{4!} \cdot \mathbf{H}^4 + \frac{\varphi^5}{5!} \cdot \mathbf{H}^5 \\
&\quad + \frac{\varphi^6}{6!} \cdot \mathbf{H}^6 + \ldots \\
&= \exp(\varphi \cdot \mathbf{H})
\end{aligned}$$

In other words:

$$\mathbf{R}_{\mathbf{n},\varphi} = \exp\left(\varphi \cdot \mathbf{Ad}\left(\mathbf{n}\right)\right) = \exp\left(\varphi \cdot \mathbf{n}\wedge\right)$$

CORRECTION FOR EXERCISE 1.6.– (Schuler oscillations)

1) The state vector is $\mathbf{x} = \left(\theta,\ \alpha,\ \dot{\theta},\ \dot{\alpha}\right)$. In order to have a horizontal movement, we need a horizontal force f. Following the fundamental principle of dynamics, we have:

$$J\left(\ddot{\theta} + \ddot{\alpha}\right) = -mg\ell_1 \sin\alpha + mg\ell_2 \sin\alpha - f\,\frac{\ell_1 - \ell_2}{2}\,\cos\alpha$$

with $J = m\left(\ell_1^2 + \ell_2^2\right)$ and $f = 2ma$. Since $\ddot{\theta} = \frac{a}{r}$, the state equations of the system are written as:

$$\begin{cases} \dot{\theta} = \dot{\theta} \\ \dot{\alpha} = \dot{\alpha} \\ \ddot{\theta} = \frac{a}{r} \\ \ddot{\alpha} = -\frac{a}{r} + \frac{\ell_2 - \ell_1}{\ell_1^2 + \ell_2^2}\left(g \sin\alpha - a\cos\alpha\right) \end{cases}$$

2) The pendulum remains horizontal if, for $\alpha = 0$, we have $\ddot{\alpha} = 0$. In other words:

$$-\frac{a}{r} + \frac{\ell_2 - \ell_1}{\ell_1^2 + \ell_2^2}\left(g \sin\alpha - a\cos\alpha\right) = 0$$

or equivalently:

$$-\frac{\ell_2 - \ell_1}{\ell_1^2 + \ell_2^2} = \frac{1}{r}$$

Therefore, we must satisfy the equation $\ell_2^2 + r\ell_2 - \ell_1 r + \ell_1^2 = 0$. Solving it gives:

$$\ell_2 = \frac{-r + \sqrt{r^2 + 4\left(\ell_1 r - \ell_1^2\right)}}{2}$$

For $\ell_1 = 1$, we obtain:

$$\ell_2 = \frac{-64{\cdot}10^5 + \sqrt{(64{\cdot}10^5)^2 + 4(64{\cdot}10^5 - 1)}}{2} = 1 - 3.1 \times 10^{-7}\text{m}$$

3) The equation describing the oscillations is:

$$\ddot{\alpha} = -\frac{a}{r} - \frac{1}{r}\left(g \sin\alpha - a\cos\alpha\right)$$

For $a = 0$, we have:

$$\ddot{\alpha} = -\frac{g}{r}\sin\alpha$$

which is the equation of a pendulum of length $\ell = r$. By linearizing this equation, we obtain the characteristic polynomial $s^2 + \frac{g}{r}$ and therefore the pulse $\omega = \sqrt{\frac{g}{r}}$. The Schuler period is, therefore, equal to:

$$T = 2\pi\sqrt{\frac{r}{g}} = 5072\,\text{sec} = 84\,\text{min}$$

4) The program, which can be found in `schuler.m`, is the following:

```
r=10; l1=1; l2=(-r+sqrt(r^2+4*(l1*r-l1^2)))/2;
dt=0.05; x=[1;0.1;0;0];
for t=0:dt:10,
a=randn(1);
x=x+dt*(0.25*f(x,a)+0.75*(f(x+dt*(2/3)*f(x,a),a)));
end
```

Note that for an initialization $\alpha = \dot{\alpha} = 0$, the pendulum always points toward the center of the Earth. Otherwise, it oscillates and conserves this oscillation at the Schuler frequency. This oscillation can be observed in modern inertial units and there are methods for compensating for it using information gathered by the other inertial sensors.

CORRECTION FOR EXERCISE 1.7.– (Brake detector)

1) We have:

$$\begin{aligned} &\mathbf{o}_0\mathbf{m} = \mathbf{o}_0\mathbf{o}_1 + \mathbf{o}_1\mathbf{m} \\ \Leftrightarrow\ & a_0\mathbf{i}_0 + b_0\mathbf{j}_0 = x\mathbf{i}_0 + y\mathbf{j}_0 + a_1\mathbf{i}_1 + b_1\mathbf{j}_1 \end{aligned}$$

Let us express this relation in the coordinate system $\mathcal{R}_0$:

$$\begin{pmatrix} a_0 \\ b_0 \end{pmatrix} = \begin{pmatrix} x \\ y \end{pmatrix} + a_1 \begin{pmatrix} \cos\theta \\ \sin\theta \end{pmatrix} + b_1 \begin{pmatrix} -\sin\theta \\ \cos\theta \end{pmatrix}$$

which leads to:

$$\begin{pmatrix} a_0 \\ b_0 \end{pmatrix} = \begin{pmatrix} x \\ y \end{pmatrix} + \underbrace{\begin{pmatrix} \cos\theta & -\sin\theta \\ \sin\theta & \cos\theta \end{pmatrix}}_{\mathbf{R}_\theta} \begin{pmatrix} a_1 \\ b_1 \end{pmatrix}$$

2) We have:

$$\mathbf{R}_\theta^\mathrm{T} \cdot \dot{\mathbf{R}}_\theta = \dot{\theta} \cdot \begin{pmatrix} \cos\theta & \sin\theta \\ -\sin\theta & \cos\theta \end{pmatrix} \begin{pmatrix} -\sin\theta & -\cos\theta \\ \cos\theta & -\sin\theta \end{pmatrix} = \dot{\theta} \begin{pmatrix} 0 & -1 \\ 1 & 0 \end{pmatrix}$$

3) In the coordinate system $\mathcal{R}_0$, this vector $\mathbf{u}$ is expressed by:

$$\mathbf{u}_{|\mathcal{R}_0} = \begin{pmatrix} \dot{a}_0 \\ \dot{b}_0 \end{pmatrix} = \begin{pmatrix} \dot{x} \\ \dot{y} \end{pmatrix} + \dot{\mathbf{R}}_\theta \begin{pmatrix} a_1 \\ b_1 \end{pmatrix} + \mathbf{R}_\theta \begin{pmatrix} \dot{a}_1 \\ \dot{b}_1 \end{pmatrix}$$

Therefore:

$$\mathbf{u}_{|\mathcal{R}_1} = \mathbf{R}_\theta^\mathrm{T} \cdot \mathbf{u}_{|\mathcal{R}_0} = \mathbf{R}_\theta^\mathrm{T} \cdot \begin{pmatrix} \dot{x} \\ \dot{y} \end{pmatrix} + \mathbf{R}_\theta^\mathrm{T} \cdot \dot{\mathbf{R}}_\theta \begin{pmatrix} a_1 \\ b_1 \end{pmatrix} + \begin{pmatrix} \dot{a}_1 \\ \dot{b}_1 \end{pmatrix}$$

in other words:

$$\begin{aligned} \mathbf{u}_{|\mathcal{R}_1} &= \mathbf{R}_\theta^\mathrm{T} \cdot \begin{pmatrix} \dot{x} \\ \dot{y} \end{pmatrix} + \dot{\theta} \begin{pmatrix} -b_1 \\ a_1 \end{pmatrix} + \begin{pmatrix} \dot{a}_1 \\ \dot{b}_1 \end{pmatrix} \\ &= \begin{pmatrix} v \\ 0 \end{pmatrix} + \dot{\theta} \cdot \begin{pmatrix} -b_1 \\ a_1 \end{pmatrix} + \begin{pmatrix} \dot{a}_1 \\ \dot{b}_1 \end{pmatrix} = \begin{pmatrix} v - \dot{\theta} b_1 + \dot{a}_1 \\ \dot{\theta} a_1 + \dot{b}_1 \end{pmatrix} \end{aligned}$$

4) We have:

$$\begin{aligned} \mathbf{a}_{|\mathcal{R}_0} &= \begin{pmatrix} \ddot{a}_0 \\ \ddot{b}_0 \end{pmatrix} = \frac{d}{dt} \left(\begin{pmatrix} \dot{x} \\ \dot{y} \end{pmatrix} + \dot{\mathbf{R}}_\theta \begin{pmatrix} a_1 \\ b_1 \end{pmatrix} + \mathbf{R}_\theta \begin{pmatrix} \dot{a}_1 \\ \dot{b}_1 \end{pmatrix} \right) \\ &= \begin{pmatrix} \ddot{x} \\ \ddot{y} \end{pmatrix} + \ddot{\mathbf{R}}_\theta \begin{pmatrix} a_1 \\ b_1 \end{pmatrix} + 2\dot{\mathbf{R}}_\theta \begin{pmatrix} \dot{a}_1 \\ \dot{b}_1 \end{pmatrix} + \mathbf{R}_\theta \begin{pmatrix} \ddot{a}_1 \\ \ddot{b}_1 \end{pmatrix} \end{aligned}$$

Therefore:

$$\begin{aligned} \mathbf{a}_{|\mathcal{R}_1} &= \mathbf{R}_\theta^\mathrm{T} \, \mathbf{a}_{|\mathcal{R}_0} \\ &= \mathbf{R}_\theta^\mathrm{T} \begin{pmatrix} \ddot{x} \\ \ddot{y} \end{pmatrix} + \mathbf{R}_\theta^\mathrm{T} \ddot{\mathbf{R}}_\theta \begin{pmatrix} a_1 \\ b_1 \end{pmatrix} + 2\mathbf{R}_\theta^\mathrm{T} \dot{\mathbf{R}}_\theta \begin{pmatrix} \dot{a}_1 \\ \dot{b}_1 \end{pmatrix} + \begin{pmatrix} \ddot{a}_1 \\ \ddot{b}_1 \end{pmatrix} \end{aligned}$$

However:

$$\mathbf{R}_\theta^{\mathrm{T}}\ddot{\mathbf{R}}_\theta = \mathbf{R}_\theta^{\mathrm{T}}\frac{d}{dt}\dot{\mathbf{R}}_\theta = \mathbf{R}_\theta^{\mathrm{T}}\frac{d}{dt}\left(\dot{\theta}\begin{pmatrix} -\sin\theta & -\cos\theta \\ \cos\theta & -\sin\theta \end{pmatrix}\right)$$

$$= \begin{pmatrix} \cos\theta & \sin\theta \\ -\sin\theta & \cos\theta \end{pmatrix}\left(\ddot{\theta}\begin{pmatrix} -\sin\theta & -\cos\theta \\ \cos\theta & -\sin\theta \end{pmatrix} + \dot{\theta}^2\begin{pmatrix} -\cos\theta & \sin\theta \\ -\sin\theta & -\cos\theta \end{pmatrix}\right)$$

$$= \ddot{\theta}\begin{pmatrix} 0 & -1 \\ 1 & 0 \end{pmatrix} - \dot{\theta}^2\begin{pmatrix} 1 & 0 \\ 0 & 1 \end{pmatrix} = \begin{pmatrix} -\dot{\theta}^2 & -\ddot{\theta} \\ \ddot{\theta} & -\dot{\theta}^2 \end{pmatrix}$$

Finally:

$$\mathbf{a}_{|\mathcal{R}_1} = \begin{pmatrix} \alpha \\ \beta \end{pmatrix} + \begin{pmatrix} -\dot{\theta}^2 & -\ddot{\theta} \\ \ddot{\theta} & -\dot{\theta}^2 \end{pmatrix}\begin{pmatrix} a_1 \\ b_1 \end{pmatrix} + 2\dot{\theta}\begin{pmatrix} -\dot{b}_1 \\ \dot{a}_1 \end{pmatrix} + \begin{pmatrix} \ddot{a}_1 \\ \ddot{b}_1 \end{pmatrix}$$

$$= \begin{pmatrix} \alpha - \dot{\theta}^2 a_1 - \ddot{\theta} b_1 - 2\dot{\theta}\dot{b}_1 + \ddot{a}_1 \\ \beta + \ddot{\theta} a_1 - \dot{\theta}^2 b_1 + 2\dot{\theta}\dot{a}_1 + \ddot{b}_1 \end{pmatrix}$$

5) The vehicle in front of us is braking if:

$$< \mathbf{a}_{|\mathcal{R}_1}, \mathbf{u}_{|\mathcal{R}_1} > \leq 0$$

in other words if:

$$\begin{pmatrix} \alpha - \dot{\theta}^2 a_1 - \ddot{\theta} b_1 - 2\dot{\theta}\dot{b}_1 + \ddot{a}_1 \\ \beta + \ddot{\theta} a_1 - \dot{\theta}^2 b_1 + 2\dot{\theta}\dot{a}_1 + \ddot{b}_1 \end{pmatrix}^{\mathrm{T}} \cdot \begin{pmatrix} v - \dot{\theta} b_1 + \dot{a}_1 \\ \dot{\theta} a_1 + \dot{b}_1 \end{pmatrix} \leq 0$$

CORRECTION FOR EXERCISE 1.8.– (Modeling an underwater robot)

The derivative of the position vector is obtained by noticing that:

$$\dot{\mathbf{p}} = v\mathbf{i}_1 \overset{[1.7]}{=} v\begin{pmatrix} \cos\theta\cos\psi \\ \cos\theta\sin\psi \\ -\sin\theta \end{pmatrix}$$

since $\mathbf{i}_1$ corresponds to the first column of matrix [1.7]. Finally, by taking into account equations [1.11], we can write the state equations for our submarine:

$$\begin{cases} \dot{p}_x &= v\cos\theta\cos\psi \\ \dot{p}_y &= v\cos\theta\sin\psi \\ \dot{p}_z &= -v\sin\theta \\ \dot{v} &= u_1 \\ \dot{\psi} &= \frac{\sin\varphi}{\cos\theta}\cdot vu_2 + \frac{\cos\varphi}{\cos\theta}\cdot vu_3 \\ \dot{\theta} &= \cos\varphi\cdot vu_2 - \sin\varphi\cdot vu_3 \\ \dot{\varphi} &= vu_4 + \tan\theta\cdot(\sin\varphi\cdot vu_2 + \cos\varphi\cdot vu_3) \end{cases}$$

Here, we have a kinematic model (in other words, it does not involve forces or torques). There are no parameters and the model can, therefore, be considered correct if the underwater robot is solid (that is it cannot be contorted), and if its trajectory is tangent to the axis of the robot. Such a model allows us to use nonlinear control methods such as feedback linearization which will be discussed in Chapter 2. These methods are indeed less robust with respect to a model error, but are formidably efficient if an accurate model for our system is known.

CORRECTION FOR EXERCISE 1.9.– (3D robot graphics)

2) We now give the MATLAB function drawing the 3D graphic:

```
function draw(x)
Auv0=[ 0 0 10 0 0 10 0 0;
-1 1 0 -1 -0.2 0 0.2 1;
0 0 0 0 1 0 1 0;
1 1 1 1 1 1 1 1 ];
E=eulermat(x(7),x(6),x(5)); %phi,theta,psi
R=[E,[x(1);x(2);x(3)];0 0 0 1];
Auv=R*Auv0;
plot3(Auv(1,:),Auv(2,:),Auv(3,:),'blue');
plot3(Auv(1,:),Auv(2,:),0*Auv(3,:),'black'); % ombre
```

4) In order to simulate our robot, we first need to write the following evolution function:

```
function xdot = f(x,u)
v=x(4); psi=x(5); theta=x(6); phi=x(7);
xdot=[v*cos(theta)*cos(psi);
```

```
v*cos(theta)*sin(psi);
-v*sin(theta); u(1);
(sin(phi)/cos(theta))*v*u(2)+(cos(phi)/cos(theta))*v*u(3);
cos(phi)*v*u(2)-sin(phi)*v*u(3);
-0.1*sin(phi)*cos(theta)+tan(theta)*(sin(phi)*v*u(2)
+cos(phi)*v*u(3))];
```

The simulation could be done by using Euler's method. This simulation will be picked up in exercise 2.4 in section 2.8 for performing control (refer to the file `auv3d.m` as well):

```
dt=0.1;
x=[0;0;10;0.1;0;0;0];
for t=0:dt:10,
u=...;
x=x+dt*f(x,u);
draw(x);
end
```

CORRECTION FOR EXERCISE 1.10.–

In order to proceed correctly, we must draw the arms one after the other and draw the associated coordinate systems. For drawing the coordinate system associated with a 4×4 homogeneous transformation matrix, we can use the following function:

```
function drawaxis(R)
A=R*[0 1;0 0; 0 0; 1 1]; plot3(A(1,:),A(2,:),A(3,:),'red');
% axes des x
A=R*[0 0;0 1; 0 0; 1 1];
plot3(A(1,:),A(2,:),A(3,:),'green'); % axes des y
A=R*[0 0;0 0; 0 1; 1 1];
plot3(A(1,:),A(2,:),A(3,:),'blue'); % axes des z
end
```

Each of the seven arms can be drawn by connecting a triangle of the plane xy of the coordinate system i to its counterpart in the coordinate system $i+1$. This can be done by the following procedure:

```
function drawArm(R1,R2)
J0=[-0.1 0.3 -0.1 -0.1; -0.1 -0.1 3*0.1 -0.1; 0 0 0 0; 1 1
1 1];
J1=R1*J0; J2=R2*J0;
```

```
J=[J1(:,1),J2(:,1),J2(:,2),J1(:,2),J1(:,3),J2(:,3),...
J2(:,1),J2(:,2),J2(:,3),J1(:,3),J1(:,1),J1(:,2)];
plot3(J(1,:),J(2,:),J(3,:),'black');
end
```

Finally, the seven arms of the robot in a configuration $\mathbf{q}$ (whose components are the coordinates of the articulations) can be drawn as follows:

```
function draw(q)
R=eye(4,4); drawaxis(R);
for j=1:7,
Rold=R;
R=R*Transl([0 0 r(j)])*Transl([d(j) 0 0])*Rot([0 a(j) 0]);
drawArm(Rold,R); R=R*Rot([0 0 q(j)]);
drawaxis(R);
end; end
```

What remains is then to configure the robot parameters in the main program and to move the robot in a loop. This may be done as in the following program:

```
a=[0,pi/2,0,-pi/2,-pi/2,-pi/2,0];
d=[0,0.5,0,-0.1,-0.3 -1,0];
r=[1,0.5,1, 0.1,1,0.2,0.2];
q=[0.3;0.3;0.3;0;1.5;0.1;1];
dt=0.05;
for i=1:length(a),
for h=0:dt:2*pi, draw(q); q(i)=q(i)+dt; end;
end
```

The whole program is given in the file `staubli.m`.

CORRECTION FOR EXERCISE 1.11.– (3D modeling of a wheel)

1) The state variables we will choose are (1) the coordinates (x, y) of the point of contact, (2) the rotation vector ω expressed in the absolute coordinate system and (3) the orientation (ψ, θ, φ) of the wheel. Thus, we have a system of order 8. Following the equation of the gyro, (or the fundamental principle of dynamics for rotating bodies), the torque $\mathbf{c}$ exerted by the external forces is proportional to the differential of the rotation vector ω following the relation:

$$\mathbf{c} = \mathbf{I}\dot{\omega}$$

where $\mathbf{I}$ is the inertia matrix. By taking as point of reference the center of gravity of the wheel, we can see that the torque $\mathbf{c}$ which makes the wheel turn is generated by the reaction $\mathbf{r}$ of the ground applied at the point of contact $\mathbf{p}$. Let us place ourselves in the plane orthogonal to the movement. Following the fundamental principle of dynamics for translations applied at point $\mathbf{g}$:

$$\underbrace{\begin{pmatrix} r_x \\ r_z - mg \end{pmatrix}}_{\text{forces}} = m \underbrace{\begin{pmatrix} v\omega_z \\ 0 \end{pmatrix}}_{\text{acceleration of } \mathbf{p}} + \underbrace{m\frac{\dot{\theta}^2}{\rho}\begin{pmatrix} -\sin\theta \\ -\cos\theta \end{pmatrix}}_{\text{acceleration of } \mathbf{g}-\mathbf{p}}$$

In other words:

$$\begin{pmatrix} r_x \\ r_y \end{pmatrix} = m \begin{pmatrix} v\omega_z - \frac{\dot{\theta}^2}{\rho}\sin\theta \\ g - \frac{\dot{\theta}^2}{\rho}\cos\theta \end{pmatrix}$$

The torque $\mathbf{c}$ exerted by $\mathbf{r}$ is, therefore, given by:

$$\mathbf{c} = \underbrace{\rho\left(r_x\cos\theta - r_z\sin\theta\right)}_{=\rho m(v\omega_z\cos\theta - g\sin\theta)} \cdot \underbrace{\begin{pmatrix} \sin\psi \\ -\cos\psi \\ 0 \end{pmatrix}}_{\mathbf{n}}$$

Moreover, following relation [1.10], we have:

$$\begin{pmatrix} \dot{\psi} \\ \dot{\theta} \\ \dot{\varphi} \end{pmatrix} = \begin{pmatrix} 0 & -\sin\psi & \cos\theta\cos\psi \\ 0 & \cos\psi & \cos\theta\sin\psi \\ 1 & 0 & -\sin\theta \end{pmatrix}^{-1} \cdot \omega$$

Finally, the speed at the point of contact is $\rho \cdot \dot{\varphi}$. This point is moving in the direction of $\mathbf{n}$, which gives us:

$$\begin{pmatrix} \dot{x} \\ \dot{y} \end{pmatrix} = \rho \cdot \dot{\varphi} \cdot \mathbf{n} = \begin{pmatrix} v\sin\psi \\ -v\cos\psi \end{pmatrix}$$

Finally, the state equations are:

$$\begin{pmatrix} \dot{x} \\ \dot{y} \\ \omega \\ \dot{\psi} \\ \dot{\theta} \\ \dot{\varphi} \end{pmatrix} = \begin{pmatrix} \rho \cdot \dot{\varphi} \sin \psi \\ -\rho \cdot \dot{\varphi} \cos \psi \\ m \left(\rho^2 \dot{\varphi} \omega_z \cos \theta - g \rho \sin \theta \right) \cdot \mathbf{I}^{-1} \cdot \begin{pmatrix} \sin \psi \\ -\cos \psi \\ 0 \end{pmatrix} \\ \begin{pmatrix} 0 & -\sin \psi & \cos \theta \cos \psi \\ 0 & \cos \psi & \cos \theta \sin \psi \\ 1 & 0 & -\sin \theta \end{pmatrix}^{-1} \cdot \omega \end{pmatrix}$$

Even though $\dot{\varphi}$ appears on the right-hand side, these are indeed explicit state equations since $\dot{\varphi}$ is expressed, by the last of these six equations, as a function of the state variables. The evolution function can be programmed in MATLAB as follows (refer to the file `wheel.m`):

```
function xdot=f(x)
w=x(3:5); m=1; g=9.81;
psi=x(6); theta=x(7);
I=diag([1 0.5 0.5])*(m/2)*rho^2
n=[sin(psi);-cos(psi);0];
deuler=[0 -sin(psi) cos(theta)*cos(psi);
0 cos(psi) cos(theta)*sin(psi);
1 0 -sin(theta) ]\w;
dphi=deuler(3);
dw=m*(rho^2*dphi*w(3)*cos(theta)-g*rho*sin(theta))*inv(I)*n;
xdot=[rho*dphi*n(1:2);dw;deuler];
end
```

2) If, for instance, we displace the masses from the center, the wheel tends to accelerate, given the principle of conservation of angular momentum. We could use mass transfers in order to control the vector ω of the wheel and therefore be able to control its trajectory.

CORRECTION FOR EXERCISE 1.12.– (Mechanical stability of an underwater robot)

1) Given that the body is homogeneous, with the same density as water, Archimedes' principle compensates for its weight. Everything is as if the body was in space. Once submerged, it does not turn if $(\mathbf{a} - \mathbf{g}) \wedge \mathbf{f} = \mathbf{0}$.

2) We will have a rotational stability if, in addition to the equilibrium relation, we have $\langle \mathbf{a} - \mathbf{g}, \mathbf{f} \rangle > 0$.

3) The center of buoyancy $\mathbf{a}$ (where Archimedes' force $\mathbf{f}$ is exerted, which is vertical and pointing upward) is behind the center of gravity. The masses must be moved back in order to have the horizontal condition of equilibrium.

4) Here, $\mathbf{f}$ corresponds to Archimedes' force. Its point of application $\mathbf{a}$ is below $\mathbf{g}$ and the condition for stability is not satisfied. We must bring $\mathbf{g}$ down by moving the masses.

5) Here, the force $\mathbf{f}$ is the drag (resistance of water). The center of drag (which is too low here) must be in the same horizontal plane as $\mathbf{g}$ (stability). Less resistance to water must be presented in the lower part of the robot. The stability also has to be taken into account. The center of drag (be careful, this concept only makes sense in the vicinity of a direction) has to be behind $\mathbf{g}$ (principle of arrow feathering). We may, for example, add ailerons to the rear of the robot.

6) The propellers must point to $\mathbf{g}$ in order to avoid rotations (except of course when the aim is to turn). In order to be more stable, we could think that the propellers should be placed at the front, in other words in the direction of the movement. However, this will not work since the propellers turn with the robot. Therefore, there is neither stability, nor instability.

7) The compass is disturbed by the metal. When the robot is close to the wall, it wants to regulate itself toward the wrong direction. This way, it moves the compass away from the wall, where it is no longer disturbed. Thus, a cycle is created. We can avoid this by merging (using a Kalman filter) the compass data with that of the gyros.

8) Configuration (a) behaves in the same way as configuration (b) since the momentums relative to the center of the hull are identical. Configuration (c) is more maneuverable than configuration (d) since its momentums caused by the propellers are greater.

2

Feedback Linearization

Due to their multiple rotation capabilities, robots are considered to be strongly nonlinear systems. In this chapter, we will look at designing nonlinear controllers in order to constrain the state vector of the robot to follow a fixed forward path or to remain within a determined area of its workspace. In contrast to the linear approach, which offers a general methodology but is limited to the neighborhood of a point of the state space [KAI 80, JAU 15], nonlinear approaches only apply to limited classes of systems, but they allow us to extend the effective operating range of the system. Indeed, there is no general method of globally stabilizing nonlinear systems. However, there is a multitude of methods that apply to particular cases [FAN 01]. The aim of this chapter is to present one of the more representative theoretical methods (whereas in the following chapter, we will be looking at more pragmatic approaches). This method is called *feedback linearization* and it requires knowledge of an accurate and reliable state machine for our robot. The robots considered here are mechanical systems whose modeling can be found in [JAU 05]. We will assume in this chapter that the state vector is entirely known. In practice, it has to be approximated from sensor measurements. We will see in Chapter 7 how such an approximation is performed.

2.1. Controlling an integrator chain

As we will show further on in this chapter, feedback linearization leads to the problem of controlling a system which is composed of several integrator chains decoupled from one another. In this section, we will therefore consider an integrator chain whose input u and output y are linked together by the differential equation:

$$y^{(n)} = u$$

2.1.1. *Proportional-derivative controller*

Let us first of all stabilize this system using a *proportional-derivative* controller of the type:

$$u = \alpha_0 (w - y) + \alpha_1 (\dot{w} - \dot{y}) + \cdots + \alpha_{n-1} \left(w^{(n-1)} - y^{(n-1)} \right) + w^{(n)}$$

where w is the desired setpoint for y. Let us note that w may depend on time. The fact that this controller requires the differentials of y is not a problem within the frame defined by the feedback linearization. Indeed, all of these derivatives can be described as analytic functions of the state $\mathbf{x}$ of the system and the input $\mathbf{u}$. Concerning the setpoint $w(t)$, it is chosen by the user and an analytic expression of $w(t)$ may be assumed to be known (for instance, $w(t) = \sin(t)$). Thus, calculating the differentials of w is done in a formal manner and no sensitivity of the differential operator with respect to the noise has to be feared.

The feedback system is described by the differential equation:

$$\begin{aligned} y^{(n)} &= u = \alpha_0 (w - y) + \alpha_1 (\dot{w} - \dot{y}) + \ldots \\ &\quad + \alpha_{n-1} \left(w^{(n-1)} - y^{(n-1)} \right) + w^{(n)} \end{aligned}$$

If we define the error e between the setpoint w and the output y as $e = w - y$, this equation becomes:

$$e^{(n)} + \alpha_{n-1} e^{(n-1)} + \cdots + \alpha_1 \dot{e} + \alpha_0 e = 0$$

This differential equation is called the *error dynamics equation*. Its characteristic polynomial, given by:

$$P(s) = s^n + \alpha_{n-1} s^{n-1} + \cdots + \alpha_1 s + \alpha_0 \qquad [2.1]$$

can thus be chosen arbitrarily among the polynomials of degree n. Of course, we will choose all roots with a negative real part, in order to ensure the stability of the system. For instance, if $n = 3$ and if we want all the poles to be equal to -1, we will take:

$$s^3 + \alpha_2 s^2 + \alpha_1 s + \alpha_0 = (s+1)^3 = s^3 + 3s^2 + 3s + 1$$

where

$$\alpha_2 = 3, \alpha_1 = 3, \alpha_0 = 1$$

The controller obtained is then given by:

$$u = (w - y) + 3(\dot{w} - \dot{y}) + 3(\ddot{w} - \ddot{y}) + \dddot{w}$$

NOTE 2.1.– In this book, we will choose, for reasons of simplicity, to position all our poles at –1. The previous reasoning, applied for various degrees n, leads us to the following controls:

$$\begin{array}{ll} n=1 & u = (w - y) + \dot{w} \\ n=2 & u = (w - y) + 2(\dot{w} - \dot{y}) + \ddot{w} \\ n=3 & u = (w - y) + 3(\dot{w} - \dot{y}) + 3(\ddot{w} - \ddot{y}) + \dddot{w} \\ n=4 & u = (w - y) + 4(\dot{w} - \dot{y}) + 6(\ddot{w} - \ddot{y}) + 4(\dddot{w} - \dddot{y}) + \ddddot{w} \end{array} \quad [2.2]$$

Note that the coefficients correspond to those of Pascal's triangle:

$$\begin{array}{l} 1 \\ 1\ 1 \\ 1\ 2\ 1 \\ 1\ 3\ 3\ 1 \\ 1\ 4\ 6\ 4\ 1 \end{array}$$

2.1.2. *Proportional-integral-derivative controller*

In order to compensate for the constant disturbances, we may decide to add an integral term. We then obtain a proportional-integral-derivative (PID) controller, which is of the form:

$$\begin{aligned} = {} & \alpha_{-1} \int_{\tau=0}^{t} (w(\tau) - y(\tau))\, d\tau \\ & + \alpha_0 (w - y) + \alpha_1 (\dot{w} - \dot{y}) + \cdots + \alpha_{n-1} \left(w^{(n-1)} - y^{(n-1)} \right) + w^{(n)} \end{aligned} \quad [2.3]$$

The feedback system is described by the differential equation:

$$y^{(n)} = \alpha_{-1} \int_{\tau=0}^{t} (w(\tau) - y(\tau))\, d\tau$$
$$+ \alpha_0 (w - y) + \alpha_1 (\dot{w} - \dot{y}) + \cdots + \alpha_{n-1}\Big(w^{(n-1)} - y^{(n-1)}\Big) + w^{(n)}$$

Hence, by differentiating once:

$$e^{(n+1)} + \alpha_{n-1}e^{(n)} + \cdots + \alpha_1\ddot{e} + \alpha_0\dot{e} + \alpha_{-1}e = 0$$

The characteristic polynomial:

$$P(s) = s^{n+1} + \alpha_{n-1}s^n + \cdots + \alpha_1 s^2 + \alpha_0 s + \alpha_{-1}$$

can be chosen arbitrarily, as with the proportional-derivative controller.

2.2. Introductory example

Before giving the principles of feedback linearization, we will consider an introductory example. Let us take the pendulum of Figure 2.1. The input of this system is the torque u exerted on the pendulum.

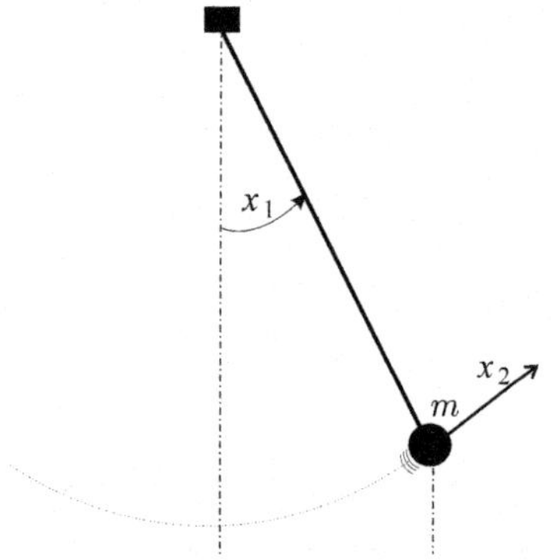

Figure 2.1. *Simple pendulum with state vector* $\mathbf{x} = (x_1, x_2)$

Its state representation is assumed to be:

$$\begin{cases} \begin{pmatrix} \dot{x}_1 \\ \dot{x}_2 \end{pmatrix} = \begin{pmatrix} x_2 \\ -\sin x_1 + u \end{pmatrix} \\ y \quad = x_1 \end{cases}$$

This is of course a normalized model in which the coefficients (mass, gravity and length) have all been set to 1. We would like the position $x_1(t)$ of the pendulum to be equal to some setpoint $w(t)$ which may vary over time. By using a feedback linearization method, explained later, we would like to have a state feedback controller such that the error $e = w - x_1$ converges towards 0 at $\exp(-t)$ (which means that we place the poles at -1). Let us differentiate y until the input u appears. We have:

$$\begin{aligned}\dot{y} &= x_2 \\ \ddot{y} &= -\sin x_1 + u\end{aligned}$$

Let us choose:

$$u = \sin x_1 + v \qquad [2.4]$$

where v corresponds to the new, so-called intermediate input. We obtain

$$\ddot{y} = v \qquad [2.5]$$

Such a feedback is called *linearizing feedback* because it transforms the nonlinear system into a linear system. The system obtained in this way can be stabilized by standard linear techniques. Let us take as an example a proportional-derivative controller:

$$\begin{aligned}v &= (w - y) + 2(\dot{w} - \dot{y}) + \ddot{w} \\ &= (w - x_1) + 2(\dot{w} - x_2) + \ddot{w}\end{aligned}$$

By injecting this expression of v into [2.5], we obtain:

$$\ddot{y} = (w - x_1) + 2(\dot{w} - x_2) + \ddot{w}$$

which yields:

$$e + 2\dot{e} + \ddot{e} = 0$$

where $e = w - x_1$ is the error between the position of the pendulum and its setpoint. The complete controller is expressed by:

$$u \overset{[2.4]}{=} \sin x_1 + (w - x_1) + 2(\dot{w} - x_2) + \ddot{w}$$

If we now want the angle x_1 of the pendulum to be equal to $\sin t$ once the transient regime has passed, we simply need to take $w(t) = \sin t$. Thus, $\dot{w}(t) = \cos t$ and $\ddot{w} = -\sin t$. Consequently, the controller is given by:

$$u = \sin x_1 + (\sin t - x_1) + 2(\cos t - x_2) - \sin t$$

In this very simple example, we can see that the proposed controller is nonlinear and depends on time. No approximation arising from linearization has been performed. Of course, a linearization has been done by a first feedback in order to make the system linear, but this linearization did not introduce any approximation.

2.3. Principle of the method

2.3.1. *Principle*

Here we will look at generalizing the method described in the previous section. Let us consider the nonlinear system described by:

$$\begin{cases} \dot{\mathbf{x}} = \mathbf{f}(\mathbf{x}) + \mathbf{g}(\mathbf{x})\mathbf{u} \\ \mathbf{y} = \mathbf{h}(\mathbf{x}) \end{cases} \qquad [2.6]$$

where the number of inputs and the number of outputs are both equal to m. The idea of the method of feedback linearization is to transform the system using a controller of the type $\mathbf{u} = \mathbf{r}(\mathbf{x}, \mathbf{v})$, where $\mathbf{v}$ is the new input, also of dimension m. This operation requires that the state is completely accessible. If this is not the case, we need to build an observer in a nonlinear context, which is a very difficult operation. Since the state is assumed to be accessible, the vector $\mathbf{y}$ must not really be considered an output, but rather as the vector of the setpoint variables.

In order to perform this transformation, we need to express the successive derivatives of each of the y_i in function of the state and of the input. We stop differentiating y_i as soon as the inputs begin to be involved in the expression of the derivative. We thus have an equation of the type:

$$\begin{pmatrix} y_1^{(k_1)} \\ \vdots \\ y_m^{(k_m)} \end{pmatrix} = \mathbf{A}(\mathbf{x})\,\mathbf{u} + \mathbf{b}(\mathbf{x}) \qquad [2.7]$$

where k_i denotes the number of times we need to differentiate y_i in order to make an input appear (refer to the examples given in the following section for a better understanding). Under the hypothesis that the matrix $\mathbf{A}(\mathbf{x})$ is invertible, the transformation:

$$\mathbf{u} = \mathbf{A}^{-1}(\mathbf{x})(\mathbf{v} - \mathbf{b}(\mathbf{x})) \qquad [2.8]$$

where $\mathbf{v}$ is our new input (see Figure 2.2), forms a linear system $\mathcal{S}_\mathrm{L}$ of m inputs to m outputs described by the differential equations:

$$\mathcal{S}_\mathrm{L}: \begin{cases} y_1^{(k_1)} &= v_1 \\ \quad\vdots &= \vdots \\ y_m^{(k_m)} &= v_m \end{cases}$$

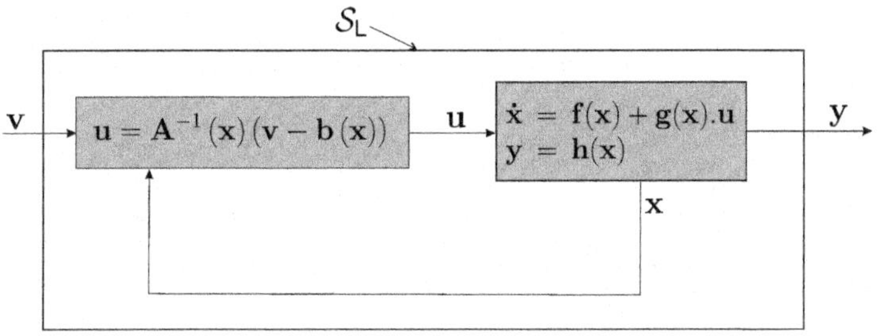

Figure 2.2. *The nonlinear system, once transformed, becomes linear and decoupled; and therefore becomes easy to control*

This system is linear and completely decoupled (in other words, each input v_i acts on one and only one output y_i). It is therefore very simple to control using standard linear techniques. Here, the system to control is composed of decoupled integrator chains; we will use m PID controllers whose principles we have already recalled in section 2.1. Let us note that in order to use such a controller, it is necessary to have the derivatives of the outputs. Since we are assumed to have access to all the state variables x_i of the system, a formal expression of these derivatives in function of x_i is easily obtained by using the state equations.

NOTE 2.2.– Robots are called *redundant* if they have more inputs than necessary, in other words, if $\dim \mathbf{u} > \dim \mathbf{y}$. In this case, the matrix $\mathbf{A}(\mathbf{x})$ is rectangular. In order to apply the transformation in [2.8], we may use a

Moore-Penrose pseudoinverse. If $\mathbf{A}$ is of full rank (in other words equal to $\dim \mathbf{y}$), this pseudoinverse is given by:

$$\mathbf{A}^{\dagger} = \mathbf{A}^{\mathrm{T}} \cdot \left(\mathbf{A} \cdot \mathbf{A}^{\mathrm{T}}\right)^{-1}$$

We will therefore have:

$$\dim \mathbf{v} = \dim \mathbf{y} < \dim \mathbf{u}$$

and we are in a situation identical to that of the square robot (i.e. non-redundant).

2.3.2. *Relative degree*

By properly analyzing the path used to obtain equation [2.7], we realize that the kth derivative of the ith output $y_i^{(k)}$ is expressed in the form:

$$\begin{aligned} y_i^{(k)} &= \hat{b}_{ik}(\mathbf{x}) \text{ if } j < k_i \\ y_i^{(k)} &= \hat{\mathbf{a}}_{ik}^{\mathrm{T}}(\mathbf{x}) \cdot \mathbf{u} + \hat{b}_{ik}(\mathbf{x}) \text{ if } k = k_i \\ y_i^{(k)} &= \hat{\mathbf{a}}_{ik}(\mathbf{x}, \mathbf{u}, \dot{\mathbf{u}}, \ddot{\mathbf{u}}, \dots) \text{ if } k > k_i \end{aligned}$$

The coefficient k_i is called the *relative degree* of the ith output. By measuring the state of the system $\mathbf{x}$ and its input $\mathbf{u}$, we can thus have all the successive derivatives of the outputs $y_i^{(k)}$, as long as k remains smaller than or equal to k_i. Indeed, given the high-frequency noise that appears in the signals, we cannot reliably obtain the derivative of the signals by using derivators. We therefore have an analytic function:

$$\Delta : \begin{array}{ll} \mathbb{R}^m & \rightarrow \mathbb{R}^{(k_1+1)\cdot\ldots\cdot(k_m+1)} \\ (\mathbf{x}, \mathbf{u}) & \mapsto \Delta(\mathbf{x}, \mathbf{u}) = \left(y_1, \dot{y}_1, \dots, y_1^{(k_1)}, y_2, \dot{y}_2, \dots, y_m^{(k_m)}\right) \end{array}$$

that allows us to have all the derivatives of the outputs (until their relative degree), and this without using digital derivators.

EXAMPLE 2.1.– Let us consider the system described by:

$$\begin{cases} \dot{x} = xu + x^3 \\ y = 2x \end{cases}$$

We have:

$$\begin{aligned} y &= 2x \\ \dot{y} &= 2\dot{x} = 2xu + 2x^3 \\ \ddot{y} &= 2\dot{x}u + 2x\dot{u} + 6\dot{x}x^2 = 2(xu + x^3)u + 2x\dot{u} + 6(xu + x^3)x^2 \end{aligned}$$

Therefore, we have a relative degree $k = 1$ for the output y. We can therefore have $\dot{y}$ without using digital derivators. This is not the case for $\ddot{y}$ because having u with a high level of precision does not mean that we have $\dot{u}$. Here, we have $\Delta(x, u) = (2x, 2xu + 2x^3)$.

2.3.3. *Differential delay matrix*

We call *differential delay* r_{ij} separating the input u_j from the output y_i the number of times we need to differentiate y_i in order to make u_j appear. The matrix $\mathbf{R}$ of the r_{ij} is called the *differential delay matrix*. When plainly reading the state equations, this matrix can be obtained without calculation, simply by counting the number of integrators each input u_j must be subjected to in order to algebraically affect the output y_i (some examples are discussed in more detail in the exercises). The relative degree for each output can be obtained by taking the minimum of each row. Let us take for example:

$$\mathbf{R} = \begin{pmatrix} \mathbf{1} & 2 & 2 \\ \mathbf{3} & 4 & \mathbf{3} \\ 4 & \infty & \mathbf{2} \end{pmatrix}$$

The associated system is composed of three inputs, three outputs and the relative degrees (components in bold in the preceding formula) are $k_1 = 1, k_2 = 3, k_3 = 2$. If there is a j such that $\forall i, r_{ij} > k_i$ (or equivalently if a column has no element in bold), the matrix $\mathbf{R}$ is called *unbalanced*. In our example, it is unbalanced since there is a j (here $j = 2$) such that $\forall i, r_{ij} > k_i$. If the matrix is unbalanced, then for all i, $y_i^{(k_i)}$ does not depend on u_j. In this case, the jth column of $\mathbf{A}(\mathbf{x})$ will be zero and $\mathbf{A}(\mathbf{x})$ will always be singular. Thus, transformation [2.8] will have no meaning. One method to overcome this is to delay some of the inputs u_j by adding one or more integrators in front of the system. Adding an integrator in front of the jth input amounts to adding 1 to the jth column of $\mathbf{R}$. In our example, if we add an integrator in front of u_1, we obtain:

$$\mathbf{R} = \begin{pmatrix} \mathbf{2} & \mathbf{2} & \mathbf{2} \\ 4 & 4 & \mathbf{3} \\ 5 & \infty & \mathbf{2} \end{pmatrix}$$

The relative degrees become $k_1 = 2, k_2 = 3, k_3 = 2$ and the matrix $\mathbf{R}$ becomes balanced.

2.3.4. *Singularities*

The matrix $\mathbf{A}(\mathbf{x})$ involved in feedback [2.8] may not be invertible. The values for $\mathbf{x}$ such that $\det(\mathbf{A}(\mathbf{x})) = 0$ are called *singularities*. Although they generally form a set of zero measures in the state space, studying singularities is fundamental since they are sometimes impossible to avoid. We call *set of acceptable outputs* of system [2.6] the quantity:

$$\mathbb{S}_y = \{\mathbf{y} \in \mathbb{R}^m \mid \exists \mathbf{x} \in \mathbb{R}^n, \exists \mathbf{u} \in \mathbb{R}^m, \mathbf{f}(\mathbf{x}) + \mathbf{g}(\mathbf{x}) \cdot \mathbf{u} = \mathbf{0}, \mathbf{y} = \mathbf{h}(\mathbf{x})\}$$

The set $\mathbb{S}_y$ is therefore composed, by the projection on $\mathbb{R}^m$, of a differentiable manifold (or surface) of dimension m (since we have $m + n$ equations for $2m + n$ variables). Thus, except for the degenerate case, $\mathbb{S}_y$ is a subset of $\mathbb{R}^m$ with a non-empty interior and an exterior.

In order to properly understand this, consider the example of a cart on rails as represented in Figure 2.3. This cart can be propelled by a horizontal ventilator whose angle of thrust can be controlled. However, be careful, the rotation speed of the ventilator is fixed.

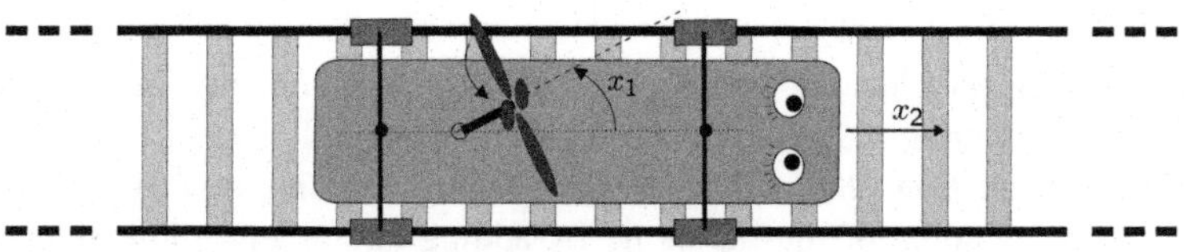

Figure 2.3. *Robot cart propelled by a ventilator*

The state equations that model this system are given by:

$$\begin{cases} \dot{x}_1 = u \\ \dot{x}_2 = \cos x_1 - x_2 \\ y \;\; = x_2 \end{cases}$$

where x_1 is the ventilator's angle of thrust, x_2 is the speed of the cart. Let us note that we have taken into account a viscous friction force. The set of acceptable outputs is:

$$\begin{aligned} \mathbb{S}_y &= \{y \mid \exists x_1, \exists x_2, \exists u, u = 0, \cos x_1 - x_2 = 0, y = x_2\} \\ &= \{y \mid \exists x_1, \cos x_1 = y\} = [-1, 1] \end{aligned}$$

This means that we will not be able to stabilize the cart at a speed that is strictly greater than 1, in absolute value. Let us apply a feedback linearizing method. We have:

$$\begin{aligned} \dot{y} &= \cos x_1 - x_2 \\ \ddot{y} &= -\left(\sin x_1\right) u - \cos x_1 + x_2 \end{aligned}$$

and therefore, the linearizing controller is given by:

$$u = \frac{-1}{\sin x_1}\left(v + \cos x_1 - x_2\right)$$

The feedback system therefore has the following equation:

$$\ddot{y} = v$$

It may appear that any value for y can be reached, since v can be chosen arbitrarily. This would be correct, if we did not have the singularity that appears when $\sin x_1 = 0$. Let us take, for example, $v(t) = 1$ and $\mathbf{x}(0) = \left(\frac{\pi}{3}, 0\right)$. We should have:

$$\begin{aligned} \ddot{y}(t) &= v\left(t\right) = 1 \\ \dot{y}(t) &= \dot{y}\left(0\right) + \int_0^t \ddot{y}(\tau) d\tau = \cos x_1(0) - x_2(0) + t = \frac{1}{2} + t \\ y(t) &= y\left(0\right) + \int_0^t \dot{y}(\tau) d\tau = x_2\left(0\right) + t^2 + \frac{1}{2}t = t^2 + \frac{1}{2}t \end{aligned}$$

which is physically impossible. This is what happens when we apply such a controller: the input u directs the angle of the ventilator toward the correct direction, and the equation $\ddot{y} = v$ is then satisfied, at least in the very beginning. Then x_1 is canceled out and the singularity is reached. The equation $\ddot{y} = v$ can no longer be satisfied. For some systems, it can happen that such a singularity can be crossed. This is not the case here.

2.4. Cart

2.4.1. *First model*

Consider a cart described by the following state equations:

$$\begin{cases} \dot{x} = v\cos\theta \\ \dot{y} = v\sin\theta \\ \dot{\theta} = u_1 \\ \dot{v} = u_2 \end{cases}$$

where v is the speed of the cart, θ is its orientation and (x, y) is the coordinates of its center. The state vector is given by $\mathbf{x} = (x, y, \theta, v)$. We would like to calculate a controller that would allow us to describe a cycloid with the equation:

$$\begin{cases} x_d(t) = R\sin(f_1 t) + R\sin(f_2 t) \\ y_d(t) = R\cos(f_1 t) + R\cos(f_2 t) \end{cases}$$

where $R = 15$, $f_1 = 0.02$ and $f_2 = 0.12$. For this, we use a feedback linearizing method. We have:

$$\begin{pmatrix} \ddot{x} \\ \ddot{y} \end{pmatrix} = \begin{pmatrix} u_2\cos\theta - u_1 v\sin\theta \\ u_2\sin\theta + u_1 v\cos\theta \end{pmatrix} = \begin{pmatrix} -v\sin\theta & \cos\theta \\ v\cos\theta & \sin\theta \end{pmatrix} \begin{pmatrix} u_1 \\ u_2 \end{pmatrix}$$

If we take as input:

$$\begin{pmatrix} u_1 \\ u_2 \end{pmatrix} = \begin{pmatrix} -v\sin\theta & \cos\theta \\ v\cos\theta & \sin\theta \end{pmatrix}^{-1} \begin{pmatrix} v_1 \\ v_2 \end{pmatrix}$$

where (v_1, v_2) is the new input vector, we obtain the linear system:

$$\begin{pmatrix} \ddot{x} \\ \ddot{y} \end{pmatrix} = \begin{pmatrix} v_1 \\ v_2 \end{pmatrix}$$

Let us transform this system so that all our poles are at -1. Following [2.2], we obtain:

$$\begin{cases} v_1 = (x_d - x) + 2(\dot{x}_d - \dot{x}) + \ddot{x}_d = (x_d - x) + 2(\dot{x}_d - v\cos\theta) + \ddot{x}_d \\ v_2 = (y_d - y) + 2(\dot{y}_d - \dot{y}) + \ddot{y}_d = (y_d - y) + 2(\dot{y}_d - v\sin\theta) + \ddot{y}_d \end{cases}$$

The transformed system then obeys the following differential equations:

$$\begin{pmatrix} (x_d - x) + 2(\dot{x}_d - \dot{x}) + (\ddot{x}_d - \ddot{x}) \\ (y_d - y) + 2(\dot{y}_d - \dot{y}) + (\ddot{y}_d - \ddot{y}) \end{pmatrix} = \begin{pmatrix} 0 \\ 0 \end{pmatrix}$$

If we define the error vector $\mathbf{e} = (e_x, e_y) = (x_d - x, y_d - y)$, the error dynamics are written as:

$$\begin{pmatrix} e_x + 2\dot{e}_x + \ddot{e}_x \\ e_y + 2\dot{e}_y + \ddot{e}_y \end{pmatrix} = \begin{pmatrix} 0 \\ 0 \end{pmatrix}$$

which is stable and quickly converges toward 0. Therefore, the controller will be:

$$\begin{pmatrix} u_1 \\ u_2 \end{pmatrix} = \begin{pmatrix} -v\sin\theta & \cos\theta \\ v\cos\theta & \sin\theta \end{pmatrix}^{-1} \begin{pmatrix} (x_d - x) + 2(\dot{x}_d - v\cos\theta) + \ddot{x}_d \\ (y_d - y) + 2(\dot{y}_d - v\sin\theta) + \ddot{y}_d \end{pmatrix} \qquad [2.9]$$

where:

$$\begin{aligned} \dot{x}_d(t) &= Rf_1\cos(f_1 t) + Rf_2\cos(f_2 t) \\ \dot{y}_d(t) &= -Rf_1\sin(f_1 t) - Rf_2\sin(f_2 t) \\ \ddot{x}_d(t) &= -Rf_1^2\sin(f_1 t) - Rf_2^2\sin(f_2 t) \\ \ddot{y}_d(t) &= -Rf_1^2\cos(f_1 t) - Rf_2^2\cos(f_2 t) \end{aligned}$$

The following MATLAB program, found in the file `cycloide.m`, simulates the behavior of the controlled system:

```
x=[10;10;0;2]; %x=[x,y,theta,v)
dt=0.1; R=15; f1=0.02;f2=0.12;
for t=0:dt:10,
w = R*[sin(f1*t)+sin(f2*t);cos(f1*t)+cos(f2*t)];
dw =
R*[f1*cos(f1*t)+f2*cos(f2*t);-f1*sin(f1*t)-f2*sin(f2*t)];
ddw=R*[-f1*f1*sin(f1*t)-f2*f2*sin(f2*t);-f1*f1*cos(f1*t)
-f2*f2*cos(f2*t)];
u=control(x,w,dw,ddw);
x=x+f(x,u)*dt;
end;
```

For the evolution of the robot, the program uses the following evolution function:

```
function xdot=f(x,u)
theta=x(3);v=x(4);
xdot=[v*cos(theta); v*sin(theta); u(1); u(2)];
end
```

As for the control, it is performed by the function:

```
  function u=control(x,w,dw,ddw)
v=0.25 *( w - [x(1);x(2)])+1*(dw -
[x(4)*cos(x(3));x(4)*sin(x(3))])+ddw;
A=[-x(4)*sin(x(3)), cos(x(3)); x(4)*cos(x(3)), sin(x(3))];
u=inv(A)*v;
end
```

2.4.2. *Second model*

Let us now assume that the cart is described by the state equations:

$$\begin{cases} \dot{x} = u_1 \cos\theta \\ \dot{y} = u_1 \sin\theta \\ \dot{\theta} = u_2 \end{cases}$$

Let us choose as output the vector $\mathbf{y} = (x, y)$. The method of feedback linearization leads to a matrix $\mathbf{A}(\mathbf{x})$ which is still singular. As was explained in section 2.3.3, this can be predicted without any calculation simply by observing the differential delay matrix:

$$\mathbf{R} = \begin{pmatrix} 1 & 2 \\ 1 & 2 \end{pmatrix}$$

This matrix contains a column whose elements never correspond to the minimum of the related row (in other words, a column without elements in bold). We will illustrate how to get out of such a situation by adding integrators in front of certain inputs. Let us, for instance, add an integrator whose state variable will be denoted by z, in front of the first input. Recall that adding an integrator in front of the jth input of the system means delaying this input and therefore adding 1 to all the elements of column j of $\mathbf{R}$. The matrix $\mathbf{R}$ is then in equilibrium. We obtain a new system described by:

$$\begin{cases} \dot{x} = z\cos\theta \\ \dot{y} = z\sin\theta \\ \dot{\theta} = u_2 \\ \dot{z} = c_1 \end{cases}$$

We have:

$$\begin{cases} \ddot{x} = \dot{z}\cos\theta - z\dot{\theta}\sin\theta = c_1\cos\theta - zu_2\sin\theta \\ \ddot{y} = \dot{z}\sin\theta + z\dot{\theta}\cos\theta = c_1\sin\theta + zu_2\cos\theta \end{cases}$$

in other words:

$$\begin{pmatrix} \ddot{x} \\ \ddot{y} \end{pmatrix} = \begin{pmatrix} \cos\theta & -z\sin\theta \\ \sin\theta & z\cos\theta \end{pmatrix} \begin{pmatrix} c_1 \\ u_2 \end{pmatrix}$$

The matrix is not singular, except in the unlikely case where the variable z is zero (here, z can be understood as the speed of the vehicle). The method of feedback linearization can therefore work. Let us take:

$$\begin{pmatrix} c_1 \\ u_2 \end{pmatrix} = \begin{pmatrix} \cos\theta & -z\sin\theta \\ \sin\theta & z\cos\theta \end{pmatrix}^{-1} \begin{pmatrix} v_1 \\ v_2 \end{pmatrix} = \begin{pmatrix} \cos\theta & \sin\theta \\ -\frac{\sin\theta}{z} & \frac{\cos\theta}{z} \end{pmatrix} \begin{pmatrix} v_1 \\ v_2 \end{pmatrix}$$

in order to have a feedback system of the form:

$$\begin{pmatrix} \ddot{x} \\ \ddot{y} \end{pmatrix} = \begin{pmatrix} v_1 \\ v_2 \end{pmatrix}$$

Figure 2.4 illustrates the feedback linearization that we have just performed.

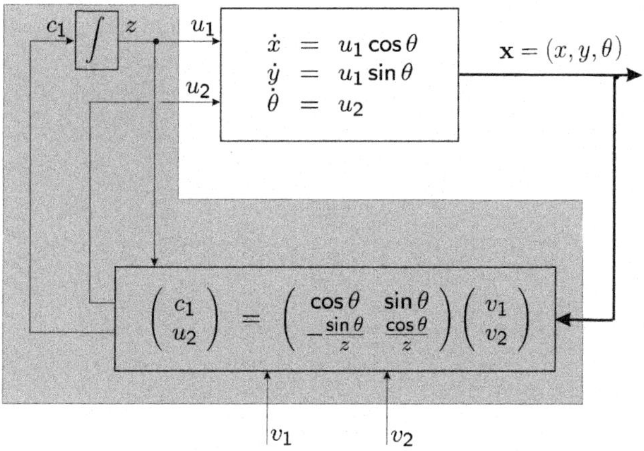

Figure 2.4. *Dynamic feedback linearization*

In order to have all the poles at -1, we need to take (see equation [2.2]):

$$\begin{pmatrix} c_1 \\ u_2 \end{pmatrix} = \begin{pmatrix} \cos\theta & \sin\theta \\ -\frac{\sin\theta}{z} & \frac{\cos\theta}{z} \end{pmatrix} \begin{pmatrix} (x_d - x) + 2(\dot{x}_d - z\cos\theta) + \ddot{x}_d \\ (y_d - y) + 2(\dot{y}_d - z\sin\theta) + \ddot{y}_d \end{pmatrix}$$

The state equations of the controller are therefore:

$$\begin{cases} \dot{z} = (\cos\theta)(x_d - x + 2(\dot{x}_d - z\cos\theta) + \ddot{x}_d) \\ \qquad + (\sin\theta)(y_d - y + 2(\dot{y}_d - z\sin\theta) + \ddot{y}_d) \\ u_1 = z \\ u_2 = -\frac{\sin\theta}{z}(x_d - x + 2(\dot{x}_d - z\cos\theta) + \ddot{x}_d) \\ \qquad + \frac{\cos\theta}{z} \cdot (y_d - y + 2(\dot{y}_d - z\sin\theta) + \ddot{y}_d) \end{cases}$$

2.5. Controlling a tricycle

2.5.1. *Speed and heading control*

Let us consider the tricycle represented in Figure 2.5. Its evolution equation is given by:

$$\begin{pmatrix} \dot{x} \\ \dot{y} \\ \dot{\theta} \\ \dot{v} \\ \dot{\delta} \end{pmatrix} = \begin{pmatrix} v\cos\delta\cos\theta \\ v\cos\delta\sin\theta \\ v\sin\delta \\ u_1 \\ u_2 \end{pmatrix}$$

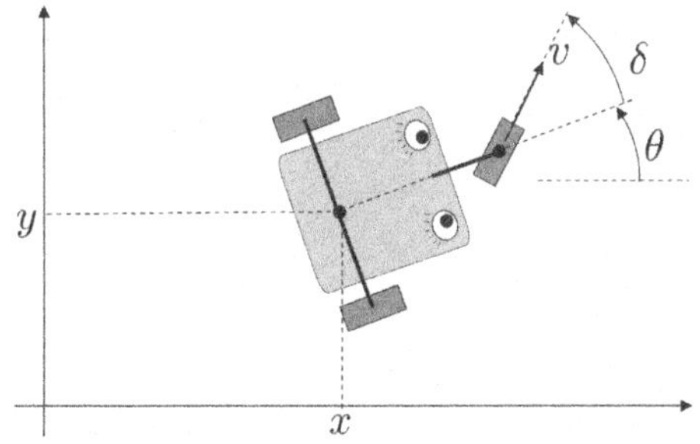

Figure 2.5. *Tricycle robot to be controlled*

We have assumed here that the distance between the center of the rear axle and the axis of the front wheel was equal to 1 m. Let us choose as output the vector $\mathbf{y} = (v, \theta)$. The first-order derivatives of the outputs y_1 and y_2 are expressed by:

$$\dot{y}_1 = \dot{v} = u_1,$$
$$\dot{y}_2 = \dot{\theta} = v\sin\delta$$

Since the derivative $\dot{y}_2$ of y_2 does not involve the input, we may differentiate it once more:

$$\ddot{y}_2 = \dot{v}\sin\delta + v\dot{\delta}\cos\delta = u_1\sin\delta + u_2 v\cos\delta$$

The expressions for $\dot{y}_1$ and $\ddot{y}_2$ can be rewritten in matrix form:

$$\begin{pmatrix} \dot{y}_1 \\ \ddot{y}_2 \end{pmatrix} = \underbrace{\begin{pmatrix} 1 & 0 \\ \sin\delta & v\cos\delta \end{pmatrix}}_{\mathbf{A}(\mathbf{x})} \begin{pmatrix} u_1 \\ u_2 \end{pmatrix}$$

By setting the feedback $\mathbf{u} = \mathbf{A}^{-1}(\mathbf{x})\ \mathbf{v}$, where $\mathbf{v}$ is the new input, our feedback system is rewritten as:

$$\mathcal{S}_{\mathrm{L}} : \begin{pmatrix} \dot{y}_1 \\ \ddot{y}_2 \end{pmatrix} = \begin{pmatrix} v_1 \\ v_2 \end{pmatrix}$$

and therefore becomes linear and decoupled. We now have two decoupled monovariate systems. The first, of order 1, can be stabilized by a proportional controller. As for the second, second-order system, a proportional-derivative controller is best adapted. If $\mathbf{w} = (w_1,\ w_2)$ represents the setpoint for $\mathbf{y}$, this controller is expressed by:

$$\begin{cases} v_1 = (w_1 - y_1) + \dot{w}_1 \\ v_2 = (w_2 - y_2) + 2(\dot{w}_2 - \dot{y}_2) + \ddot{w}_2 \end{cases}$$

if we want all our poles to be equal to -1 (refer to equation [2.2]). Therefore the equations of a state feedback controller for our nonlinear system are given by:

$$\mathbf{u} = \begin{pmatrix} 1 & 0 \\ \sin\delta & v\cos\delta \end{pmatrix}^{-1} \begin{pmatrix} (w_1 - v) + \dot{w}_1 \\ w_2 - \theta + 2\left(\dot{w}_2 - \frac{v\sin\delta}{L}\right) + \ddot{w}_2 \end{pmatrix} \qquad [2.10]$$

Let us note that this controller does not have a state variable. It is therefore a *static* controller.

NOTE 2.3.– Since:

$$\det(\mathbf{A}(\mathbf{x})) = \frac{v\cos\delta}{L}$$

can be zero, there are singularities for which the control $\mathbf{u}$ is not defined. Appropriate processing has to be provided when such singularities are encountered by the system.

2.5.2. *Position control*

Let us now try to make our tricycle follow a desired trajectory (x_d, y_d). For this, let us choose as output the vector $\mathbf{y} = (x, y)$. We have:

$$\begin{cases} \dot{x} = v \cos \delta \cos \theta \\ \ddot{x} = \dot{v} \cos \delta \cos \theta - v\dot{\delta} \sin \delta \cos \theta - v\dot{\theta} \cos \delta \sin \theta \\ \quad = u_1 \cos \delta \cos \theta - v u_2 \sin \delta \cos \theta - v^2 \sin \delta \cos \delta \sin \theta \\ \dot{y} = v \cos \delta \sin \theta \\ \ddot{y} = \dot{v} \cos \delta \sin \theta - v\dot{\delta} \sin \delta \sin \theta + v\dot{\theta} \cos \delta \cos \theta \\ \quad\; u_1 \cos \delta \sin \theta - v u_2 \sin \delta \sin \theta + v^2 \sin \delta \cos \delta \cos \theta \end{cases}$$

Thus:

$$\begin{pmatrix} \ddot{x} \\ \ddot{y} \end{pmatrix} = \underbrace{\begin{pmatrix} \cos \delta \cos \theta & -v \sin \delta \cos \theta \\ \cos \delta \sin \theta & -v \sin \delta \sin \theta \end{pmatrix}}_{\mathbf{A}(\mathbf{x})} \begin{pmatrix} u_1 \\ u_2 \end{pmatrix} + \underbrace{\begin{pmatrix} -v^2 \sin \delta \cos \delta \sin \theta \\ v^2 \sin \delta \cos \delta \cos \theta \end{pmatrix}}_{\mathbf{b}(\mathbf{x})}$$

However, the determinant of $\mathbf{A}(\mathbf{x})$ is zero since the two columns of the matrix $\mathbf{A}(\mathbf{x})$ are collinear to the vector $(\cos \theta, \sin \theta)$. This means that the controllable part of the acceleration is forcibly in the vehicle heading direction. Thus, $\ddot{x}$ and $\ddot{y}$ will not be independently controllable. The method of feedback linearization can therefore not be applied.

2.5.3. *Choosing another output*

In order to avoid having a singular matrix $\mathbf{A}(\mathbf{x})$, let us now choose the center of the front wheel as output. We have:

$$\mathbf{y} = \begin{pmatrix} x + \cos \theta \\ y + \sin \theta \end{pmatrix}$$

By differentiating once, we have:

$$\begin{pmatrix} \dot{y}_1 \\ \dot{y}_2 \end{pmatrix} = \begin{pmatrix} \dot{x} - \dot{\theta} \sin \theta \\ \dot{y} + \dot{\theta} \cos \theta \end{pmatrix} = v \begin{pmatrix} \cos \delta \cos \theta - \sin \delta \sin \theta \\ \cos \delta \sin \theta + \sin \delta \cos \theta \end{pmatrix} = v \begin{pmatrix} \cos (\delta + \theta) \\ \sin (\delta + \theta) \end{pmatrix}$$

Differentiating again, we obtain:

$$\begin{pmatrix} \ddot{y}_1 \\ \ddot{y}_2 \end{pmatrix} = \begin{pmatrix} \dot{v}\cos(\delta+\theta) - v\left(\dot{\delta}+\dot{\theta}\right)\sin(\delta+\theta) \\ \dot{v}\sin(\delta+\theta) + v\cos(\delta+\theta) \end{pmatrix}$$
$$= \begin{pmatrix} u_1\cos(\delta+\theta) - v\,(u_2 + v\sin\delta)\sin(\delta+\theta) \\ u_1\sin(\delta+\theta) + v\,(u_2 + v\sin\delta)\cos(\delta+\theta) \end{pmatrix}$$

and therefore:

$$\begin{pmatrix} \ddot{y}_1 \\ \ddot{y}_2 \end{pmatrix} = \underbrace{\begin{pmatrix} \cos(\delta+\theta) & -v\sin(\delta+\theta) \\ \sin(\delta+\theta) & v\cos(\delta+\theta) \end{pmatrix}}_{\mathbf{A}(\mathbf{x})} \begin{pmatrix} u_1 \\ u_2 \end{pmatrix}$$
$$\underbrace{+\, v^2\sin\delta \begin{pmatrix} -\sin(\delta+\theta) \\ \cos(\delta+\theta) \end{pmatrix}}_{\mathbf{b}(\mathbf{x})}$$

The determinant of $\mathbf{A}(\mathbf{x})$ is never equal to zero, except when $v = 0$. The linearizing control is therefore $\mathbf{u} = \mathbf{A}^{-1}(\mathbf{x}) \cdot (\mathbf{v} - \mathbf{b}(\mathbf{x}))$. Therefore, the tricycle control (the one that places all the poles at -1) is expressed by:

$$\mathbf{u} = \mathbf{A}^{-1}(\mathbf{x}) \left(\left(\begin{pmatrix} x_d \\ y_d \end{pmatrix} - \begin{pmatrix} x+\cos\theta \\ y+\sin\theta \end{pmatrix} \right) + 2\left(\begin{pmatrix} \dot{x}_d \\ \dot{y}_d \end{pmatrix} - \begin{pmatrix} v\cos(\delta+\theta) \\ v\sin(\delta+\theta) \end{pmatrix} \right) \right.$$
$$\left. + \begin{pmatrix} \ddot{x}_d \\ \ddot{y}_d \end{pmatrix} - \mathbf{b}(\mathbf{x}) \right)$$

where $\mathbf{w} = (x_d, y_d)$ is the desired trajectory for the output $\mathbf{y}$.

2.6. Sailboat

Automatic control for sailing robots [PET 11] is a complex problem given the strong nonlinearities implied in the evolution of the system. Here we will consider the sailboat of Figure 2.6 whose state equations [JAU 04] are given by:

$$\begin{cases} \dot{x} &= v\cos\theta \\ \dot{y} &= v\sin\theta - 1 \\ \dot{\theta} &= \omega \\ \dot{\delta}_s &= u_1 \\ \dot{\delta}_r &= u_2 \\ \dot{v} &= f_s \sin\delta_s - f_r \sin\delta_r - v \\ \dot{\omega} &= (1-\cos\delta_s)\, f_s - \cos\delta_r \cdot f_r - \omega \\ f_s &= \cos(\theta+\delta_s) - v\sin\delta_s \\ f_r &= v\sin\delta_r \end{cases} \qquad [2.11]$$

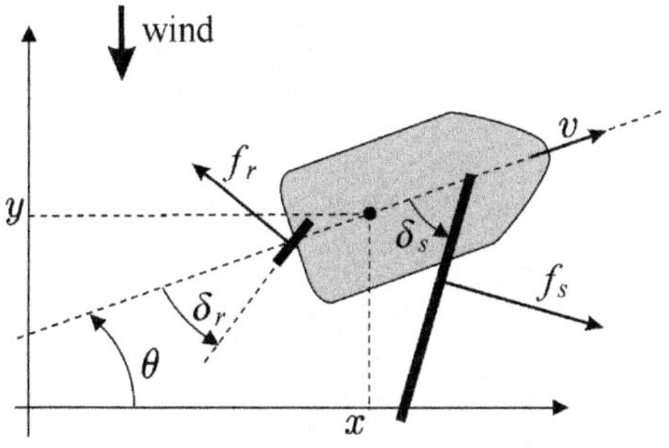

Figure 2.6. *Sailing robot to be controlled*

This is of course a normalized model in which many coefficients (masses, lengths, etc.) have been set to 1 in order to simplify the following developments. The state vector $\mathbf{x} = (x, y, \theta, \delta_s, \delta_r, v, \omega)$, of dimension 7, is composed of:

– position coordinates; in other words, the x, y coordinates of the sailboat's center of gravity, the orientation θ, and the angles δ_s and δ_r of the sail and the rudder;

– kinematic coordinates v and ω representing respectively the speed of the center of gravity and the angular speed of the boat.

The inputs u_1 and u_2 of the system are the differentials of the angles δ_s and δ_r. The indices s and r refer respectively to the sail and the rudder.

2.6.1. *Polar curve*

Let us take as outputs $\mathbf{y} = (\theta, v)$. The polar curve is the set of acceptable outputs (refer to section 2.3.4), in other words, the set $\mathbb{S}_y$ of all pairs (θ, v) over which we are able to stabilize. In stationary regime, we have:

$$\dot{\theta} = 0, \dot{\delta}_s = 0, \dot{\delta}_r = 0, \dot{v} = 0, \dot{\omega} = 0$$

Thus, following the state equations in [2.11], we obtain:

$$\begin{aligned} \mathbb{S}_y = \{(\theta, v) \mid \; & f_s \sin \delta_s - f_r \sin \delta_r - v = 0 \\ & (1 - \cos \delta_s) f_s - \cos \delta_r f_r = 0 \\ & f_s = \cos(\theta + \delta_s) - v \sin \delta_s \\ & f_r = v \sin \delta_r \; \} \end{aligned}$$

An interval calculation method [HER 10] allows us to obtain the estimation of Figure 2.7.

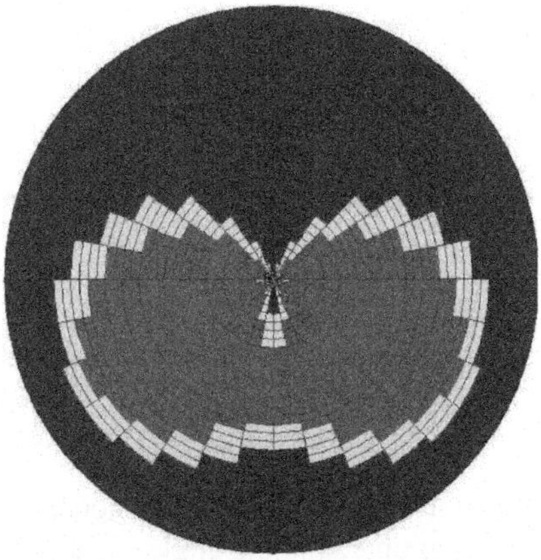

Figure 2.7. *Internal frame (in light gray) and external frame (in dark gray) of the polar curve*

2.6.2. *Differential delay*

We may associate with the state equations of our sailboat a *differential delay graph* between the variables (see Figure 2.8). Within this graph, a solid arrow can be interpreted, depending on the reader, either as a cause and effect relationship, a differential delay, or as a state equation. A dotted arrow represents an algebraic (and not a differential) dependency. On the graph, we can distinguish two types of variables: the state variables, pointed at by solid arrows, and link variables (in gray), pointed at by dotted arrows. The derivative of a state variable is expressed as an algebraic function of all the variables which are directly before it. Likewise, a link variable is an algebraic function of the variables which are directly before it.

The differential delay between a variable and an input u_j is thus the minimum number of solid arrows to traverse in order to reach this variable from u_j. Just as in [JAU 04], let us take as output the vector $\mathbf{y} = (\delta_s, \theta)$. The differential delay matrix is:

$$\mathbf{R} = \begin{pmatrix} 1 & \infty \\ 3 & 3 \end{pmatrix}$$

The infinity here can be interpreted as the fact that there is no causal connection that links u_2 to δ_s. The relative degrees are therefore $k_1 = 1$ and $k_2 = 3$.

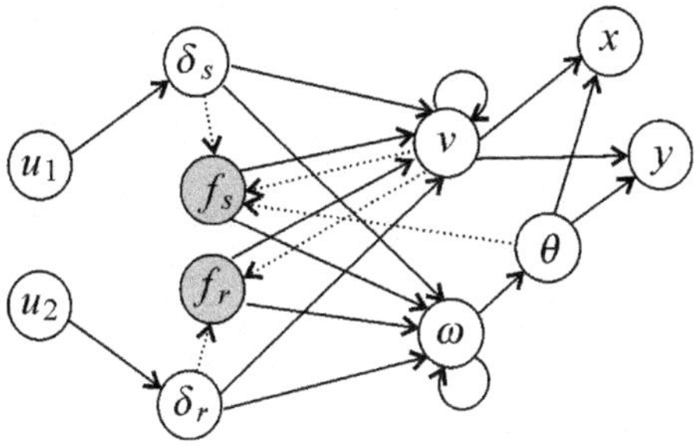

Figure 2.8. *Graph of the differential delays for the sailing robot*

2.6.3. *The method of feedback linearization*

Let us recall that the outputs (these are actually setpoint variables) chosen are the sail opening $y_1 = \delta_s$ and the heading $y_2 = \theta$. In order to apply a feedback linearization method, we first of all need to differentiate the outputs as many times as the relative degree requires it, in other words three times for θ and once for δ_s. By looking at the differential dependency graph, we can observe that in order to express $\dddot{\theta}$ in function of $\mathbf{x}$ and $\mathbf{u}$, we need to do the same with $\ddot{\omega}, \dot{\omega}, \dot{\delta}_r, \dot{\delta}_s, \dot{f}_r, \dot{f}_s, \dot{v}$. This yields:

$$\begin{cases} \dot{v} &= f_s \sin\delta_s - f_r \sin\delta_r - v \\ \dot{f}_s &= -(\omega + u_1)\sin(\theta+\delta_s) - \dot{v}\sin\delta_s - vu_1\cos\delta_s \\ \dot{f}_r &= \dot{v}\sin\delta_r + vu_2\cos\delta_r \\ \dot{\omega} &= (1-\cos\delta_s)\cdot f_s - \cos\delta_r \cdot f_r - \omega \\ \ddot{\omega} &= u_1\sin\delta_s \cdot f_s + (1-\cos\delta_s)\cdot \dot{f}_s + u_2 \sin\delta_r \cdot f_r - \cos\delta_r \cdot \dot{f}_r - \dot{\omega} \\ \dddot{\theta} &= \ddot{\omega} \end{cases}$$

We have:

$$\begin{pmatrix} \dot{y}_1 \\ \dddot{y}_2 \end{pmatrix} = \begin{pmatrix} \dot{\delta}_s \\ \dddot{\theta} \end{pmatrix} = \underbrace{\begin{pmatrix} 1 & 0 \\ f_s\sin\delta_s & f_r\sin\delta_r \end{pmatrix}}_{\mathbf{A}_1(\mathbf{x})} \begin{pmatrix} u_1 \\ u_2 \end{pmatrix}$$

$$+ \underbrace{\begin{pmatrix} 0 & 0 \\ 1-\cos\delta_s & -\cos\delta_r \end{pmatrix}}_{\mathbf{A}_2(\mathbf{x})} \begin{pmatrix} \dot{f}_s \\ \dot{f}_r \end{pmatrix} + \underbrace{\begin{pmatrix} 0 \\ -\dot{\omega} \end{pmatrix}}_{\mathbf{b}_1(\mathbf{x})}$$

However:

$$\begin{pmatrix} \dot{f}_s \\ \dot{f}_r \end{pmatrix} = \underbrace{\begin{pmatrix} -(\sin(\theta+\delta_s) + v\cos\delta_s) & 0 \\ 0 & v\cos\delta_r \end{pmatrix}}_{\mathbf{A}_3(\mathbf{x})} \begin{pmatrix} u_1 \\ u_2 \end{pmatrix}$$

$$+ \underbrace{\begin{pmatrix} -\omega\sin(\theta+\delta_s) + \dot{v}\sin\delta_s \\ \dot{v}\sin\delta_r \end{pmatrix}}_{\mathbf{b}_2(\mathbf{x})}$$

and thus we have a relation of the form:

$$\begin{pmatrix} \dot{y}_1 \\ \dddot{y}_2 \end{pmatrix} = \mathbf{A}_1\mathbf{u} + \mathbf{A}_2(\mathbf{A}_3\mathbf{u} + \mathbf{b}_2) + \mathbf{b}_1$$

$$= (\mathbf{A}_1 + \mathbf{A}_2\mathbf{A}_3)\mathbf{u} + \mathbf{A}_2\mathbf{b}_2 + \mathbf{b}_1 = \mathbf{A}\mathbf{u} + \mathbf{b}$$

In order to set $(\dot{y}_1, \dddot{y}_2)$ to a certain setpoint $\mathbf{v} = (v_1, v_2)$, we need to take:

$$\mathbf{u} = \mathbf{A}^{-1}(\mathbf{x})\,(\mathbf{v} - \mathbf{b}\,(\mathbf{x}))$$

The system looped in this manner is governed by the differential equations:

$$\mathcal{S}_{\mathrm{L}} : \begin{cases} \dot{y}_1 &= v_1, \\ \dddot{y}_2 &= v_2 \end{cases} \qquad [2.12]$$

which are linear and decoupled. The linearized system is of order 4 instead of 7. We have thus lost control over three variables which happen to be x, y and v. The loss of control over x and y was predictable (we want the boat to advance and therefore it is only natural that this corresponds to an instability for these two variables x and y). As for the loss of control over v, this is without consequence since the associated dynamics are stable. How indeed would it be possible to design a boat that would be able to keep a fixed heading and sail opening, without its speed converging toward a finite value?

Let us now determine the singularities of our linearizing feedback loop. By calculating the expression of $\mathbf{A}(\mathbf{x})$, we can show that:

$$\det(\mathbf{A}(\mathbf{x})) = f_r \sin\delta_r - v\cos^2\delta_r \overset{[2.11]}{=} v\left(2\sin^2\delta_r - 1\right)$$

We have a singularity when this quantity is equal to zero, in other words if:

$$v = 0 \text{ or } \delta_r = \frac{\pi}{4} + k\frac{\pi}{2} \qquad [2.13]$$

The singularity corresponding to $v = 0$ is relatively simple to understand: when the boat is not advancing, we can no longer control it. The condition on the rudder angle δ_r is more delicate to interpret. Indeed, the condition $\delta_r = \pm\frac{\pi}{4}$ translates to a maximal rotation. Any action on the rudder when $\delta_r = \pm\frac{\pi}{4}$ translates to a slower rotation. This is what this singularity means.

We are dealing with two decoupled monovariate systems here. Let us denote by $\mathbf{w} = (w_1,\ w_2)$ the setpoint for $\mathbf{y}$. We will sometimes write $\mathbf{w} = (\hat{\delta}_s, \hat{\theta})$ in order to recall that w_1 and w_2 are the setpoints corresponding to the sail opening angle and the heading. Let us choose the proportional and derivative controller given by:

$$\begin{cases} v_1 = (w_1 - y_1) + \dot{w}_1 \\ v_2 = (w_2 - y_2) + 3\,(\dot{w}_2 - \dot{y}_2) + 3\,(\ddot{w}_2 - \ddot{y}_2) + \dddot{w}_2 \end{cases}$$

which allows all poles of the feedback system to be equal to -1 (refer to equation [2.2]). By assuming that the setpoint $\mathbf{w}$ is constant, the state equations of the state feedback controller for our nonlinear system are given by:

$$\mathbf{u} = \mathbf{A}^{-1}(\mathbf{x}) \left(\begin{pmatrix} w_1 - \delta_s \\ w_2 - \theta - 3\dot{\theta} - 3\ddot{\theta} \end{pmatrix} - \mathbf{b}(\mathbf{x}) \right) \qquad [2.14]$$

However, $\dot{\theta}$ and $\ddot{\theta}$ are analytic functions of the state $\mathbf{x}$. Indeed, we have:

$$\begin{aligned} \dot{\theta} &= \omega \\ \ddot{\theta} &= (1 - \cos\delta_s) f_s - \cos\delta_r f_r - \omega \end{aligned}$$

Equation [2.14] can therefore be written in the form:

$$\mathbf{u} = \mathbf{r}(\mathbf{x}, \mathbf{w}) = \mathbf{r}\left(\mathbf{x}, \hat{\delta}_s, \hat{\theta}\right) \qquad [2.15]$$

This controller is *static* since it does not have a state variable.

2.6.4. *Polar curve control*

In some situations, the boater does not want complete autonomy of his boat, only steering assistance. He does not wish to decide the angle of the sails, but simply its speed and heading. In summary, he would like to choose a point on the polar curve and it is up to the controller to perform low-level control. In cruising regime, we have:

$$\begin{cases} 0 &= \bar{f}_s \sin\bar{\delta}_s - \bar{f}_r \sin\bar{\delta}_r - \bar{v} \\ 0 &= \left(1 - \cos\bar{\delta}_s\right) \cdot \bar{f}_s - \cos\bar{\delta}_r \cdot \bar{f}_r \\ \bar{f}_s &= \cos\left(\bar{\theta} + \bar{\delta}_s\right) - \bar{v}\sin\bar{\delta}_s \\ \bar{f}_r &= \bar{v}\sin\bar{\delta}_r \end{cases}$$

If $\left(\bar{\theta}, \bar{v}\right)$ is in the polar curve, we can calculate $\left(\bar{f}_r, \bar{\delta}_r, \bar{f}_s, \bar{\delta}_s\right)$ (there is at least one solution, by definition of the polar curve). Thus, it is sufficient to inject $\left(\bar{\theta}, \bar{\delta}_s\right)$ in controller [2.15] in order to preform our control. Figure 2.9 illustrates the docking of a sailboat in a harbor using this approach [HER 10].

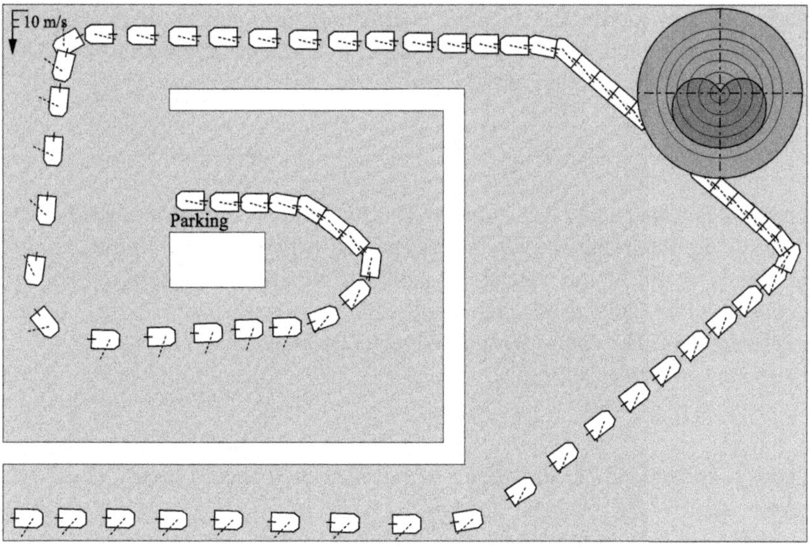

Figure 2.9. *By using a linearizing controller, the robot docks in its place in the harbor; the polar curve is represented on the top right corner*

2.7. Kinematic model and dynamic model

2.7.1. *Principle*

The dynamic models for the robots are of the form:

$$\dot{\mathbf{x}} = \mathbf{f}(\mathbf{x}, \mathbf{u})$$

where $\mathbf{u}$ is the vector of the external forces (that are under our control). The function $\mathbf{f}$ involves dynamic coefficients (such as masses, inertial moments, coefficients of friction, etc.) as well as geometric coefficients (such as lengths). The dynamic coefficients are generally not well known and can change over time with wear or usage. If we now take as a new input the vector $\mathbf{a}$ of the desired accelerations at the application points of the forces (in the direction of the forces), we obtain a new model, referred to as *kinematic*, of the form:

$$\dot{\mathbf{x}} = \varphi(\mathbf{x}, \mathbf{a})$$

but in this new model, the dynamic coefficients have almost all vanished. It is possible to switch from a dynamic model to a kinematic model using a so-called *high-gain* controller, of the form:

$$\mathbf{u} = K(\mathbf{a} - \mathbf{a}(\mathbf{x}, \mathbf{u}))$$

where K is a very large real number. The function $\mathbf{a}(\mathbf{x}, \mathbf{u})$ is a function that allows us to obtain the acceleration associated with the forces, in function of the forces $\mathbf{u}$ and of the state $\mathbf{x}$. In practice, we are not looking to express $\mathbf{a}(\mathbf{x}, \mathbf{u})$ within the controller, but rather to measure $\mathbf{a}(\mathbf{x}, \mathbf{u})$ using accelerometers. The controller that will actually be implemented is:

$$\mathbf{u} = K(\mathbf{a} - \widetilde{\mathbf{a}})$$

where $\widetilde{\mathbf{a}}$ corresponds to the vector of the measured accelerations. Thus, we have:

$$\dot{\mathbf{x}} = \mathbf{f}(\mathbf{x}, K(\mathbf{a} - \mathbf{a}(\mathbf{x}, \mathbf{u}))) \Leftrightarrow \dot{\mathbf{x}} = \varphi(\mathbf{x}, \mathbf{a})$$

The simple high-gain feedback has allowed us to get rid of numerous dynamic parameters and switch from an uncertain system to a reliable system, with well-known geometric coefficients. This high-gain feedback is known in electronics as an operational amplifier, where it is used with the same idea of robustness.

Switching from a dynamic model to a kinematic system has the following advantages:

– the linearizing controller developed in this chapter requires using a reliable model such as a kinematic model. If the coefficients (which are not measured) are not well known (such as in the case of dynamic systems), the linearizing controller will not work in practice;

– the kinematic model is easier to put into equations. It is not necessary to have a dynamic model to obtain the latter;

– the servo-motors (see section 2.7.3) incorporate this high-gain controller. We may therefore see them as mechanical operational amplifiers.

We will illustrate the concept in the following section, through the example of the inverted rod pendulum.

2.7.2. *Example of the inverted rod pendulum*

Let us consider the inverted rod pendulum, composed of a pendulum in an unstable equilibrium on top of a moving cart, as represented in Figure 2.10.

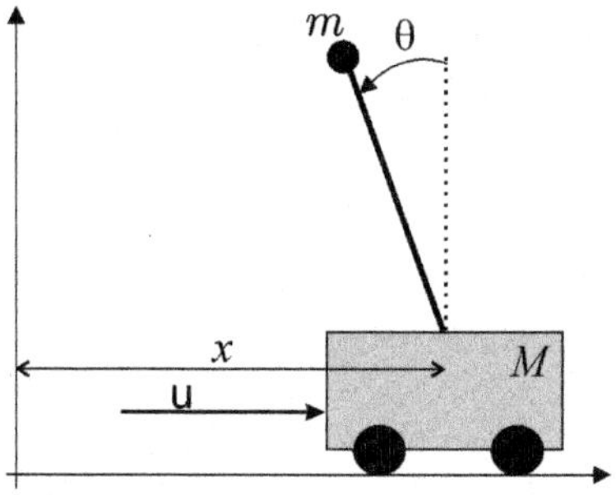

Figure 2.10. *Inverted rod pendulum to be modeled and controlled*

2.7.2.1. *Dynamic model*

The quantity u is the force exerted on the cart of mass M, x indicates the position of the cart, θ is the angle between the pendulum and the vertical direction. The state equations are written in the form:

$$\frac{d}{dt}\begin{pmatrix} x \\ \theta \\ \dot{x} \\ \dot{\theta} \end{pmatrix} = \begin{pmatrix} \dot{x} \\ \dot{\theta} \\ \frac{-m\sin\theta(\ell\dot{\theta}^2 - g\cos\theta)}{M+m\sin^2\theta} \\ \frac{\sin\theta((M+m)g - m\ell\dot{\theta}^2\cos\theta)}{\ell(M+m\sin^2\theta)} \end{pmatrix} + \begin{pmatrix} 0 \\ 0 \\ \frac{1}{M+m\sin^2\theta} \\ \frac{\cos\theta}{\ell(M+m\sin^2\theta)} \end{pmatrix} u \qquad [2.16]$$

2.7.2.2. *Kinematic model*

Let us recall the the state equations of the inverted rod pendulum, but instead of taking the force as input, we take the acceleration $a = \ddot{x}$. We obtain, following [2.16]:

$$a = \frac{1}{M + m\sin^2\theta}\left(-m\sin\theta(\ell\dot{\theta}^2 - g\cos\theta) + u\right) \qquad [2.17]$$

Therefore:

$$\begin{aligned}
\ddot{\theta} &\overset{[2.16]}{=} \frac{\sin\theta((M+m)g - m\ell\dot{\theta}^2\cos\theta)}{\ell(M+m\sin^2\theta)} + \frac{\cos\theta}{\ell(M+m\sin^2\theta)}u \\
&\overset{[2.17]}{=} \frac{\sin\theta((M+m)g - m\ell\dot{\theta}^2\cos\theta)}{\ell(M+m\sin^2\theta)} \\
&+ \frac{\cos\theta}{\ell(M+m\sin^2\theta)}\left(m\sin\theta\left(\ell\dot{\theta}^2 - g\cos\theta\right)\right) + \left(M+m\sin^2\theta\right)a\right) \\
&= \frac{1}{\ell(M+m\sin^2\theta)}\left((M+m)g\sin\theta - gm\sin\theta\cos^2\theta\right. \\
&+ \left.\left(M+m\sin^2\theta\right)\cos\theta\cdot a\right) \\
&= \frac{g\sin\theta}{\ell} + \frac{\cos\theta}{\ell}a
\end{aligned}$$

Let us note that this relation could have been obtained directly by noting that:

$$\ell\ddot{\theta} = \underbrace{a\cdot\cos\theta}_{\text{acceleration of } A \text{ that contributes to the rotation}} + \underbrace{g\cdot\sin\theta}_{\text{acceleration of } B}$$

NOTE 2.4.– In order to obtain this relation in a more rigorous manner, we need to write the temporal derivative of the speed composition formula. In other words:

$$\dot{\mathbf{v}}_A = \dot{\mathbf{v}}_B + \overrightarrow{AB} \wedge \overrightarrow{\dot{\omega}}$$

and write this formula in the coordinate system of the pendulum. We obtain:

$$\begin{pmatrix} a\cos\theta \\ -a\sin\theta \\ 0 \end{pmatrix} = \begin{pmatrix} -g\sin\theta \\ n \\ 0 \end{pmatrix} + \begin{pmatrix} 0 \\ \ell \\ 0 \end{pmatrix} \wedge \begin{pmatrix} 0 \\ 0 \\ \dot{\omega} \end{pmatrix}$$

where n corresponds to the normal acceleration of the mass m. We thus obtain the desired relation as well as the normal acceleration $n = -a\sin\theta$ which will not be used.

Finally, the kinematic model is written as:

$$\frac{d}{dt}\begin{pmatrix} x \\ \theta \\ \dot{x} \\ \dot{\theta} \end{pmatrix} = \begin{pmatrix} \dot{x} \\ \dot{\theta} \\ 0 \\ \frac{g\sin\theta}{\ell} \end{pmatrix} + \begin{pmatrix} 0 \\ 0 \\ 1 \\ \frac{\cos\theta}{\ell} \end{pmatrix} a \qquad [2.18]$$

This model, referred to as the *kinematic model*, only involves positions, speeds and accelerations. It is a lot more simple than the dynamic model and involves fewer coefficients. However, it corresponds less to reality since the correct input is a force and not an acceleration. In practice, we may switch from dynamic model [2.16] with input u to kinematic model [2.18] with input a by calculating u using a *high-gain proportional controller* of the form:

$$u = K(a - \ddot{x}) \qquad [2.19]$$

with K very large and where a is a new input.

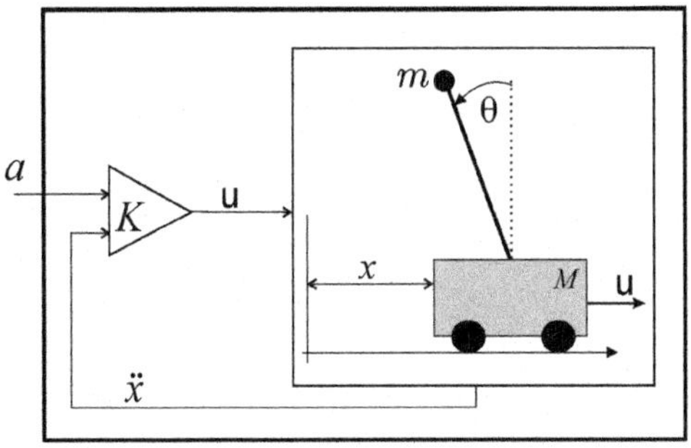

Figure 2.11. *The inverted rod pendulum, looped by a high gain K, behaves like a kinematic model*

The acceleration $\ddot{x}$ can be measured using an accelerometer. If K is sufficiently large, we will of course have the controller u that will give us the desired acceleration a; in other words, we will have $\ddot{x} = a$. Thus, system [2.16] can be described by the state equations in [2.18] which do not involve any of the inertial parameters of the system. A controller designed over the kinematic model will therefore be more robust than the controller designed over the dynamic system since the controller will function for any values of m, M, the inertial momentums, friction, etc. Let us recall that this high-gain controller is very close to the principle of the operational amplifier. In addition to being more robust, such an approach allows us to have a simpler model that is easier to obtain. For the implementation of the controller in [2.19], we of course do not need to use state equations [2.16] in order to express $\ddot{x}$, but measure $\ddot{x}$ instead. It is this measurement that allows us to have a controller that is independent of the dynamic parameters.

Let us recall our inverted rod pendulum and try to make the pendulum oscillate from left to right with a desired angle of the form $\theta_d = \sin t$. Let us apply a linearizing controller for this. We have:

$$\ddot{\theta} = \frac{g \sin \theta}{\ell} + \frac{\cos \theta}{\ell} a$$

We will therefore take:

$$a = \frac{\ell}{\cos \theta} \left(v - \frac{g \sin \theta}{\ell} \right)$$

where v is the new input. We will then choose:

$$v = (\theta_d - \theta) + 2 \left(\dot{\theta}_d - \dot{\theta} \right) + \ddot{\theta}_d = \sin t - \theta + 2 \cos t - 2\dot{\theta} - \sin t$$

and finally:

$$\begin{aligned} u &= K (a - \ddot{x}) \\ &= K \left(\frac{\ell}{\cos \theta} \left(\sin t - \theta + 2 \cos t - 2\dot{\theta} - \sin t - \frac{g \sin \theta}{\ell} \right) - \ddot{x} \right) \end{aligned}$$

Note that the inertial parameters are not taken into account in this controller. This controller ensures that the system will respect its setpoint angle. However, the position of the cart can diverge, since u does not depend on x. The dynamics of x are hidden and moreover unstable here. These hidden dynamics are conventionally referred to as *zero dynamics*.

2.7.3. *Servo-motors*

A mechanical system is controlled by forces or torques and obeys a dynamic system that depends on numerous little-known coefficients. This same mechanical system represented by a kinematic model is controlled by positions, speeds or accelerations. The kinematic model depends on well-known geometric coefficients and is much simpler to put into equations. In practice, we switch from a dynamic model to its kinematic equivalent by adding servo-motors. In summary, a servo-motor is a DC motor with an electrical control circuit and a sensor (of position, speed or acceleration). The control circuit calculates the voltage u to give the motor in order for the quantity measured by the sensor to correspond to the setpoint w. There are three types of servo-motors:

– the *position servo*. The sensor measures the position (or the angle) x of the motor and the control law is expressed by $u = K(x - w)$. If K is large, we may conclude that $x \simeq w$;

– the *speed servo*. The sensor measures the speed (or the angular speed) $\dot{x}$ of the motor and the control law is expressed by $u = K(\dot{x} - w)$. If K is large, we have $\dot{x} \simeq w$;

– the *acceleration servo*. The sensor measures the acceleration (tangential or angular) $\ddot{x}$ of the motor and the control law is expressed by $u = K(\ddot{x} - w)$. If K is large, we have $\ddot{x} \simeq w$. It is this type of servo-motor that we have chosen for the inverted rod pendulum.

Thus, when we wish to control a mechanical system, the use of servo-motors allows us (1) to have a model that is easier to obtain, (2) to have a model with fewer coefficients that is closer to reality, and (iii) to have a more robust controller with respect to any modification of the dynamic coefficients of the system.

2.8. Exercises

EXERCISE 2.1.– Crank

Let us consider the manipulator robot, or *crank* of Figure 2.12 (on the left). This robot is composed of two arms of length ℓ_1 and ℓ_2. Its two degrees of freedom denoted by x_1 and x_2 are represented in the figure. The inputs u_1, u_2 of the system are the angular speeds of the arms (in other words, $u_1 = \dot{x}_1 and u_2 = \dot{x}_2$). We will take as output the vector $\mathbf{y} = (y_1, y_2)$ corresponding to the end of the second arm.

1) Give the state equations of the robot. We will take the state vector $\mathbf{x} = (x_1, x_2)$.

2) We would like $\mathbf{y}$ to follow a setpoint $\mathbf{w}$ describing a target circle (on the right of Figure 2.12). This setpoint satisfies:

$$\mathbf{w} = \mathbf{c} + r \cdot \begin{pmatrix} \cos t \\ \sin t \end{pmatrix}$$

Give the expression of a control law that allows us to perform this task. We will use a feedback linearization method and we will place the poles at -1.

3) Study the singularities of the control.

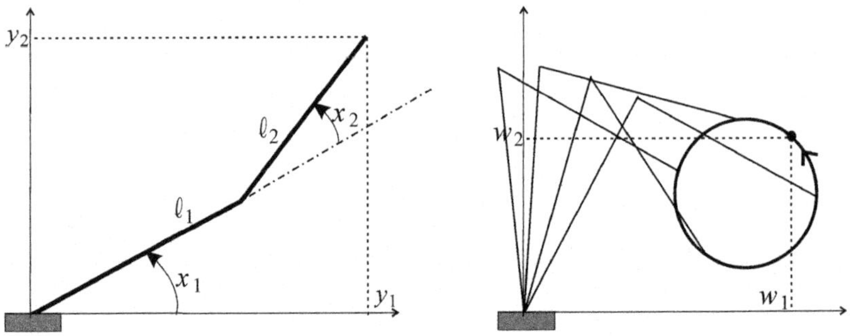

Figure 2.12. *Manipulator robot whose end effector must follow a circle*

4) Let us consider the case $\ell_1 = \ell_2$, $\mathbf{c} = (3, 4)$ and $r = 1$. For which values of ℓ_1 are we certain to be able to move freely on the target circle, without encountering singularities?

5) Write a MATLAB program illustrating this control law.

EXERCISE 2.2.– The three pools

Let us consider a flow system with three pools as represented in Figure 2.13.

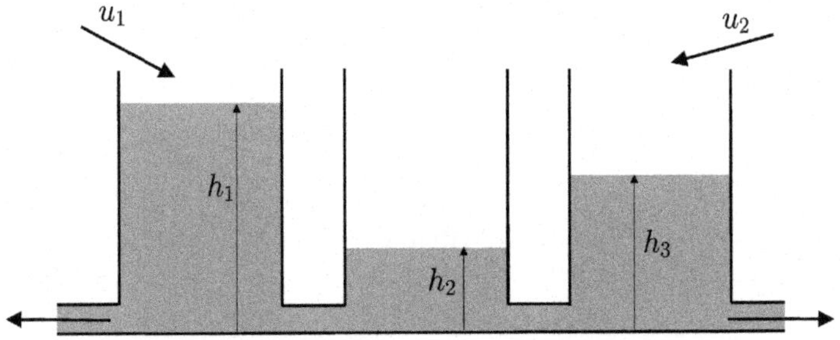

Figure 2.13. *System composed of three pools containing water and connected by two channels*

This system is described by the following state equations:

$$\begin{cases} \dot{h}_1 = -\alpha\left(h_1\right) - \alpha\left(h_1 - h_2\right) + u_1 \\ \dot{h}_2 = \alpha\left(h_1 - h_2\right) - \alpha\left(h_2 - h_3\right) \\ \dot{h}_3 = -\alpha\left(h_3\right) + \alpha\left(h_2 - h_3\right) + u_2 \\ y_1 = h_1 \\ y_2 = h_3 \end{cases}$$

where $\alpha(h) = a{\cdot}\text{sign}(h)\sqrt{2g|h|}$. We have chosen our outputs here to be the water levels in the first and third pools.

1) Propose a feedback that would make the system linear and decoupled.

2) Propose a proportional-integral controller for the linearized system.

3) Give the state equations of the obtained controller.

4) Write a MATLAB program that simulates the system and its control law.

EXERCISE 2.3.– Train robot

Let us consider a robot A (on the left of Figure 2.14) described by the following state equations (tank model):

$$\begin{cases} \dot{x}_a = v_a \cos\theta_a \\ \dot{y}_a = v_a \sin\theta_a \\ \dot{\theta}_a = u_{a1} \\ \dot{v}_a = u_{a2} \end{cases}$$

where v_a is the speed of the robot, θ_a its orientation and (x_a, y_a) the coordinates of its center. We assume to be able to measure the state variables of our robot with very high precision.

1) Calculate $\ddot{x}_a, \ddot{y}_a$ in function of $x_a, y_a, v_a, \theta_a, u_{a1}, u_{a2}$.

2) Propose a controller that allows us to follow the trajectory:

$$\begin{cases} \hat{x}_a(t) = L_x \sin(\omega t) \\ \hat{y}_a(t) = L_y \cos(\omega t) \end{cases}$$

with $\omega = 0.1$, $L_x = 15$ and $L_y = 7$. A feedback linearization method must be used for this.

3) A second robot B of the same type as A wishes to follow robot A (see Figure 2.15). We define a virtual attachment point with coordinates $(\hat{x}_b, \hat{y}_b)$ in order for vehicle B (on the left of the figure) to be able to attach itself to our robot A. We can send it the information associated with the attachment point wirelessly.

This point will be positioned at the rear of robot A at a distance ℓ of our reference point (x_a, y_a). Give the expression of these quantities in function of the state of our vehicle A.

4) Simulate this second vehicle B in MATLAB together with its controller following robot A.

5) Add a third robot C that follows B with the same principle. Simulate the entire system in MATLAB.

6) Given that, in this exercise, the reference path is precisely known, propose a controller that will allow robots B and C to precisely follow robot A.

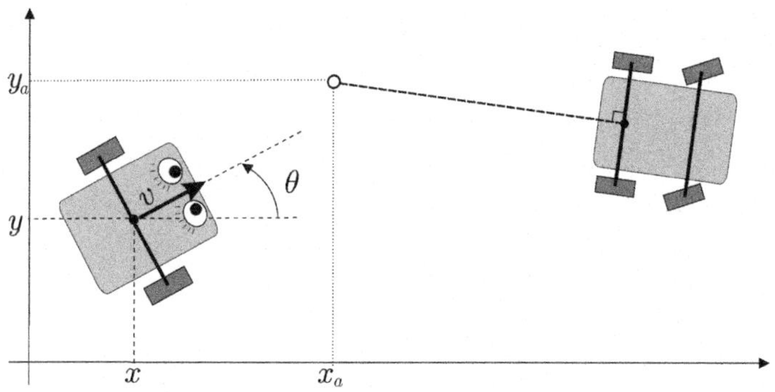

Figure 2.14. *Our robot (with eyes) following a vehicle (here a car) whose state equations are unknown. This car has an imaginary attachment point (small white circle) that we must attach to*

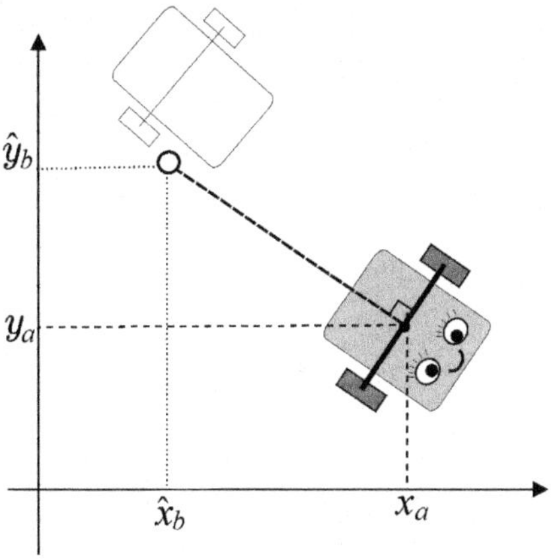

Figure 2.15. *Robot B (dotted) has to follow robot A*

EXERCISE 2.4.– Controlling a 3D underwater robot

Let us consider the underwater robot already discussed in exercise 1.9. This robot is described by the following state equations:

$$\begin{cases} \dot{p}_x &= v\cos\theta\cos\psi \\ \dot{p}_y &= v\cos\theta\sin\psi \\ \dot{p}_z &= -v\sin\theta \\ \dot{v} &= u_1 \\ \dot{\psi} &= \frac{\sin\varphi}{\cos\theta}\cdot v\cdot u_2 + \frac{\cos\varphi}{\cos\theta}\cdot v\cdot u_3 \\ \dot{\theta} &= \cos\varphi\cdot v\cdot u_2 - \sin\varphi\cdot v\cdot u_3 \\ \dot{\varphi} &= -0.1\ \sin\varphi + \tan\theta\cdot v\cdot(\sin\varphi\cdot u_2 + \cos\varphi\cdot u_3) \end{cases}$$

where (p_x, p_y, p_z) is the position of its center and (ψ, θ, φ) are the three Euler angles. Its inputs are the tangential acceleration u_1, the pitch u_2 and the yaw u_3. Suggest a controller capable of controlling the robot around the cycloid of equation:

$$\begin{pmatrix} x_d \\ y_d \\ z_d \end{pmatrix} = \begin{pmatrix} R\cdot\sin(f_1 t) + R\cdot\sin(f_2 t) \\ R\cdot\cos(f_1 t) + R\cdot\cos(f_2 t) \\ R\cdot\sin(f_3 t) \end{pmatrix}$$

where $f_1 = 0.01$, $f_2 = 6f_1$, $f_3 = 3f_1$ and $R = 20$. For the control, we will choose a time constant of 5 s. Simulate the behavior of the controller in MATLAB.

EXERCISE 2.5.– Flat system

Consider the system described by the state equations:

$$\begin{cases} \dot{x}_1 = x_1 + x_2 \\ \dot{x}_2 = x_2^2 + u \\ y \;\; = x_1 \end{cases} \qquad [2.20]$$

1) A system is called *flat* if there are two functions ϕ, ψ such that:

$$\begin{cases} \mathbf{x} = \phi\left(y, \dot{y}, \dots, y^{(r-1)}\right) \\ u = \psi\left(y, \dot{y}, \dots, y^{(r-1)}, y^{(r)}\right) \end{cases}$$

Show that our system is flat. Give the expressions of ϕ and ψ.

2) Give the expression of a linearizing feedback for the system of the form $u = \gamma\left(v, y, \dot{y}\right)$, where v is a new input. This feedback transforms our system into a system described by $\ddot{y} = v$.

3) We would like to control the system $\ddot{y} = v$ using a proportional-derivative controller. Give the expression of the controller $v = \eta\left(w, \dot{w}, \ddot{w}, y, \dot{y}\right)$, where w is a setpoint that varies with time and that allows us to have an error $e = w - y$ that converges toward zero. All the poles are equal to -1.

4) Deduce from the above equations an output feedback controller for the initial system of the form $u = \rho\left(w, \dot{w}, \ddot{w}, y, \dot{y}\right)$ which is such that the error $e = w - y$ converges toward 0.

EXERCISE 2.6.– Pursuit

Let us consider two robots described by the following state equations:

$$\begin{cases} \dot{x}_1 = u_1 \cos\theta_1 \\ \dot{y}_1 = u_1 \sin\theta_1 \\ \dot{\theta}_1 = u_2 \end{cases} \text{ and } \begin{cases} \dot{x}_2 = v_1 \cos\theta_2 \\ \dot{y}_2 = v_1 \sin\theta_2 \\ \dot{\theta}_2 = v_2 \end{cases}$$

In this exercise, robot 1 tries to follow robot 2 (see Figure 2.16).

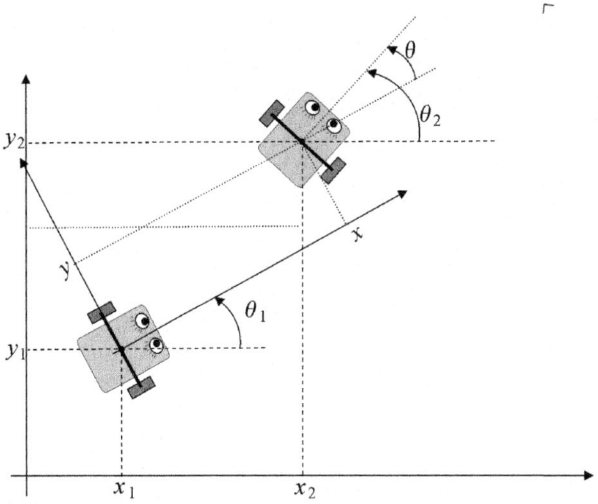

Figure 2.16. *Robot 1 is in pursuit of robot 2*

1) Let $\mathbf{x} = (x, y, \theta)$ be the position vector of robot 2 in the coordinate system of robot 1. Show that $\mathbf{x}$ satisfies a state equation of the form

$$\dot{\mathbf{x}} = \mathbf{f}(\mathbf{x}, \mathbf{v}, \mathbf{u})$$

2) We assume that the control variables v_1 and v_2 of robot 2 are known (a polynomial in t, for example). Suggest a controller that generates us $\mathbf{u}$ in order to have $x = w_1$ and $y = w_2$, where $\mathbf{w} = (w_1, w_2)$ corresponds to a setpoint in relative position. The poles for the error are fixed at -1.

3) Study the singularities of this controller.

4) Illustrate this control law with MATLAB in the situation where robot 1 would like to point toward robot 2 while keeping a distance of 10 m.

EXERCISE 2.7.– Controlling the SAUCISSE robot

Consider the underwater robot represented in Figure 2.17.

Figure 2.17. *The* SAUCISSE *robot in a pool*

This is the SAUCISSE robot, built by students of the ENSTA Bretagne for the SAUC'E competition (*Student Autonomous Underwater Challenge Europe*). It includes three propellers. Propellers 1 and 2 on the left and the right are able to act on the speed of the robot and its angular speed. Propeller 3 acts on the depth of the robot. This robot is stable in roll and pitch and we will assume that its angles of bank φ and elevation θ are always zero. The state equations of the robot are the following:

$$\begin{cases} \dot{x} &= v_x \\ \dot{y} &= v_y \\ \dot{z} &= v_z \\ \dot{\psi} &= \omega \\ \dot{v}_x &= u_1 \cos\psi \\ \dot{v}_y &= u_1 \sin\psi \\ \dot{v}_z &= u_3 \\ \dot{\omega} &= u_2 \end{cases}$$

Let us note that no nonholonomic constraint has been assumed in this model. The speed vector of the robot (v_x, v_y) is not necessarily in its axis, in contrast to the case of the cart model. The robot can therefore operate in crab steering mode. However, the propulsion is necessarily in the direction of the robot axis. If we are limited to the horizontal plane, this model is known as a *hovercraft*.

1) Give the differential dependency graph associated with this system.

2) Let us choose as output the vector $\mathbf{y} = (x, y, z)$. Give the differential delay matrix and deduce the relative degrees from it. What can we conclude?

3) In order to balance the differential delays by delaying u_1, we add two integrators in front of u_1. Our new system will admit as new inputs $\mathbf{a} = (a_1, a_2, a_3)$ with:

$$\begin{cases} \ddot{u}_1 = a_1 \\ u_2 = a_2 \\ u_3 = a_3 \end{cases} \qquad [2.21]$$

What are the new state equations of the delayed system? Give the differential dependency graph as well as the associated differential delay matrix.

4) Perform a feedback linearization of the delayed system.

5) Deduce from the above the controller corresponding to our robot. We will place all the poles at -1.

2.9. Corrections

CORRECTION FOR EXERCISE 2.1.– (Crank)

1) The state equations of the crank are:

$$\begin{cases} \dot{x}_1 = u_1 \\ \dot{x}_2 = u_2 \\ y_1 = \ell_1 \cos x_1 + \ell_2 \cos(x_1 + x_2) \\ y_2 = \ell_1 \sin x_1 + \ell_2 \sin(x_1 + x_2) \end{cases}$$

2) By differentiating the output, we obtain:

$$\begin{aligned} \dot{y}_1 &= -\ell_1 \dot{x}_1 \sin x_1 - \ell_2 (\dot{x}_1 + \dot{x}_2) \sin(x_1 + x_2) \\ &= -\ell_1 u_1 \sin x_1 - \ell_2 (u_1 + u_2) \sin(x_1 + x_2) \\ \dot{y}_2 &= \ell_1 \dot{x}_1 \cos x_1 + \ell_2 (\dot{x}_1 + \dot{x}_2) \cos(x_1 + x_2) \\ &= \ell_1 u_1 \cos x_1 + \ell_2 (u_1 + u_2) \cos(x_1 + x_2) \end{aligned}$$

Thus:

$$\dot{\mathbf{y}} = \underbrace{\begin{pmatrix} -\ell_1 \sin x_1 - \ell_2 \sin(x_1 + x_2) & -\ell_2 \sin(x_1 + x_2) \\ \ell_1 \cos x_1 + \ell_2 \cos(x_1 + x_2) & \ell_2 \cos(x_1 + x_2) \end{pmatrix}}_{\mathbf{A}(\mathbf{x})} \mathbf{u}$$

We take $\mathbf{u} = \mathbf{A}^{-1}(\mathbf{x}) \cdot \mathbf{v}$ to have two decoupled integrators. We then choose the proportional controller:

$$\mathbf{v} = (\mathbf{w} - \mathbf{y}) + \dot{\mathbf{w}} = \mathbf{c} + r \cdot \begin{pmatrix} \cos t \\ \sin t \end{pmatrix} - \begin{pmatrix} \ell_1 \cos x_1 + \ell_2 \cos(x_1 + x_2) \\ \ell_1 \sin x_1 + \ell_2 \sin(x_1 + x_2) \end{pmatrix} + r \begin{pmatrix} -\sin t \\ \cos t \end{pmatrix}$$

which places all of our poles at -1.

3) By multilinearity of the determinant, we have:

$$\det \mathbf{A}(\mathbf{x}) = -\ell_1 \ell_2 \underbrace{\det \begin{pmatrix} \sin x_1 & \sin(x_1 + x_2) \\ \cos x_1 & \cos(x_1 + x_2) \end{pmatrix}}_{= \sin x_2} + \ell_2^2 \underbrace{\det \begin{pmatrix} -\sin(x_1 + x_2) & -\sin(x_1 + x_2) \\ \cos(x_1 + x_2) & \cos(x_1 + x_2) \end{pmatrix}}_{=0}$$

This determinant is equal to zero if $\ell_1 \ell_2 \sin x_2 = 0$. In other words, if, $x_2 = k\pi,\ k\mathbb{Z}$ or if one of the two arms is of length zero.

4) If $\ell_1 = \ell_2 \neq 0$, we have a singularity if $\sin x_2 = 0$. Thus, either both arms are folded up (and therefore $\mathbf{y}$ is not on the circle) or both arms are stretched out (see Figure 2.18). In the latter case (which is of interest to us), the point $\mathbf{y}$ is on the circle of radius $2\ell_1$ that intersects the target circle if:

$$\ell_1 + \ell_2 \in \sqrt{4^2 + 3^2} \pm 1 = 5 \pm 1 = [4, 6]$$

where $\sqrt{4^2 + 3^2}$ corresponds to the distance of the center of the circle to the origin. We will have a singularity on the circle if $\ell_1 = \ell_2 \in [2, 3]$. If we wish to move freely on the circle, we need to choose $\ell_1 = \ell_2 > 3$.

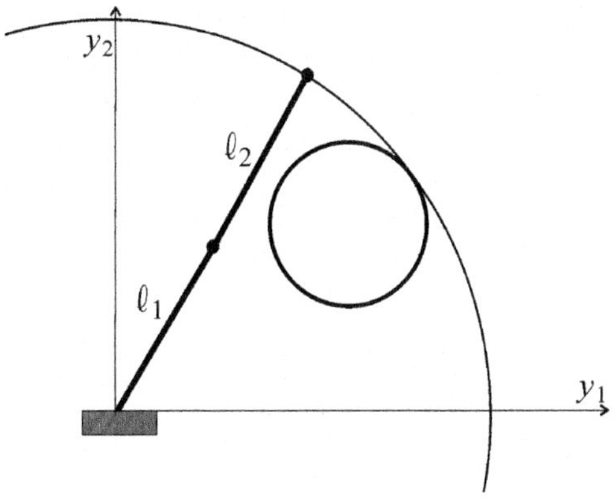

Figure 2.18. *Illustration of a singularity, when both arms are stretched out. The end effector might not be able to follow the circle*

5) The program can be found in the file `crank.m`.

CORRECTION FOR EXERCISE 2.2.– (The three pools)

1) The derivatives of the outputs y_1 and y_2 are expressed by:

$$\begin{aligned} \dot{y}_1 &= \dot{h}_1 = -\alpha(h_1) - \alpha(h_1 - h_2) + u_1 \\ \dot{y}_2 &= \dot{h}_3 = -\alpha(h_3) + \alpha(h_2 - h_3) + u_2 \end{aligned}$$

or, in vector form:

$$\begin{pmatrix} \dot{y}_1 \\ \dot{y}_2 \end{pmatrix} = \underbrace{\begin{pmatrix} 1 & 0 \\ 0 & 1 \end{pmatrix}}_{\mathbf{A}(\mathbf{x})} \mathbf{u} + \underbrace{\begin{pmatrix} -\alpha(h_1) - \alpha(h_1 - h_2) \\ -\alpha(h_3) + \alpha(h_2 - h_3) \end{pmatrix}}_{\mathbf{b}(\mathbf{x})}$$

The feedback:

$$\mathbf{u} = \mathbf{A}^{-1}(\mathbf{x})(\mathbf{v} - \mathbf{b}(\mathbf{x})) = \mathbf{v} - \begin{pmatrix} -\alpha(h_1) - \alpha(h_1 - h_2) \\ -\alpha(h_3) + \alpha(h_2 - h_3) \end{pmatrix}$$

where **v** is our new input, makes our system linear. More precisely, the system looped in such a way has the form:

$$\begin{cases} \dot{y}_1 = v_1 \\ \dot{y}_2 = v_2 \end{cases}$$

2) Let us now try to control this linear system by a system composed of two proportional-integral (PI) controllers of the form:

$$\begin{cases} v_1(t) = \alpha_0\left(w_1(t) - y_1(t)\right) + \alpha_{-1}\int_0^t \left(w_1(\tau) - y_1(\tau)\right) d\tau + \dot{w}_1 \\ v_2(t) = \beta_0\left(w_2(t) - y_2(t)\right) + \beta_{-1}\int_0^t \left(w_2(\tau) - y_2(\tau)\right) d\tau + \dot{w}_2 \end{cases}$$

where w_1 and w_2 are the new setpoints for y_1 and y_2. If we want all our poles to be equal to -1, we need:

$$\begin{cases} s^2 + \alpha_0 s + \alpha_{-1} = (s+1)^2 = s^2 + 2s + 1 \\ s^2 + \beta_0 s + \beta_{-1} = (s+1)^2 = s^2 + 2s + 1 \end{cases}$$

in other words, $\alpha_{-1} = \beta_{-1} = 1, \alpha_0 = \beta_0 = 2$.

3) The state equations for the controller are therefore:

$$\begin{cases} \dot{z}_1 = w_1 - y_1 \\ \dot{z}_2 = w_2 - y_2 \\ v_1 = z_1 + 2\left(w_1 - y_1\right) + \dot{w}_1 \\ v_2 = z_2 + 2\left(w_2 - y_2\right) + \dot{w}_2 \end{cases}$$

The state equations of a state feedback controller for our nonlinear system are therefore:

$$\begin{cases} \dot{z}_1 = w_1 - h_1 \\ \dot{z}_2 = w_2 - h_3 \\ u_1 = z_1 + 2\left(w_1 - h_1\right) + \dot{w}_1 + \alpha\left(h_1\right) + \alpha\left(h_1 - h_2\right) \\ u_2 = z_2 + 2\left(w_2 - h_3\right) + \dot{w}_2 + \alpha\left(h_3\right) - \alpha\left(h_2 - h_3\right) \end{cases}$$

4) The program can be found in the file `pools.m`.

CORRECTION FOR EXERCISE 2.3.– (Train robot)

1) We have:

$$\begin{cases} \ddot{x}_a = \dot{v}_a \cos\theta_a - v_a\dot{\theta}_a \sin\theta_a = u_{a2}\cos\theta_a - v_a u_{a1}\sin\theta_a \\ \ddot{y}_a = \dot{v}_a \sin\theta_a + v_a\dot{\theta}_a \cos\theta_a = u_{a2}\sin\theta_a + v_a u_{a1}\cos\theta_a \end{cases}$$

Therefore:

$$\begin{pmatrix} \ddot{x}_a \\ \ddot{y}_a \end{pmatrix} = \underbrace{\begin{pmatrix} -v_a\sin\theta_a & \cos\theta_a \\ v_a\cos\theta_a & \sin\theta_a \end{pmatrix}}_{\mathbf{A}(v_a,\theta_a)} \begin{pmatrix} u_{a1} \\ u_{a2} \end{pmatrix}$$

2) By using a feedback linearization method, we obtain:

$$\begin{pmatrix} u_{a1} \\ u_{a2} \end{pmatrix} = \mathbf{A}^{-1}(v_a,\theta_a)\cdot\begin{pmatrix} (\hat{x}_a - x_a) + 2\left(\frac{d\hat{x}_a}{dt} - v_a\cos\theta_a\right) + \frac{d^2}{dt^2}\hat{x}_a \\ (\hat{y}_a - y_a) + 2\left(\frac{d\hat{y}_a}{dt} - v_a\sin\theta_a\right) + \frac{d^2}{dt^2}\hat{y}_a \end{pmatrix}$$

$$= \mathbf{A}^{-1}(v_a,\theta_a)\cdot\begin{pmatrix} L_x\sin\omega t - x_a + 2\omega L_x\cos\omega t - 2v_a\cos\theta_a - \omega^2 L_x\sin\omega t \\ L_y\cos\omega t - y_a - 2\omega L_y\sin\omega t + 2v_a\sin\theta_a - \omega^2 L_y\cos\omega t \end{pmatrix}$$

3) Let us apply the controller obtained in the previous question. It is given by:

$$\begin{pmatrix} u_{b1} \\ u_{b2} \end{pmatrix} = \mathbf{A}^{-1}(v_b,\theta_b)\cdot\begin{pmatrix} (\hat{x}_b - x_b) + 2\left(\frac{d\hat{x}_b}{dt} - v_b\cos\theta_b\right) + \frac{d^2\hat{x}_b}{dt^2} \\ (\hat{y}_b - y_b) + 2\left(\frac{d\hat{y}_b}{dt} - v_b\sin\theta_b\right) + \frac{d^2\hat{y}_b}{dt^2} \end{pmatrix}$$

We have:

$$\begin{aligned} \hat{x}_b &= x_a - \ell\cos\theta_a \\ \hat{y}_b &= y_a - \ell\sin\theta_a \\ \frac{d}{dt}\hat{x}_b &= \dot{x}_a + \ell\dot{\theta}_a\sin\theta_a = v_a\cos\theta_a + \ell u_{1a}\sin\theta_a \end{aligned}$$

$$\frac{d}{dt}\hat{y}_b = \dot{y}_a - \ell\dot{\theta}_a \cos\theta_a = v_a \sin\theta_a - \ell u_{1a} \cos\theta_a$$

In order to have $\frac{d^2}{dt^2}\hat{x}_b, \frac{d^2}{dt^2}\hat{y}_b$, we would need to have $\dot{u}_{1a}$, which is not the case. Here, we will simply assume that these two quantities are equal to zero and hope that this approximation will not lead to the instability of the system. We are therefore no longer assured of an exponential convergence of the error toward 0, however we will note that in practice, the behavior remains acceptable.

4) We simply need to recall the same controller as in the previous question. The program can be found in the file `train.m`. Figure 2.19 illustrates the behavior of our train. Cart A is turning in an ellipse, cart B follows cart A and cart C follows cart B.

5) We simply need to take completely independent controllers with a simple temporal delay, for example:

$$\begin{pmatrix} \hat{x}_a \\ \hat{y}_a \end{pmatrix} = \begin{pmatrix} L_x \sin\omega t \\ L_y \cos\omega t \end{pmatrix}, \begin{pmatrix} \hat{x}_b \\ \hat{y}_b \end{pmatrix} = \begin{pmatrix} L_x \sin\omega\,(t-1) \\ L_y \cos\omega\,(t-1) \end{pmatrix}$$
$$\text{and} \begin{pmatrix} \hat{x}_c \\ \hat{y}_c \end{pmatrix} = \begin{pmatrix} L_x \sin\omega\,(t-2) \\ L_y \cos\omega\,(t-2) \end{pmatrix}$$

This control law can be considered centralized since it requires a supervisor capable of sending coherent setpoints to all the robots.

CORRECTION FOR EXERCISE 2.4.– (Controlling a 3D underwater robot)

Let us choose as output the position vector of the robot $\mathbf{p}$. We have:

$$\ddot{\mathbf{p}}(t) = \begin{pmatrix} \dot{v}\cos\theta\cos\psi - v\dot{\theta}\sin\theta\cos\psi - v\dot{\psi}\cos\theta\sin\psi \\ \dot{v}\cos\theta\sin\psi - v\dot{\theta}\sin\theta\sin\psi + v\dot{\psi}\cos\theta\cos\psi \\ -\dot{v}\sin\theta - v\dot{\theta}\cos\theta \end{pmatrix}$$

$$= \begin{pmatrix} \cos\theta\cos\psi & -v\cos\theta\sin\psi & -v\sin\theta\cos\psi \\ \cos\theta\sin\psi & v\cos\theta\cos\psi & -v\sin\theta\sin\psi \\ -\sin\theta & 0 & -v\cos\theta \end{pmatrix} \begin{pmatrix} \dot{v} \\ \dot{\psi} \\ \dot{\theta} \end{pmatrix}$$

$$= \underbrace{\begin{pmatrix} \cos\theta\cos\psi & -v\cos\theta\sin\psi & -v\sin\theta\cos\psi \\ \cos\theta\sin\psi & v\cos\theta\cos\psi & -v\sin\theta\sin\psi \\ -\sin\theta & 0 & -v\cos\theta \end{pmatrix} \begin{pmatrix} 1 & 0 & 0 \\ 0 & \frac{v\sin\varphi}{\cos\theta} & v\frac{\cos\varphi}{\cos\theta} \\ 0 & v\cos\varphi & -v\sin\varphi \end{pmatrix}}_{\mathbf{A}(\mathbf{x})} \begin{pmatrix} u_1 \\ u_2 \\ u_3 \end{pmatrix}$$

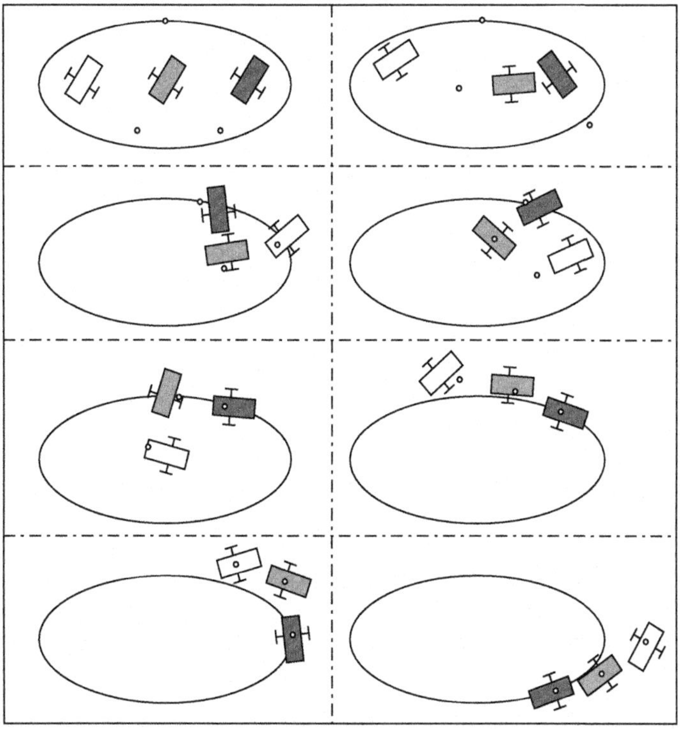

Figure 2.19. *Illustration of the execution of the program; cart A is in bold gray, cart B is in light gray and cart C is in white; the setpoints and attachment points are represented by small circles*

By looping the system using the controller $\mathbf{u} = \mathbf{A}^{-1}(\mathbf{x})\mathbf{v}$, we obtain the decoupled linear system $\ddot{\mathbf{p}} = \mathbf{v}$. This system has three inputs and three outputs. If $\mathbf{w} = (w_1, w_2, w_3)$ represents the path to follow for $\mathbf{p}$, we may consider the following controller:

$$\mathbf{v} = 0.04 \cdot (\mathbf{w} - \mathbf{p}) + 0.4 \cdot (\dot{\mathbf{w}} - \dot{\mathbf{p}}) + \ddot{\mathbf{w}}$$

in order for all the poles to be equal to -0.2 (which corresponds to a characteristic polynomial $P(s) = (s+0.2)^2$). The state feedback controller for our robot is therefore:

$$\mathbf{u} = \mathbf{A}^{-1}(\mathbf{x}) \cdot \left(0.04 \cdot (\mathbf{w} - \mathbf{p}) + 0.4 \cdot \left(\dot{\mathbf{w}} - \begin{pmatrix} v\cos\theta\cos\psi \\ v\cos\theta\sin\psi \\ -v\sin\theta \end{pmatrix} \right) + \ddot{\mathbf{w}} \right)$$

with:

$$\mathbf{w} = \begin{pmatrix} R\sin(f_1 t) + R\sin(f_2 t) \\ R\cos(f_1 t) + R\cos(f_2 t) \\ R\sin(f_3 t) \end{pmatrix}$$

$$\dot{\mathbf{w}} = \begin{pmatrix} Rf_1\cos(f_1 t) + Rf_2\cos(f_2 t) \\ -Rf_1\sin(f_1 t) - Rf_2\sin(f_2 t) \\ Rf_3\cos(f_3 t) \end{pmatrix}$$

$$\ddot{\mathbf{w}} = \begin{pmatrix} -Rf_1^2\sin(f_1 t) - Rf_2^2\sin(f_2 t) \\ -Rf_1^2\cos(f_1 t) - Rf_2^2\cos(f_2 t) \\ -Rf_3^2\sin(f_3 t) \end{pmatrix}$$

The graphical simulation of the robot is given in the file `auv3d.m`. The program contains parts of the results of exercise 1.9.

CORRECTION FOR EXERCISE 2.5.– (Flat system)

1) We have:

$$y = x_1,\ \dot{y} = x_1 + x_2$$
$$\ddot{y} = \dot{x}_1 + \dot{x}_2 = x_1 + x_2 + x_2^2 + u$$

Therefore:

$$x_1 = y,\ \ x_2 = \dot{y} - x_1 = \dot{y} - y$$
$$\begin{aligned} u &= \ddot{y} - \left(x_1 + x_2 + x_2^2\right) = \ddot{y} - \left(y + \dot{y} - y + (\dot{y} - y)^2\right) \\ &= \ddot{y} - \dot{y} - \dot{y}^2 - y^2 + 2y\dot{y} \end{aligned}$$

Thus:

$$\phi(y,\dot{y}) = \begin{pmatrix} y \\ \dot{y} - y \end{pmatrix}$$

$$\psi(y,\dot{y},\ddot{y}) = \ddot{y} - \dot{y} - \dot{y}^2 - y^2 + 2y\dot{y}$$

2) Since $u = \ddot{y} - \dot{y} - \dot{y}^2 + y^2 + 2y\dot{y}$, by taking:

$$u = v - \dot{y} - \dot{y}^2 - y^2 + 2y\dot{y}$$

we obtain $\ddot{y} = v$.

3) By taking $v = (w - y) + 2(\dot{w} - \dot{y}) + \ddot{w}$, the differential equation on the error is $\ddot{e} + 2\dot{e} + e = 0$. The characteristic polynomial is $s^2 + 2s + 1 = (s+1)^2$. The error therefore converges toward zero at e^{-t}.

4) We have:

$$\begin{aligned} u &= v - \dot{y} - \dot{y}^2 + y^2 + 2y\dot{y} \\ &= (w - y) + 2(\dot{w} - \dot{y}) + \ddot{w} - \dot{y} - \dot{y}^2 + y^2 + 2y\dot{y} \\ &= w + 2\dot{w} + \ddot{w} - y - y^2 - 3\dot{y} - \dot{y}^2 + 2y\dot{y} \end{aligned}$$

CORRECTION FOR EXERCISE 2.6.– (Pursuit)

1) The relative distance between the two robots (i.e. the position of robot 2 expressed in the coordinate system of robot 1) is written as:

$$\begin{pmatrix} x \\ y \\ \theta \end{pmatrix} = \begin{pmatrix} \cos\theta_1 & \sin\theta_1 & 0 \\ -\sin\theta_1 & \cos\theta_1 & 0 \\ 0 & 0 & 1 \end{pmatrix} \begin{pmatrix} x_2 - x_1 \\ y_2 - y_1 \\ \theta_2 - \theta_1 \end{pmatrix}$$

Let us differentiate the first two components of this relation with respect to t. We obtain:

$$\begin{pmatrix} \dot{x} \\ \dot{y} \end{pmatrix} = \begin{pmatrix} \cos\theta_1 & \sin\theta_1 \\ -\sin\theta_1 & \cos\theta_1 \end{pmatrix} \begin{pmatrix} \dot{x}_2 - \dot{x}_1 \\ \dot{y}_2 - \dot{y}_1 \end{pmatrix} + \dot{\theta}_1 \begin{pmatrix} -\sin\theta_1 & \cos\theta_1 \\ -\cos\theta_1 & -\sin\theta_1 \end{pmatrix} \begin{pmatrix} x_2 - x_1 \\ y_2 - y_1 \end{pmatrix}$$

However:

$$\begin{cases} \dot{\theta}_1 & = u_2 \\ \begin{pmatrix} \dot{x}_2 \\ \dot{y}_2 \end{pmatrix} & = v_1 \begin{pmatrix} \cos\theta_2 \\ \sin\theta_2 \end{pmatrix} \\ \begin{pmatrix} \dot{x}_1 \\ \dot{y}_1 \end{pmatrix} & = u_1 \begin{pmatrix} \cos\theta_1 \\ \sin\theta_1 \end{pmatrix} \\ \begin{pmatrix} x_2 - x_1 \\ y_2 - y_1 \end{pmatrix} & = \begin{pmatrix} \cos\theta_1 & -\sin\theta_1 \\ \sin\theta_1 & \cos\theta_1 \end{pmatrix} \begin{pmatrix} x \\ y \end{pmatrix} \end{cases}$$

Therefore:

$$\begin{aligned} \begin{pmatrix} \dot{x} \\ \dot{y} \end{pmatrix} &= v_1 \begin{pmatrix} \cos\theta_1 & \sin\theta_1 \\ -\sin\theta_1 & \cos\theta_1 \end{pmatrix} \begin{pmatrix} \cos\theta_2 \\ \sin\theta_2 \end{pmatrix} - u_1 \begin{pmatrix} \cos\theta_1 & \sin\theta_1 \\ -\sin\theta_1 & \cos\theta_1 \end{pmatrix} \begin{pmatrix} \cos\theta_1 \\ \sin\theta_1 \end{pmatrix} \\ &+ u_2 \begin{pmatrix} 0 & 1 \\ -1 & 0 \end{pmatrix} \begin{pmatrix} x \\ y \end{pmatrix} \\ &= v_1 \begin{pmatrix} \cos(\theta_2 - \theta_1) \\ \sin(\theta_2 - \theta_1) \end{pmatrix} - u_1 \begin{pmatrix} 1 \\ 0 \end{pmatrix} + u_2 \begin{pmatrix} y \\ -x \end{pmatrix} \\ &= \begin{pmatrix} -u_1 + v_1 \cos(\theta_2 - \theta_1) + u_2 y \\ v_1 \sin(\theta_2 - \theta_1) - u_2 x \end{pmatrix} \end{aligned}$$

Moreover:

$$\dot{\theta} = \dot{\theta}_2 - \dot{\theta}_1 = v_2 - u_2$$

Thus:

$$\begin{pmatrix} \dot{x} \\ \dot{y} \\ \dot{\theta} \end{pmatrix} = \begin{pmatrix} -u_1 + v_1 \cos\theta + u_2 y \\ v_1 \sin\theta - u_2 x \\ v_2 - u_2 \end{pmatrix}$$

2) We have:

$$\begin{pmatrix} \dot{x} \\ \dot{y} \end{pmatrix} = \begin{pmatrix} v_1 \cos\theta \\ v_1 \sin\theta \end{pmatrix} + \begin{pmatrix} -1 & y \\ 0 & -x \end{pmatrix} \begin{pmatrix} u_1 \\ u_2 \end{pmatrix}$$

We propose the following controller:

$$\begin{pmatrix} u_1 \\ u_2 \end{pmatrix} = \begin{pmatrix} -1 & y \\ 0 & -x \end{pmatrix}^{-1} \left(\begin{pmatrix} w_1 - x \\ w_2 - y \end{pmatrix} + \begin{pmatrix} \dot{w}_1 \\ \dot{w}_2 \end{pmatrix} - \begin{pmatrix} v_1 \cos\theta \\ v_1 \sin\theta \end{pmatrix} \right)$$

Thus, by combining the two previous equations, we obtain the equations on the error at x and y:

$$\begin{pmatrix} w_1 - x \\ w_2 - y \end{pmatrix} + \begin{pmatrix} \dot{w}_1 - \dot{x} \\ \dot{w}_2 - \dot{y} \end{pmatrix} = 0$$

The characteristic polynomial of these two errors is $P(s) = s + 1$, which tells us that the error converges toward 0 at e^{-t}.

3) We have a singularity of our control law when:

$$\det \begin{pmatrix} -1 & y \\ 0 & -x \end{pmatrix} = 0$$

in other words when $x = 0$, and also when robot 2 is on the left or the right of robot 1.

4) The program (refer to the file `pursuit.m`) is the following:

```
xa=[-10;-10;0]; xb=[-5;-5;0]; dt=0.02;
for t=0:dt:50,
v=[3;sin(0.2*t)];
x=[cos(xa(3)),sin(xa(3)),0;-sin(xa(3)),
cos(xa(3)),0;0,0,1]*(xb-xa);
w=[10;0]; dw=[0;0];
u=inv([-1 x(2);0 -x(1)])*(w-x(1:2)+dw-v(1)*[cos(x(3));
sin(x(3))]);
xa=xa+f(xa,u)*dt; xb=xb+f(xb,v)*dt;
end;
```

This program uses the following evolution function:

```
function xdot = f(x,u)
xdot=[u(1)*cos(x(3)); u(1)*sin(x(3)); u(2)];
end
```

CORRECTION FOR EXERCISE 2.7.– (Controlling the SAUCISSE robot)

1) The differential dependency graph is given in Figure 2.20.

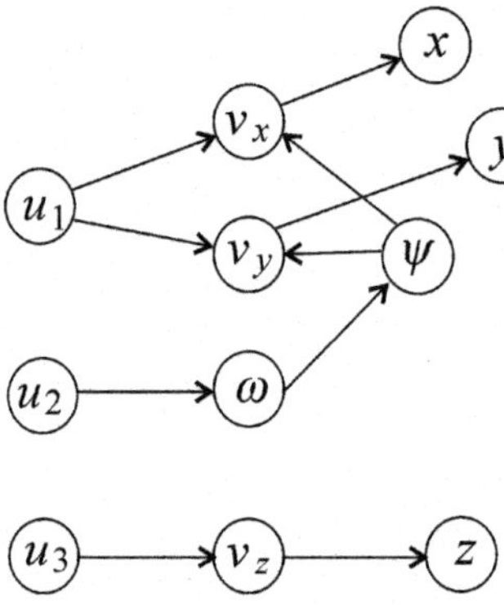

Figure 2.20. *Differential dependency graph of the underwater robot*

2) The differential delay matrix is:

$$\mathbf{R} = \begin{pmatrix} 2 & 4 & \infty \\ 2 & 4 & \infty \\ \infty & \infty & 2 \end{pmatrix}$$

The relative degrees are $k_1 = 2, k_2 = 2$ and $k_3 = 2$. However, since the second column is such that $\forall i,\ r_{i2} > k_i$, the method of linearizing feedback will necessarily lead to a matrix $\mathbf{A}(\mathbf{x})$ that is singular for any state vector $\mathbf{x}$ (see section 2.3.3).

3) We obtain the following state equations for our new system:

$$\begin{cases} \dot{x} &= v_x \\ \dot{y} &= v_y \\ \dot{z} &= v_z \\ \dot{\psi} &= \omega \\ \dot{v}_x &= c_1 \cos \psi \\ \dot{v}_y &= c_1 \sin \psi \\ \dot{v}_z &= a_3 \\ \dot{\omega} &= a_2 \\ \dot{c}_1 &= b_1 \\ \dot{b}_1 &= a_1 \end{cases}$$

where b_1 and c_1 are the new state variables attached to our integrators. The differential dependency graph is given in Figure 2.21. The differential delay matrix is given by:

$$\mathbf{R} = \begin{pmatrix} 4 & 4 & \infty \\ 4 & 4 & \infty \\ \infty & \infty & 2 \end{pmatrix}$$

The relative degrees are $k_1 = 4, k_2 = 4$ and $k_3 = 2$.

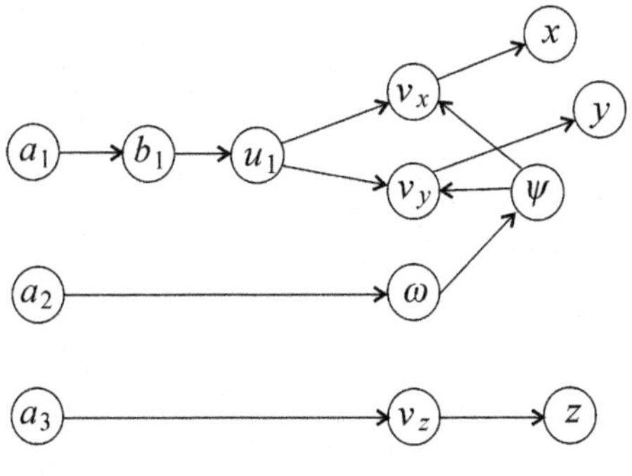

Figure 2.21. *Differential dependency graph after adding the integrators*

4) In order to linearize the delayed system by feedback, we need to calculate $\dddot{x}$ and $\dddot{y}$ in function of $\mathbf{x}$ and $\mathbf{u}$. We have:

$$\begin{cases} \ddot{x} &= c_1 \cos\psi \\ \dddot{x} &= b_1 \cos\psi - c_1\omega \sin\psi \\ \ddddot{x} &= a_1 \cos\psi - 2b_1\omega \sin\psi - c_1 a_2 \sin\psi - c_1\omega^2 \cos\psi \\ \ddot{y} &= c_1 \sin\psi \\ \dddot{y} &= b_1 \sin\psi + c_1\omega \cos\psi \\ \ddddot{y} &= a_1 \sin\psi + 2b_1\omega \cos\psi + c_1 a_2 \cos\psi - c_1\omega^2 \sin\psi \\ \ddot{z} &= a_3 \end{cases}$$

Therefore:

$$\begin{pmatrix} \ddddot{x} \\ \ddddot{y} \\ \ddot{z} \end{pmatrix} = \underbrace{\begin{pmatrix} \cos\psi & -c_1 \sin\psi & 0 \\ \sin\psi & c_1 \cos\psi & 0 \\ 0 & 0 & 1 \end{pmatrix}}_{\mathbf{A}(\mathbf{x},c_1)} \begin{pmatrix} a_1 \\ a_2 \\ a_3 \end{pmatrix} + \underbrace{\begin{pmatrix} -2b_1\omega \sin\psi - c_1\omega^2 \cos\psi \\ 2b_1\omega \cos\psi - c_1\omega^2 \sin\psi \\ 0 \end{pmatrix}}_{\mathbf{b}(\mathbf{x},b_1,c_1)}$$

In order to have a feedback system of the form:

$$\begin{pmatrix} \ddddot{x} \\ \ddddot{y} \\ \ddot{z} \end{pmatrix} = \begin{pmatrix} v_1 \\ v_2 \\ v_3 \end{pmatrix}$$

We will take:

$$\begin{aligned} \mathbf{a} &= \mathbf{A}^{-1}(\mathbf{x}, c_1)\,(\mathbf{v} - \mathbf{b}(\mathbf{x}, b_1, c_1)) \\ &= \begin{pmatrix} \cos\psi & -c_1 \sin\psi & 0 \\ \sin\psi & c_1 \cos\psi & 0 \\ 0 & 0 & 1 \end{pmatrix}^{-1} \left(\begin{pmatrix} v_1 \\ v_2 \\ v_3 \end{pmatrix} - \begin{pmatrix} -2b_1\omega \sin\psi - c_1\omega^2 \cos\psi \\ 2b_1\omega \cos\psi - c_1\omega^2 \sin\psi \\ 0 \end{pmatrix} \right) \\ &= \begin{pmatrix} \cos\psi & \sin\psi & 0 \\ -\frac{\sin\psi}{c_1} & \frac{\cos\psi}{c_1} & 0 \\ 0 & 0 & 1 \end{pmatrix} \begin{pmatrix} v_1 \\ v_2 \\ v_3 \end{pmatrix} + \begin{pmatrix} \omega^2 c_1 \\ \frac{-2\omega b_1}{c_1} \\ 0 \end{pmatrix} \end{aligned} \quad [2.22]$$

5) By noting that $\dot{b}_1 = a_1$ and $u_2 = a_2$, we deduce that the state equations of our dynamic linearizer are:

$$\left\{\begin{array}{ll} \dot{c}_1 & = b_1 \\ \dot{b}_1 & = v_1 \cos\psi + v_2 \sin\psi + \omega^2 c_1 \\ \begin{pmatrix} u_1 \\ u_2 \\ u_3 \end{pmatrix} & = \begin{pmatrix} c_1 \\ \frac{1}{c_1}\left(v_2 \cos\psi - v_1 \sin\psi - 2\omega b_1\right) \\ v_3 \end{pmatrix} \end{array}\right. \quad [2.23]$$

This linearizing feedback has $\mathbf{v}$ as input vector, $\mathbf{u}$ as output and (c_1, b_1) as state vector. In order to have all the poles at -1, we choose (refer to equation [2.2]):

$$\left\{\begin{array}{l} v_1 = (w_1 - y_1) + 4(\dot{w}_1 - \dot{y}_1) + 6(\ddot{w}_1 - \ddot{y}_1) + 4(\dddot{w}_1 - \dddot{y}_1) + \ddddot{w}_1 \\ v_2 = (w_2 - y_2) + 4(\dot{w}_2 - \dot{y}_2) + 6(\ddot{w}_2 - \ddot{y}_2) + 4(\dddot{w}_2 - \dddot{y}_2) + \ddddot{w}_2 \\ v_3 = (w_3 - y_3) + 2(\dot{w}_3 - \dot{y}_3) + \ddot{w}_3 \end{array}\right.$$

in other words:

$$\left\{\begin{array}{l} v_1 = (x_d - x) + 4(\dot{x}_d - v_x) + 6(\ddot{x}_d - c_1 \cos\psi) \\ \qquad\qquad + 4(\dddot{x}_d - (b_1 \cos\psi - c_1\omega \sin\psi)) + \ddddot{x}_d \\ v_2 = (y_d - y) + 4(\dot{y}_d - v_y) + 6(\ddot{y}_d - c_1 \sin\psi) \\ \qquad\qquad + 4(\dddot{y}_d - (b_1 \sin\psi + c_1\omega \cos\psi)) + \ddddot{y}_d \\ v_3 = (z_d - z) + 2(\dot{z}_d - v_z) + \ddot{z}_d \end{array}\right. \quad [2.24]$$

where (x_d, y_d, z_d) corresponds to the desired trajectory for the center of the robot. By combining equations [2.23] and [2.24], we obtain the requested controller, which is of the form:

$$\left\{\begin{array}{l} \dot{\mathbf{x}}_r = \mathbf{f}_r(\mathbf{x}, \mathbf{x}_r, t) \\ \mathbf{u} = \mathbf{g}_r(\mathbf{x}, \mathbf{x}_r, t) \end{array}\right.$$

where $\mathbf{x} = (x, y, z, \psi, v_x, v_y, v_z, \omega)$ is the state vector of the robot and where $\mathbf{x}_r = (c_1, b_1)$ is the state vector of the controller. The latter corresponds to two integrators added in front of the system with the aim of eliminating the singularity in the linearization. The approach presented in this example is called *dynamic feedback linearization*. Indeed, the fact of having added integrators forces a state representation for our controller, in contrast to the situation without singularity which leads to a static relation for our controller.

3

Model-free Control

When we implement a controller for a robot and perform the initial tests we rarely succeed on the first try, which leads us to the problem of debugging. It might be that the compass is subject to electromagnetic disturbances, that it is placed upside-down, that there is a unit conversion problem in the sensors, that the motors are saturated or that there is a sign problem in the equations of the controller. The problem of debugging is a complex one and it is wise to respect the *continuity principle*: each step in the construction of the robot must be of reasonable size and has to be validated before pursuing construction. Thus, for a robot, it is desirable to implement a simple intuitive controller that is easy to debug before setting up a more advanced one. This principle cannot always be applied. However, if we have a good *a priori* understanding of the control law to apply, then such a continuity principle can be followed. Among mobile robots for which a pragmatic controller can be imagined, we can distinguish at least two subclasses:

– *vehicle-robots*: these are systems built by man to be controlled by man such as the bicycle, sailboat, car, etc. We will try to copy the control law used by humans and transform it into an algorithm;

– *biomimetic robots*: these robots are inspired by the movement of human beings. We have been able to observe them for long periods and deduce the strategy developed by nature to design its control law. This is the *biomimetic* approach (see, for example, [BOY 06]). We do not include walking robots in this category because, even though we all know how to walk, it is near to impossible to know which control law we use for it. Thus, designing a control law for walking robots [CHE 07] cannot be done without a complete mechanical modeling of walking and without using any theoretical automatic control methods such as those evoked in Chapter 2.

For these two classes, we often do not have simple and reliable models available (this is the case, for example, of the sailboat or bicycle). However, the strong understanding we have of them will allow us to build a robust control law.

The aim of this chapter is to show, using several examples, how to design such control laws. These will be referred to as *mimetic control* (we are trying to imitate humans or animals) or *model-free control* (we do not use the state equations of the robot to design the controller). Although model-free approaches have been largely explored in theory (see, for instance, [FLI 13]), here we will use the intuition we have of the functioning of our robot as much as possible.

3.1. Model-free control of a robot cart

In order to illustrate the principle of model-free control, let us consider the case of a robot cart described by the equations:

$$\begin{cases} \dot{x} = v\cos\theta \\ \dot{y} = v\sin\theta \\ \dot{\theta} = u_2 \\ \dot{v} = u_1 - v \end{cases}$$

This model can be used for simulation, but not for obtaining the controller.

3.1.1. *Proportional heading and speed controller*

We will now propose a simple controller for this system by using our intuition about the system. Let us take $\widetilde{\theta} = \theta_d - \theta$ where θ_d is the desired heading and $\widetilde{v} = v_d - v$ where v_d is the desired speed.

– For speed control, we take:

$$u_1 = a_1 \tanh \widetilde{v}$$

where a_1 is a constant representing the maximum acceleration (in absolute value) that the motor is able to deliver. The hyperbolic tangent *tanh* (see Figure 3.1) is used as saturation function. Let us recall that:

$$\tanh x = \frac{e^x - e^{-x}}{e^x + e^{-x}} \qquad [3.1]$$

– For the heading control, we take:

$$u_2 = a_2 \cdot \text{sawtooth}(\widetilde{\theta})$$

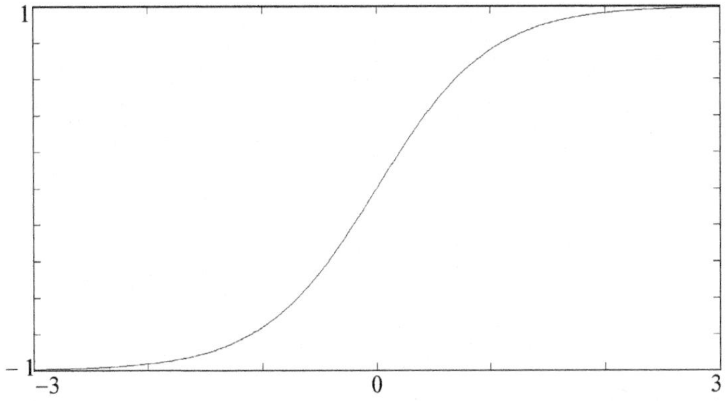

Figure 3.1. *Hyperbolic tangent function used as saturation function*

In this last formula, *sawtooth* corresponds to the sawtooth function defined by:

$$\text{sawtooth}(\widetilde{\theta}) = 2\text{atan}\left(\tan\frac{\widetilde{\theta}}{2}\right) = \text{mod}(\widetilde{\theta} + \pi, 2\pi) - \pi \qquad [3.2]$$

Let us note that for numerical reasons, it is preferable to use the expression containing the *modulus* function (`mod` in MATLAB). As shown in Figure 3.2, the function corresponds to an error in heading. The interest in taking an error $\widetilde{\theta}$ filtered by the *sawtooth* function is to avoid the problem of the $2k\pi$ modulus: we would like a $2k\pi$ to be considered non-zero.

We may summarize this controller by:

$$\begin{pmatrix} u_1 \\ u_2 \end{pmatrix} = \begin{pmatrix} a_1 \cdot \tanh(v_d - v) \\ a_2 \cdot \text{sawtooth}(\theta_d - \theta) \end{pmatrix}$$

Thus, we will perform heading control. This model-free control, which works very well in practice, does not need to use the state equations of the robot. It is based on the understanding that we have of the dynamics of the system and recalls our wireless operation method of the cart robot. It has two parameters a_1 and a_2 that are easy to set (a_1 represents the propelling power and a_2 represents the directional disturbances). Finally, this controller is easy to implement and debug.

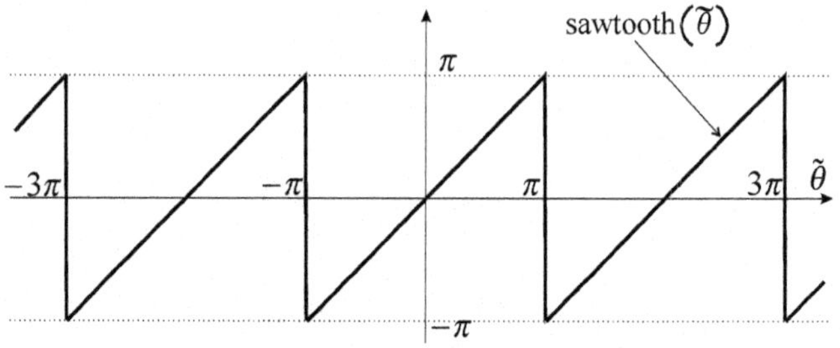

Figure 3.2. Sawtooth *function used to avoid the jumps in the heading control*

3.1.2. *Proportional-derivative heading controller*

For many robots, a proportional controller creates oscillations and it might prove to be necessary to add a damping or derivative term. This is the case for underwater exploration robots (of the type *remotely operated vehicle* (ROV)) which are meant to stabilize above the zone of interest. Underwater torpedo robots do not have this oscillation problem given their control surfaces that stabilize the heading while in movement. If the heading is constant, such a proportional-derivative controller is given by:

$$u_2 = a_2 \cdot \text{sawtooth}\,(\theta_d - \theta) + b_2\dot{\theta}$$

The quantity θ may be obtained by a compass, for example. As for $\dot{\theta}$, it is generally obtained by a gyro. Low-cost robots do not always have a gyro available and we must try to approximate $\dot{\theta}$ from measurements of θ. However, a compass might jump by 2π for small variations in heading. This is the case, for instance, when a compass returns an angle within the interval $[-\pi, \pi]$ and the heading varies around $(2k+1)\,\pi$. In this case, an approximation of $\dot{\theta}$ must be obtained and this approximation has to be insensitive to these jumps. Let us denote by:

$$\mathbf{R}_t = \begin{pmatrix} \cos\theta\,(t) & -\sin\theta\,(t) \\ \sin\theta\,(t) & \cos\theta\,(t) \end{pmatrix}$$

the rotation matrix corresponding to the heading $\theta(t)$ of the robot (let us note that this matrix is insensitive to jumps of $2k\pi$). Note that:

$$\mathbf{R}_t^{\mathrm{T}}\dot{\mathbf{R}}_t = \begin{pmatrix} 0 & -\dot{\theta}(t) \\ \dot{\theta}(t) & 0 \end{pmatrix}$$

This relation can be seen as a two-dimensional (2D) version of relation [1.1] in section1.1.2 , and can also be directly obtained by using the expression of $\mathbf{R}_t$. Thus, an Euler integration of the rotation matrix:

$$\mathbf{R}_{t+dt} = \mathbf{R}_t + dt\dot{\mathbf{R}}_t$$

translates to:

$$\mathbf{R}_{t+dt} = \mathbf{R}_t + dt \cdot \mathbf{R}_t \begin{pmatrix} 0 & -\dot{\theta}(t) \\ \dot{\theta}(t) & 0 \end{pmatrix} = \mathbf{R}_t \left(\mathbf{I} + dt \begin{pmatrix} 0 & -\dot{\theta}(t) \\ \dot{\theta}(t) & 0 \end{pmatrix} \right)$$

Therefore:

$$\begin{aligned} dt \begin{pmatrix} 0 & -\dot{\theta}(t) \\ \dot{\theta}(t) & 0 \end{pmatrix} &= \mathbf{R}_t^{\mathrm{T}}\mathbf{R}_{t+dt} - \mathbf{I} \\ &= \begin{pmatrix} \cos(\theta(t+dt) - \theta(t)) & -\sin(\theta(t+dt) - \theta(t)) \\ \sin(\theta(t+dt) - \theta(t)) & \cos(\theta(t+dt) - \theta(t)) \end{pmatrix} \\ &\quad - \begin{pmatrix} 1 & 0 \\ 0 & 1 \end{pmatrix} \end{aligned}$$

Let us take in this matrix equation the scalar equation corresponding to the second row and first column. We obtain:

$$\dot{\theta}(t) = \frac{\sin(\theta(t+dt) - \theta(t))}{dt}$$

The proportional-derivative heading controller can, therefore, be written as:

$$u_2(t) = a_2 \cdot \mathrm{sawtooth}(\theta_d - \theta(t)) + b_2 \frac{\sin(\theta(t) - \theta(t-dt))}{dt}$$

which will be insensitive to jumps of 2π.

3.2. Skate car

Let us consider the skating vehicle [JAU 10] represented in Figure 3.3.

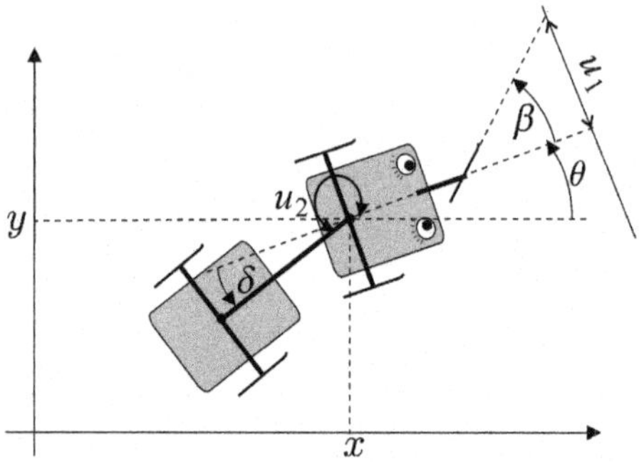

Figure 3.3. *Skating robot moving like a snake*

This vehicle that we will refer to as *a skate car* is purely imaginary. It is designed such that it moves on a frozen lake and stands on five ice skates. This system has two inputs: the tangent u_1 of the angle β of the front skate (we have chosen the tangent as input in order to avoid the singularities) and u_2 the torque exerted at the articulation between the two carts and corresponding to the angle δ. The thrust, therefore, only comes from the torque u_2 and recalls the propulsion mode of a snake or an eel [BOY 06]. Any control over u_1 will, therefore, not bring any energy to the system, but indirectly participates in the propulsion by generating waves. In this section, we will propose a model in the form of a state for simulating the system. Concerning the control law, the existing general methods cannot deal with this kind of system and it is necessary to take into account the physics of the problem. Therefore, we will propose a mimetic control law that allows us to obtain an efficient controller.

3.2.1. *Model*

Let us try to obtain state equations capable of representing the dynamics of the system in order to simulate our system. The state variables are chosen to be $\mathbf{x} = (x, y, \theta, v, \delta)$, where x, y, θ correspond to the position of the front cart, v represents the speed of the center of the front sled axle and δ is the angle between the two carts. The angular speed of the front sled is given by:

$$\dot{\theta} = \frac{v_1 \sin \beta}{L_1} \qquad [3.3]$$

where v_1 is the speed of the front skate and L_1 is the distance between the front skate and the center of the front sled axle. However:

$$v = v_1 \cos \beta$$

and therefore:

$$\dot{\theta} = \frac{v \tan \beta}{L_1} = \frac{v u_1}{L_1} \qquad [3.4]$$

Viewed from the rear sled, everything is as if there was a virtual skate in the middle of the front sled axle, moving together with it. Thus, by recalling formula [3.3], the angular speed of the rear sled is:

$$\dot{\theta} + \dot{\delta} = -\frac{v \sin \delta}{L_2}$$

where L_2 is the distance between the centers of the axles, and therefore:

$$\dot{\delta} = -\frac{v \sin \delta}{L_2} - \dot{\theta} \overset{[3.4]}{=} -\frac{v \sin \delta}{L_2} - \frac{v u_1}{L_1} \qquad [3.5]$$

Following the theorem of kinetic energy, the temporal derivative of kinetic energy is equal to the sum of the powers supplied to the system, in other words:

$$\frac{d}{dt}\left(\frac{1}{2} m v^2\right) = \underbrace{u_2 \cdot \dot{\delta}}_{\text{engine power}} - \underbrace{(\alpha v) \cdot v}_{\text{dissipated power}} \qquad [3.6]$$

where α is the coefficient of viscous friction. For reasons of simplicity, we will assume here that the force of friction is equal to αv, which is the same as assuming that only the front sled is braking. Therefore, we have:

$$m v \dot{v} \overset{[3.6]}{=} u_2 \cdot \dot{\delta} - \alpha v^2 \overset{[3.5]}{=} u_2 \cdot \left(-\frac{v \sin \delta}{L_2} - \frac{v u_1}{L_1}\right) - \alpha v^2$$

and:

$$m \dot{v} = u_2 \cdot \left(-\frac{\sin \delta}{L_2} - \frac{u_1}{L_1}\right) - \alpha v \qquad [3.7]$$

The system can be described by the following state equations:

$$\begin{cases} \dot{x} = & v\cos\theta \\ \dot{y} = & v\sin\theta \\ \dot{\theta} \overset{[3.4]}{=} & vu_1 \\ \dot{v} \overset{[3.7]}{=} & -(u_1+\sin\delta)\,u_2 - v \\ \dot{\delta} \overset{[3.5]}{=} & -v\,(u_1+\sin\delta) \end{cases} \qquad [3.8]$$

where, for reasons of simplicity, the coefficients (mass m, coefficient of viscous friction α, interaxle distances L_1, L_2, etc.) have been given unit values. This system could be made control-affine (refer to equation [2.6]) by adding an integrator in front of u_1, however, the feedback linearization method cannot be applied due to the numerous singularities. Indeed, it can be easily shown that when the speed v is zero (easy to avoid) or when $\dot{\delta} = 0$ (which necessarily happens regularly), we have a singularity. A *biomimetic* controller that imitates the propulsion of the snake or eel might be feasible.

3.2.2. *Sinusoidal control*

By trying to imitate the control strategy of an undulating snake's movement, we choose u_1 of the form:

$$u_1 = p_1\cos(p_2 t) + p_3$$

where p_1 is the amplitude, p_2 is the pulse and p_3 is the bias. We choose u_2 such that the propelling torque is a motor torque, in other words $\dot{\delta}u_2 \geq 0$. Indeed, $\dot{\delta}u_2$ corresponds to the power supplied to the robot that is transformed into kinetic energy. If u_2 is bounded by the interval $[-p_4, p_4]$, we choose a bang-bang-type controller for u_2 of the form:

$$u_2 = p_4 \cdot \text{sign}(\dot{\delta})$$

which is equivalent to exerting maximum propulsion. The chosen state feedback controller is therefore:

$$\mathbf{u} = \begin{pmatrix} p_1\cos(p_2 t) + p_3 \\ p_4\,\text{sign}\left(-v\,(u_1+\sin\delta)\right) \end{pmatrix}$$

The parameters of the controller remain to be determined. The bias parameter p_3 allows it to direct its heading. The power of the motor torque gives us p_4. The parameter p_1 is directly linked to the amplitude of the oscillation created during movement. Finally, the parameter p_2 gives the frequency of the oscillations. The simulations can help us to set the parameters p_1 and p_2 correctly. Figure 3.4 illustrates two simulations in which the robot begins with an almost zero speed. In the simulation on top, the bias p_3 is equal to zero. In the bottom simulation, $p_3 > 0$.

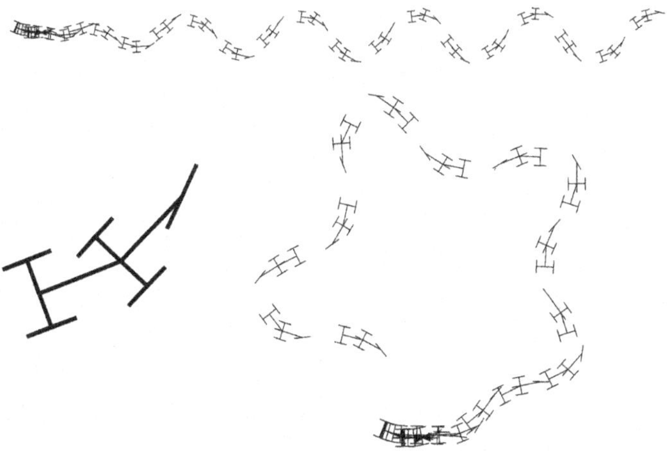

Figure 3.4. *Various simulations illustrate the control law for the skating robot*

Figure 3.5 represents the advance as a function of time. It is clear that the power supplied by the engine is very strong at startup whereas in cruising regime, it is underutilized. Such a controller forces us to oversize our engine. It would be to our advantage to have a thrust as constant as possible. The script in `snake.m` contains an implementation of this control law.

3.2.3. *Maximum thrust control*

The propulsion of the robot is done by the thrust $u_2 \cdot \dot{\delta} = -v\,(u_1 + \sin\delta) \cdot u_2$ and therefore by the engine that generates the torque u_2. In order to move as fast as possible, for a given motor, the engine should supply a maximum amount of power denoted by $\bar{p}$ which will be transformed into kinetic energy. Thus:

$$\underbrace{-v\left(u_1+\sin\delta\right)}_{\dot{\delta}}\cdot u_2=\bar{p}$$

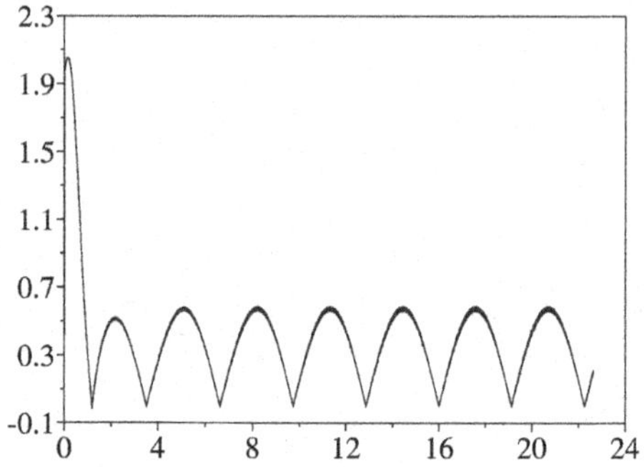

Figure 3.5. *Thrust supplied by the engine* u_2

Therefore, there are several torques (u_1, u_2) capable of supplying the desired power $\bar{p}$. Therefore, we will choose for u_2 the form:

$$u_2=\varepsilon\cdot\bar{u}_2 \text{ with } \varepsilon=\pm 1$$

where $\varepsilon(t)$ is a square wave and $\bar{u}_2$ is a constant. This choice for u_2 may be bound to the engine torque and thereby limit the mechanical load. If we choose the frequency of ε too low, the power supplied will be respected, but the front cart will collide with the rear cart. In the borderline case where ε is constant, we can observe, through the simulation, the first cart roll up to the second (which means that δ increases to infinity). We obtain, by isolating the orientation u_1 of the front skate:

$$u_1=-\left(\frac{\bar{p}}{v\varepsilon\bar{u}_2}+\sin\delta\right)$$

The maximum thrust controller is, therefore, given by:

$$\mathbf{u}=\begin{pmatrix}-\left(\frac{\bar{p}}{v\varepsilon\bar{u}_2}+\sin\delta\right)\\ \varepsilon\bar{u}_2\end{pmatrix} \quad [3.9]$$

Thus, with this controller, not only do we always thrust in the correct direction through u_2 but we can also adjust the direction u_1 in order for the torque supplied by u_2 to translate into a maximum thrust $\bar{p}$. Now, we only need to act on ε (which, as we recall, is a square wave equal to ± 1) and on the power $\bar{p}$. The duty cycle of the signal $\varepsilon(t)$ will allow us to direct our orientation and its frequency will give us the amplitude of the oscillations for the robot's path. As for $\bar{u}_2$, it allows us to control the average speed of the robot. In the simulation, this controller turns out indeed to be more efficient than the sinusoidal controller. Figure 3.6 shows the angle of the skate β as a function of time, once the cruising regime has been reached. Let us note that the angle of the front skate β =atan(u_1) makes discontinuities appear.

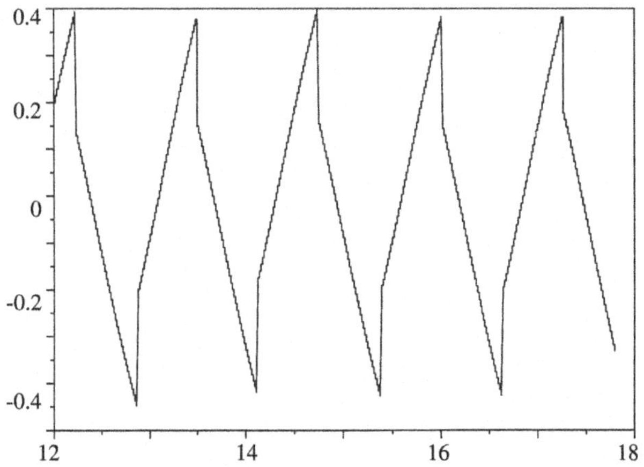

Figure 3.6. *Evolution of the front skate angle β in cruising regime*

3.2.4. *Simplification of the fast dynamics*

The state equations for the *snakeboard* contain numerous singularities and we would like to simplify them here. However, in our system, we have two interfering dynamics: one that is slow (representing the smooth evolution of the state variables) and the other that is fast (rated by ε) which creates the undulation. The idea, relatively standard in automatic control, is to average these values in a way to make the fast dynamics disappear. We can find this idea in *pulse width modulation* (PWM)-controlled Direct Current (DC) engines.

Let us consider a high-frequency square wave signal $\varepsilon(t)$. Its temporal average $\bar{\varepsilon}$ is called the *duty cycle*. This duty cycle is set to vary very slowly in time. The *temporal average* operator is linear (just like the mathematical expectation). For example:

$$\overline{2\varepsilon_1(t) - 3\varepsilon_2(t)} = 2\bar{\varepsilon}_1(t) - 3\bar{\varepsilon}_2(t)$$

However, for a nonlinear function f, we cannot write $\overline{f(\varepsilon)} = f(\bar{\varepsilon})$. For instance, $\overline{\varepsilon^{-1}} \neq \bar{\varepsilon}^{-1}$. However, we will have $\overline{\varepsilon^{-1}} = \bar{\varepsilon}$ if $\varepsilon(t) \in \{-1, 1\}$. This comes from the fact that the signals ε and ε^{-1} are equal in such a case. If $a(t)$ and $b(t)$ are slowly varying signals in time, we will also have:

$$\overline{a(t)\ \varepsilon_1(t) + b(t)\ \varepsilon_2(t)} = a(t)\ \bar{\varepsilon}_1(t) + b(t)\ \bar{\varepsilon}_2(t)$$

We will try to apply these approximations to the case of the *snakeboard* in order to eliminate the fast dynamics. By recalling the state equations in [3.8] and by injecting control law [3.9], we obtain a feedback system described by the following state equations:

$$\begin{cases} \dot{x} = v\cos\theta \\ \dot{y} = v\sin\theta \\ \dot{\theta} = -\frac{\bar{p}}{\varepsilon\bar{u}_2} - v\sin\delta \\ \dot{v} = \frac{\bar{p}}{v} - v \\ \dot{\delta} = \frac{\bar{p}}{\varepsilon\bar{u}_2} \end{cases}$$

Recall that we can act on the constants $\bar{p}$, $\bar{u}_2$ and on the square wave signal $\varepsilon = \pm 1$ that we will here consider to be high-frequency and with a duty cycle of $\bar{\varepsilon}$. We can approximate:

$$\frac{\bar{p}}{\varepsilon\bar{u}_2} = \frac{\varepsilon\bar{p}}{\bar{u}_2} \overset{\text{(by linearization)}}{\simeq} \bar{\varepsilon}\frac{\bar{p}}{\bar{u}_2}$$

Thus, the system becomes:

$$\begin{cases} \dot{x} = v\cos\theta \\ \dot{y} = v\sin\theta \\ \dot{\theta} = -\bar{p}\bar{q} - v\sin\delta \\ \dot{v} = \frac{\bar{p}}{v} - v \\ \dot{\delta} = \bar{p}\bar{q} \end{cases}$$

which now only has two inputs: $\bar{p}$ and $\bar{q} = \frac{\bar{\varepsilon}}{\bar{u}_2}$. Let us try to control the inputs θ and v by using feedback linearization method. Note that although this system is not affine in its inputs ($\bar{p}$ and $\bar{q}$), the method can be applied because, as we will see below, the necessary inversion is possible here. For this, let us define two new inputs v_1, v_2 such that:

$$\begin{cases} v_1 = -\bar{p}\bar{q} - v\sin\delta \\ v_2 = \frac{\bar{p}}{v} - v \end{cases}$$

By inverting this system relative to the inputs, we obtain:

$$\bar{p} = v\,(v_2 + v)$$
$$\bar{q} = -\frac{v_1 + v\sin\delta}{v(v_2 + v)}$$

Thus, the feedback linearized system is:

$$\begin{cases} \dot{\theta} = v_1 \\ \dot{v} = v_2 \end{cases}$$

A proportional controller is, therefore, sufficient. We will take one that places the poles at -1, in other words:

$$\begin{cases} v_1 = w_1 - \theta + \dot{w}_1 \\ v_2 = w_2 - v + \dot{w}_2 \end{cases}$$

Let us summarize the control law in its entirety. Its setpoints are w_1, w_2 which correspond to the desired heading and speed. It is given by the following table:

$$\bar{p} = v\,(w_2 + \dot{w}_2)$$

$$\bar{q} = -\frac{w_1 - \theta + \dot{w}_1 + v\sin\delta}{v\,(w_2 + \dot{w}_2)}$$

$\bar{\varepsilon} = \bar{q} \cdot \bar{u}_2$ (choose $\bar{\varepsilon} \in [-1, 1]$)

ε : duty cycle $\bar{\varepsilon}$ and frequency slot ∞

$$\mathbf{u} = \begin{pmatrix} -\left(\frac{\bar{p}}{v\varepsilon\bar{u}_2} + \sin\delta\right) \\ \varepsilon\bar{u}_2 \end{pmatrix}$$

The adjustment parameter $\bar{u}_2$ involved in this controller is quite delicate to set. We need to choose $\bar{u}_2$ small enough to have $\bar{\varepsilon} \in [-1, 1]$. However, it must not be too close to zero in order for u_1 to not be too large (which would cause too significant front skate movements). This variable $\bar{u}_2$ influences the necessary distribution between the torque (through u_2) and the movement (through u_1) for generating power. Let us finally note that this latter control law, supposed to be more efficient, uses state equations of the system in its design and it is, therefore, difficult to call it *model-free*.

3.3. Sailboat

3.3.1. *Problem*

Let us recall the principles of model-free control and try to adapt it to a line-tracking controller for our sailboat. Here, we will consider a sailboat whose sheet length is variable, but not directly the angle of the sail as was the case until now (see [2.11] in section 2.6). This robot has two inputs which are the angle of the rudder $u_1 = \delta_r$ and the maximum angle of the sail $u_2 = \delta_s^{\max}$ (equivalently u_2 corresponds to the length of the sheet). We will try to make the robot follow a line which passes through points $\mathbf{a}$ and $\mathbf{b}$ (see Figure 3.7).

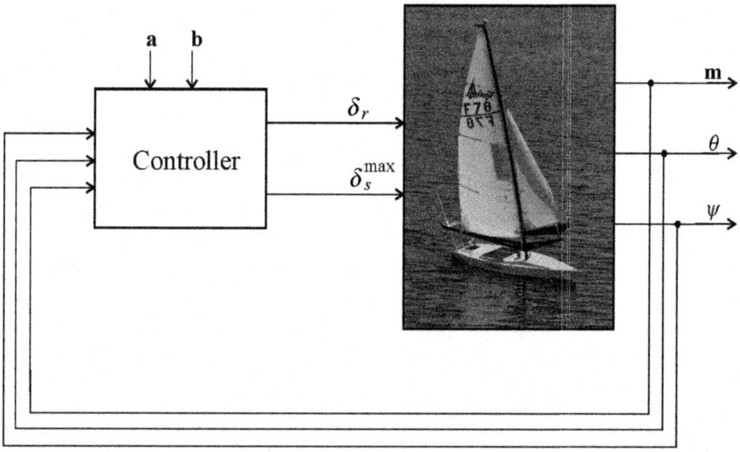

Figure 3.7. *Feedback control of the sailing robot*

This problem is influenced by the control strategies of the VAIMOS robot [GOR 11] of IFREMER, the sailing boat of the ERWAN naval school and the Optimousse robot from the ENSTA-Bretagne (see Figure 3.8).

Figure 3.8. *a) The* VAIMOS *sailing robot of* IFREMER *(in the background) and the Optimousse robot from the* ENSTA *Bretagne; b) robot of the Ecole Navale (Naval School) following a line*

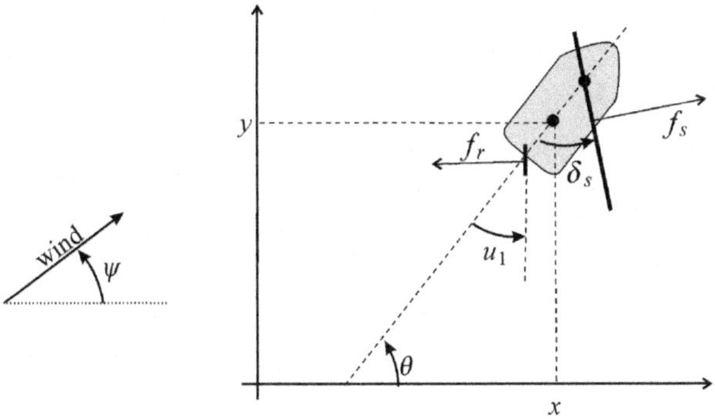

Figure 3.9. *Variables used in the state equations of the robot*

As shown in Figure 3.9, we will denote by (x, y, θ) the posture of the boat, by v its advancing speed, by ω its angular speed, by f_s the force of the wind on the sail, by f_r the force of the water on the rudder, by δ_s the angle of the sail and by ψ the angle of the wind.

3.3.2. *Controller*

We will now try to find a controller that will enable the robot to track a line. The robot will be equipped with three sensors: a compass that gives us the heading θ, a weathervane that measures the angle of the wind ψ and a Global Positionning System (GPS) that returns the position $\mathbf{m}$ of the boat. The robot will also be equipped with two actuators: a servo-motor that controls the angle of the rudder δ_r and a stepper motor that sets the length of the sheet and therefore the maximum angle $\delta_s^{\max}$ of the sail (i.e. $|\delta_s| \leq \delta_s^{\max}$). As for the controller, its setpoint is the line $\mathbf{ab}$ to track and it has a binary variable $q \in \{-1, 1\}$ called the *hysteresis* which will be used for close hauled sailing. This controller will have few parameters which will also be easy to control. Among these parameters, we find the maximum rudder angle $\delta_r^{\max}$ (typically, $\delta_r^{\max} = \frac{\pi}{4}$), the cutting distance r (i.e. we would like the distance to the line to be always smaller than r), the close haul angle ζ (typically, $\zeta = \frac{\pi}{4}$) and the angle of the sail in crosswind β (typically, $\beta = 0.3$ rad). We propose the following controller, taken from the article [JAU 12], which we will explain later:

Controller in: $\mathbf{m}, \theta, \psi, \mathbf{a}, \mathbf{b}$; out: $\delta_r, \delta_s^{\max}$; inout: q
1 $e = \det\left(\frac{\mathbf{b}-\mathbf{a}}{\|\mathbf{b}-\mathbf{a}\|}, \mathbf{m}-\mathbf{a}\right)$
2 if $\lvert e\rvert > r$ then $q = \text{sign}(e)$
3 $\varphi = \text{angle}(\mathbf{b}-\mathbf{a})$
4 $\bar{\theta} = \varphi -$ atan$\left(\frac{e}{r}\right)$
5 if $\cos\left(\psi - \bar{\theta}\right) + \cos\zeta < 0$
6 or $(\lvert e\rvert < r$ and $(\cos(\psi - \varphi) + \cos\zeta < 0))$
7 then $\bar{\theta} = \pi + \psi - q\zeta$.
8 $\delta_r = \frac{\delta_r^{\max}}{\pi}\,\text{sawtooth}(\theta - \bar{\theta})$
9 $\delta_s^{\max} = \frac{\pi}{2}\left(\frac{\cos\left(\psi - \bar{\theta}\right)+1}{2}\right)^{\frac{\log\left(\frac{\pi}{2\beta}\right)}{\log(2)}}$

The controller has a single state variable which is the binary variable $q \in \{-1, 1\}$. It is for this reason that it appears at the same time as input and output of the algorithm. Let us comment on this algorithm.

Line 1 (calculation of the algebraic distance). We calculate the algebraic distance between the robot and its line. If $e > 0$, the robot is on the left of its line, and if $e < 0$, it is on the right. In the formula, the determinant is to be understood in the following way:

$$\det(\mathbf{u}, \mathbf{v}) = u_1 v_2 - v_1 u_2$$

Line 2 (update of the hysteresis variable). When $|e| > r$, the robot is far from its line and the hysteresis variable q (that memorizes the starboard tack) is allowed to change value. If, for instance $e > r$, then q will take the value 1 and will keep it until $e < -r$.

Line 3 (calculation of the line angle). We calculate the angle φ of the line to track (see Figure 3.10). In the instruction, angle($\mathbf{u}$) represents the angle made by the vector $\mathbf{u} \in \mathbb{R}^2$ relative to the Ox axis (toward the east). The corresponding function can be found in `angle.m`.

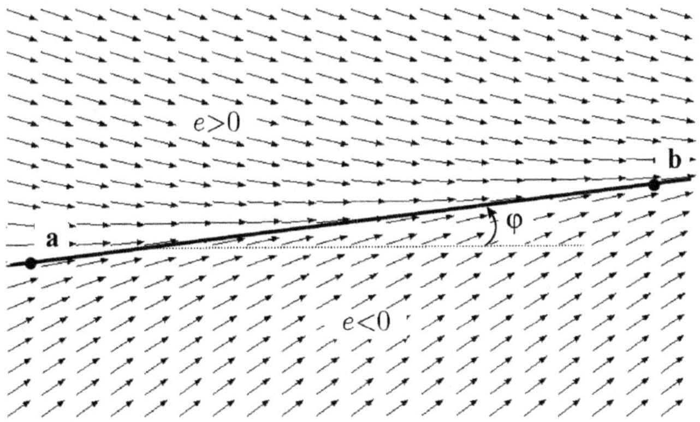

Figure 3.10. *Nominal vector field that the robot tries to track, when possible*

Line 4 (calculation of the nominal heading). We calculate the nominal angle $\bar{\theta}$ (see Figure 3.10), in other words the one that we would like to have without worrying about the wind. We take:

$$\bar{\theta} = \varphi - \text{atan}\left(\frac{e}{r}\right)$$

This expression for $\bar{\theta}$ translates to an attractive line. When $e = \pm\infty$, we have $\bar{\theta} = \varphi \pm \frac{\pi}{2}$, which means that the robot has a heading that forms an angle of $\frac{\pi}{2}$ with the line. For a distance e corresponding to the cutting distance r, i.e. $e = \pm r$, we have $\bar{\theta} = \varphi \pm \frac{\pi}{4}$. Finally, on the line we have $e = 0$ and therefore $\bar{\theta} = \varphi$, which corresponds to a heading with the direction of the line. As shown in Figure 3.11(a), some directions $\bar{\theta}$ may be incompatible with that of the wind.

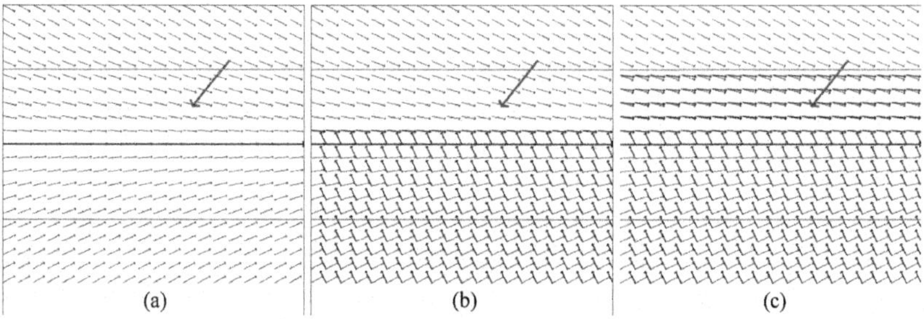

Figure 3.11. *a) The nominal vector field may be incompatible with the wind (here represented by the large arrow); b) vector field generated by the controller if we remove line 6. The thin arrows correspond to the nominal paths and the bold arrows correspond to the corrected paths; c) vector field generated by the controller with line 6 included*

Line 5. When $\cos\left(\psi - \bar{\theta}\right) + \cos\zeta < 0$, the path $\bar{\theta}$ corresponds to a direction that is too close to the wind that the robot is incapable of following (see Figure 3.12).

The heading $\bar{\theta}$ is then impossible to keep. In this case, we need to switch to *close haul* mode, which means that the robot will do everything it can to face the wind, or more formally, the new direction becomes $\bar{\theta} = \pi + \psi \pm \zeta$ (see line 7). Figure 3.11(b) represents the corresponding vector field. The thin arrows correspond to the nominal field and the bold arrows represent the corrected field when necessary. In this representation, we have removed the hysteresis effect induced by the variable q (which means that we always have $q = \text{sign}(e)$).

Line 6 (keep close hauled strategy). This instruction implements the so-called *keep close hauled strategy*. If $|e| < r$ or if $\cos(\psi - \varphi) + \cos\zeta < 0$, then the boat is forced to move upwind, even when the heading $\bar{\theta}$ is admissible, for reasons of efficiency. This strategy is shown in Figure 3.11(c). In this figure, we have chosen a close haul angle of $\zeta = \frac{\pi}{3}$ (which corresponds to difficulties moving upwind) and given this, the line is considered to be against the wind.

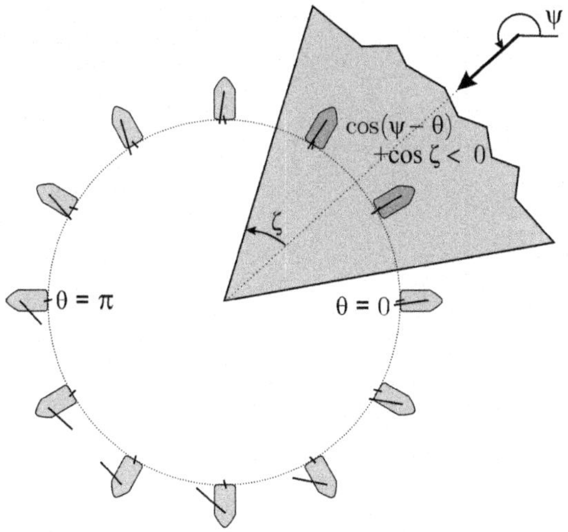

Figure 3.12. *Some directions are not possible for the sailboat. These unfeasible directions form the* no-go *zone, represented in gray*

Line 7 (close hauled heading). The boat is in close haul and we choose $\bar{\theta} = \pi + \psi - q\zeta$ (the wind direction plus or minus the close haul angle ζ). The hysteresis variable q is forced to keep the same point of sail as long as the distance of r to the line is not reached. An illustration of the resulting behavior is represented in Figure 3.13. If the nominal heading can be kept, then it is followed.

Line 8 (rudder control). At this level, the heading to maintain $\bar{\theta}$ has already been chosen and we are trying to follow it using the rudder. We perform a proportional control relative to the error $\theta - \bar{\theta}$. In order to filter out the modulus-2π problem, we use the *sawtooth* function (see formula [3.2]). Thus, we obtain:

$$\delta_r = \frac{\delta_r^{\max}}{\pi} \cdot \text{sawtooth}(\theta - \bar{\theta})$$

where $\delta_r^{\max}$ is the maximum angle of the rudder (for example, $\delta_r^{\max} = 0.5$ rad). The resulting controller is shown in Figure 3.14.

Line 9 (sail control). We choose a sail angle β (half-open sail) that the sail needs to have in crosswind. This parameter is determined experimentally depending on the sailboat and steering mode we would like to use. The maximum angle of the sail $\delta_s^{\max}$ is a function of $\psi - \theta$ which is periodic with period 2π. One possible model [JAU 12] is that of the cardioid:

$$\delta_s^{\max} = \frac{\pi}{2} \cdot \left(\frac{\cos(\psi - \theta) + 1}{2}\right)^{\eta}$$

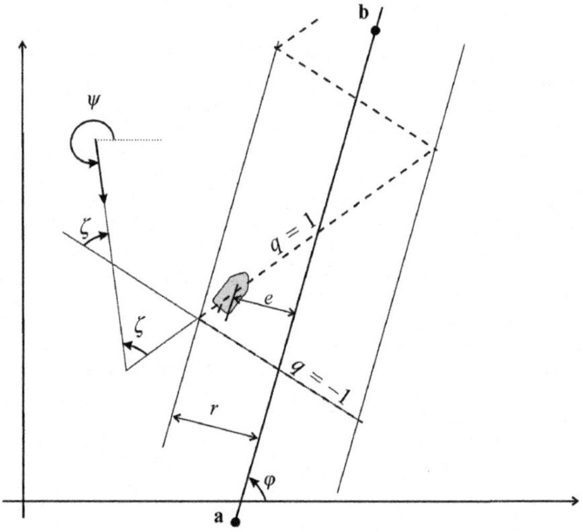

Figure 3.13. *Keep close hauled strategy by remaining within the strip centered on the line* ab *with diameter* r

where the parameter η is positive. When $\psi = \theta + \pi$, the boat is facing the wind and the model gives us $\delta_s^{\max} = 0$. When $\psi = \theta$, we have $\delta_s^{\max} = \frac{\pi}{2}$, which means that the sail is wide open when the robot is with the wind. The choice of the parameter η will be based on the angle of the sail in crosswind, in other words for $\psi = \theta \pm \frac{\pi}{2}$. The equation $\delta_s^{\max} = \beta$ for $\psi = \theta \pm \frac{\pi}{2}$ is translated by:

$$\frac{\pi}{2} \cdot \left(\frac{1}{2}\right)^{\eta} = \beta$$

i.e.:

$$\eta = \frac{\log\left(\frac{\pi}{2\beta}\right)}{\log(2)}$$

The function $\delta_s^{\max}$ is represented in Figure 3.15.

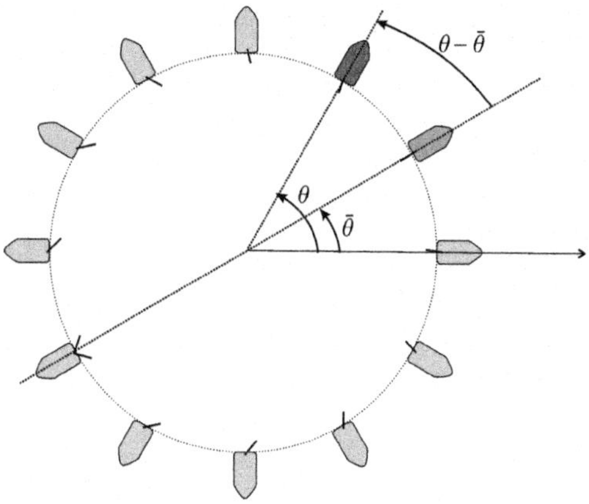

Figure 3.14. *Rudder control for the sailing robot*

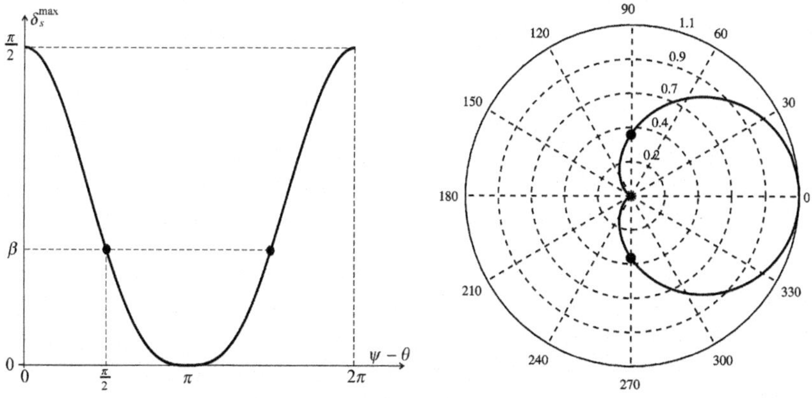

Figure 3.15. *Adjusting the maximum sail angle (or the sheet length); a) Cartesian representation and; b) polar representation*

On the tests that have been carried out, this adjustment of the sail was shown to be efficient and easy to control, given the few number of parameters.

3.3.3. *Navigation*

Once the line tracking has been correctly implemented and validated, a number of lines should be chained together with the aim of performing complex missions (such as connecting two points of the globe). In such a context, a Petri net strategy is well adapted for representing the discrete state changes [MUR 89]. Figure 3.16 shows a Petri net allowing us to manage the mission. Before the robot is launched, it is in an initial state represented by the place p_0. The transition t_1 is crossed at the start of the mission. If everything goes well, the robot is in state p_1 and is tracking its first line $\mathbf{a}_1\mathbf{b}_1$. The line $\mathbf{a}_j\mathbf{b}_j$ is validated as soon as point $\mathbf{b}_j$ is surpassed, in other words if $\langle \mathbf{b}_j - \mathbf{a}_j, \mathbf{m} - \mathbf{b}_j \rangle > 0$. This stopping criterion coupled with the path can be interpreted as a sort of validation. Once this is validated, we proceed to the next line. When the list of lines to track is empty, the mission ends (place p_3).

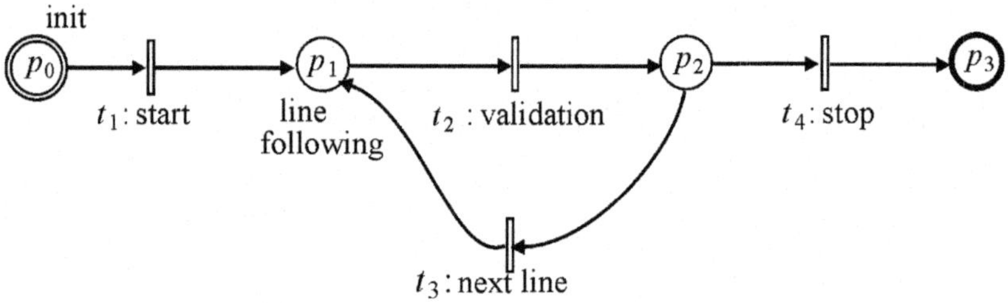

Figure 3.16. *Petri net supervising the navigation of the robot*

3.3.4. *Experiment*

Beginning in September 2011, we carried out a series of experiments with the VAIMOS sailboat in autonomous mode. We will describe one of these experiments which is simultaneously simple and representative, which took place on Thursday 28 June 2012 in the Brest bay, close to the Moulin Blanc harbor. The trajectory performed by the robot is represented in Figure 3.17. The wind comes from South-South-East, as indicated by the arrow that represents the average wind on the robot during the entire mission, deduced from the sensor. In this zone of heavy maritime traffic, interruptions in the mission can be anticipated and a permanent WiFi link is necessary between the robot and tracking boat in order to be able to cancel the mission at any time and avoid collision with other boats. Apart from these security interruptions (which by the way were not necessary during the mission), the robot is entirely autonomous. The mission is broken down into five submissions. First, the robot begins with a triangle (in the circle) in order to check whether everything is working properly. Then, it proceeds South-East against the wind by tracking the required path. It then describes a spiral. Once it is in the center of the spiral, the robot anchors virtually, in other words it maneuvers in order to remain around its attachment point. Finally, the robot returns to the harbor with the wind.

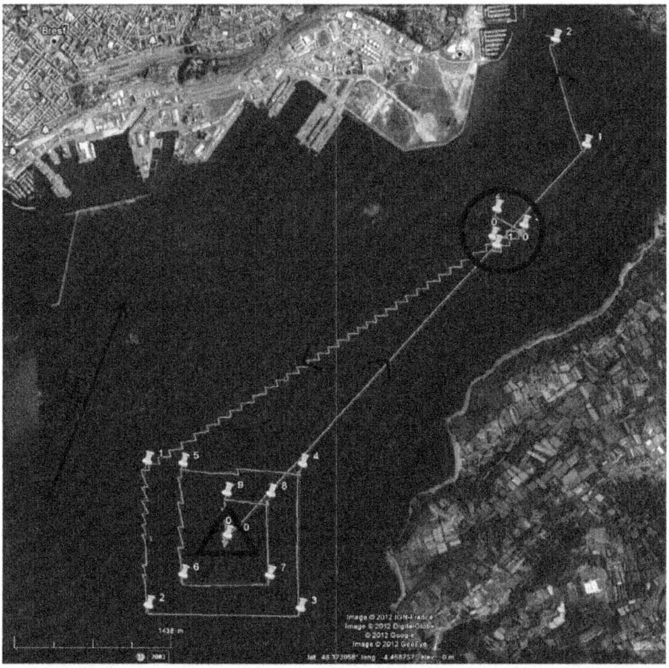

Figure 3.17. *Experiment of the spiral composed of five stages: a) the robot begins with a triangle (in the circle); b) it goes upwind following a line; c) it describes a spiral; d) it anchors virtually at the center of the spiral for several minutes; e) and goes back to the harbor*

Other large-scale experiments were also carried out, such as the journey from Brest to Douarnenez (see Figure 3.18) undertaken on the 17th–18th January 2012, thus completing a path of more than 100 km. From very high up (as in the figure), the lines seem to be tracked perfectly. Upon closer inspection, things are revealed to be less idealistic: the sailboat tacks in order to go upwind, recalibrates itself or is subjected to large waves otherwise. However, in both previously described experiments, the robot is never more than 50 m away from its track (except of course in the situations of avoidance in which it is being hauled).

NOTE 3.1.– Even though the robot never surpasses its line by more than 50 m, we could do better and improve this precision when the robot follows the nominal heading (i.e. the angle φ of the line corresponds to a sustainable heading). Indeed, in our experiments, a 10 m bias can be observed in nominal mode, which means that the distance to the line does not converge toward zero (with GPS precision). The role of the integrator is to remove such a bias. In order to implement such an integrator, we simply replace line 4 of the controller with the following two instructions:

$$\begin{cases} z = z + \alpha \, dt \, e \\ \theta = \varphi - \operatorname{atan}\left(\frac{e+z}{r}\right) \end{cases}$$

where dt is the sampling period. The variable z corresponds to the value of the integrator and naturally converges toward the constant bias that we had without the integrator and which we would like to remove. The coefficient α has to be sufficiently small to avoid a change in the behavior of our robot (which could appear in transient regime). For instance, if $e = 10$ m for 100 s, we may want a correction of 1 m of the bias. For this, we need to take $\alpha = 0.001$. Let us note that as soon as the distance on the line is greater than r (this is the case, for instance, during initialization), when the robot validates a line and continues on to the next, or when the robot is in large mode, then the integrator has to be forced to zero. Indeed, an integrator must not take up its function unless the permanent regime has been established.

Figure 3.18. *Journey from Brest to Douarnenez made by* VAIMOS*: a) the robot leaves the Moulin-Blanc harbor (in the circle); b) it avoids a submarine (in the square); c) it avoids a cargo ship (triangle)*

3.4. Exercises

EXERCISE 3.1.– Robot tank on a line

Let us consider a robot moving on a plane and described by the following state equations:

$$\begin{cases} \dot{x} = \cos\theta \\ \dot{y} = \sin\theta \\ \dot{\theta} = u \end{cases}$$

where θ is the heading of the robot and (x, y) are the coordinates of its center. This model corresponds to the Dubins car [DUB 57]. The state vector is given by $\mathbf{x} = (x, y, \theta)$.

1) Simulate this system graphically with MATLAB in various situations.

2) Propose a heading controller for the robot.

3) Propose a controller tracking a line $\mathbf{ab}$. This line must be attractive. Stop the program when point $\mathbf{b}$ is overtaken, in other words when $(\mathbf{b} - \mathbf{a})^{\mathrm{T}} (\mathbf{b} - \mathbf{m}) < 0$.

4) Make the robot track a closed path composed of a sequence of lines $\mathbf{a}_j\mathbf{b}_j$, $j \in \{1, \ldots, j_{\max}\}$.

5) Make several identical robots track the same circuit, but with different speeds. Modify the control laws in order to avoid collisions.

EXERCISE 3.2.– Van der Pol car

Consider the car represented in Figure 3.19.

2) Perform a first high-gain proportional feedback $\mathbf{u} = \rho(\mathbf{x}, \bar{\mathbf{u}})$ that would allow switching to a cart model of the from:

$$\begin{pmatrix} \dot{x} \\ \dot{y} \\ \dot{\theta} \end{pmatrix} = \begin{pmatrix} \bar{u}_1 \cos\theta \\ \bar{u}_1 \sin\theta \\ \bar{u}_2 \end{pmatrix}$$

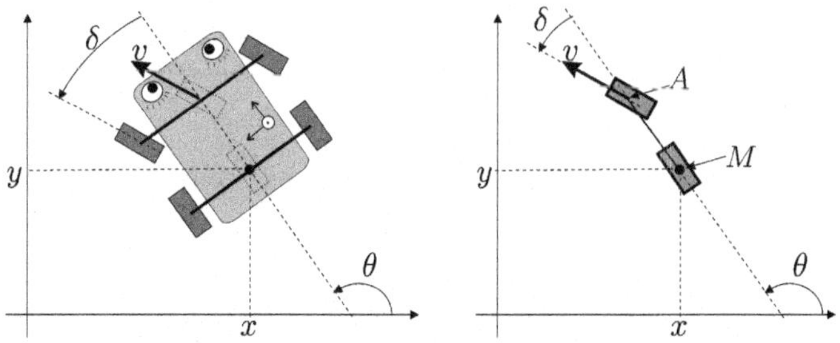

Figure 3.19. *Car moving on a plane (view from above)*

The new inputs of the feedback system $\bar{u}_1$ and $\bar{u}_2$ correspond, respectively, to the speed v and to the angular speed $\dot{\theta}$.

This car has two controls: the acceleration of the front wheels and the rotation speed of the steering wheel. The state variables of our system are composed of the position coordinates (the coordinates x, y of the center of the rear axle, the heading θ of the car and the angle δ of the front wheels) and the speed v of the center of the front axle. The evolution equation of the car is assumed to be identical to that of a tricycle (see section 2.5.1). It is written as:

$$\begin{pmatrix} \dot{x} \\ \dot{y} \\ \dot{\theta} \\ \dot{v} \\ \dot{\delta} \end{pmatrix} = \begin{pmatrix} v \cos \delta \cos \theta \\ v \cos \delta \sin \theta \\ \frac{v \sin \delta}{L} \\ u_1 \\ u_2 \end{pmatrix}$$

1) Simulate this system in MATLAB by using Euler's method.

3) Perform a second feedback $\bar{\mathbf{u}} = \sigma(\mathbf{x}, \mathbf{w})$ that would allow us to control this car's heading and speed. The new input will be $\mathbf{w} = (w_1, w_2)$ where w_1, w_2 correspond, respectively, to the desired speed and heading.

4) We would like the car to follow a path that obeys the Van der Pol equation:

$$\begin{cases} \dot{x}_1 = x_2 \\ \dot{x}_2 = -\left(0.01\, x_1^2 - 1\right) x_2 - x_1 \end{cases}$$

Propose a third controller $\mathbf{w} = \tau(\mathbf{x})$ that would allow us to perform this. Validate it using a simulation by superimposing the vector field by using the `quiver` instruction.

EXERCISE 3.3.– Anchoring

Let us consider the robot described by the following state equations:

$$\begin{cases} \dot{x} = \cos\theta \\ \dot{y} = \sin\theta \\ \dot{\theta} = u \end{cases}$$

The aim of *anchoring* the robot is to remain within the neighborhood of zero.

1) Does this system have a point of equilibrium?

2) Given the fact that the problem of remaining around zero admits a rotational symmetry, we suggest switching from a Cartesian representation (x, y, θ) toward a polar representation (α, d, φ), as shown in Figure 3.20. Give the state equations in the polar representation.

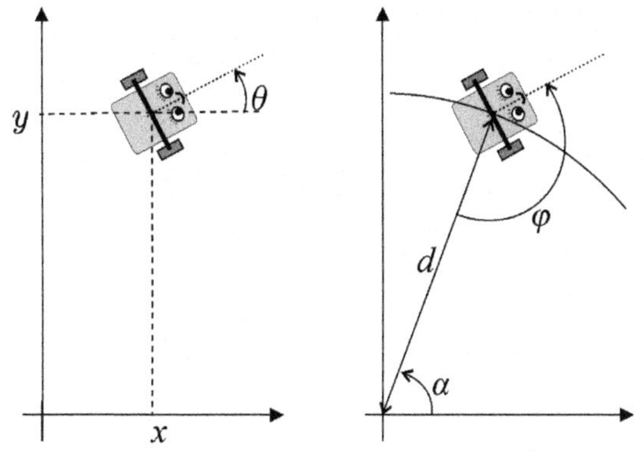

Figure 3.20. *Coordinate system change allowing us to take advantage of the rotational symmetry*

3) How is this new representation of interest for the graphical representation of the system dynamics?

4) In order to solve the anchoring problem, we propose the control law:

$$u = \begin{cases} +1 & \text{if } \cos\varphi \leq \frac{1}{\sqrt{2}} \text{ (the robot turns to the left)} \\ -\sin\varphi & \text{otherwise} \quad \text{(proportional control)} \end{cases}$$

An illustration of this control law is given in Figure 3.21. Explain how this control solves the anchoring problem. Is there a configuration that allows the robot to move arbitrarily far away from 0?

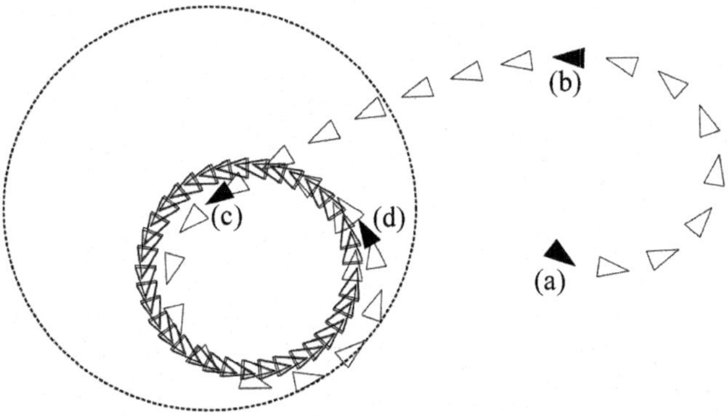

Figure 3.21. *Path of the controlled system in order to remain within the disk*

5) Simulate this control law in MATLAB with various initial conditions.

EXERCISE 3.4.– Discretization of the state space

Let us consider the following system (which arises from the previous exercise):

$$\begin{cases} \text{(i) } \dot{\varphi} = \begin{cases} \frac{\sin\varphi}{d} + 1 & \text{if } \cos\varphi \leq \frac{1}{\sqrt{2}} \\ \left(\frac{1}{d} - 1\right)\sin\varphi & \text{otherwise} \end{cases} \\ \text{(ii) } \dot{d} = -\cos\varphi \end{cases}$$

The associated vector field is represented in Figure 3.22, where $\varphi \in [\pi, \pi]$ and $d \in [0, 10]$. A path $\phi(t, \mathbf{x}_0)$ which corresponds to the simulation visualized in the previous exercise is also drawn.

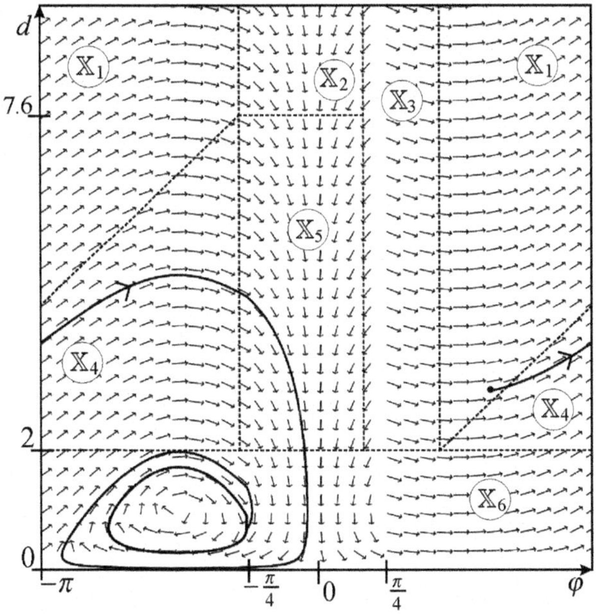

Figure 3.22. *Vector field associated with our system*

In this same figure, the state space is cut into six areas. Indeed, given the adjunction of the line $\varphi = \pi$ with the line $\varphi = -\pi$, the state space has a cylindrical nature and we have six areas, whereas we can see eight areas in the figure.

Succession relation. We define the rotation, denoted by $\hookrightarrow$ between zones $\mathbb{A}, \mathbb{B}$ of the state space as follows:

$$(\mathbb{A} \hookrightarrow \mathbb{B}) \Leftrightarrow \exists \mathbf{x}_0 \in \mathbb{A} \in \phi(\eta(\mathbf{x}_0), \mathbf{x}_0) \in \mathbb{B}$$

where $\eta(\mathbf{x}_0)$ is the time the system exits $\mathbb{A}$.

Convention. If there is $\mathbf{x}_0 \in \mathbb{A}$, such that $\forall t > 0, \phi(t, \mathbf{x}_0) \subset \mathbb{A}$, then $\eta(\mathbf{x}_0) = \infty$. Thus, $\phi(\eta(\mathbf{x}_0), \mathbf{x}_0) \in \mathbb{A}$ and therefore will we have $(\mathbb{A} \hookrightarrow \mathbb{A})$.

1) Draw the graph associated with this relation.

2) From this, deduce a superset of the state space in which the system will remain trapped.

EXERCISE 3.5.– Sailing robot

Consider the sailboat described by the following state equations:

$$\left\{\begin{array}{ll} \dot{x} & = v\cos\theta + p_1 a\cos\psi \\ \dot{y} & = v\sin\theta + p_1 a\sin\psi \\ \dot{\theta} & = \omega \\ \dot{v} & = \frac{f_s\sin\delta_s - f_r\sin u_1 - p_2 v^2}{p_9} \\ \dot{\omega} & = \frac{f_s(p_6 - p_7\cos\delta_s) - p_8 f_r\cos u_1 - p_3\omega v}{p_{10}} \\ f_s & = p_4\left\|\mathbf{w}_{\text{ap}}\right\|\sin\left(\delta_s - \psi_{\text{ap}}\right) \\ f_r & = p_5 v\sin u_1 \\ \sigma & = \cos\psi_{\text{ap}} + \cos u_2 \\ \delta_s & = \left\{\begin{array}{ll} \pi + \psi_{\text{ap}} & \text{if } \sigma \leq 0 \\ -\text{sign}\left(\sin\psi_{\text{ap}}\right)\cdot u_2 & \text{otherwise} \end{array}\right. \\ \mathbf{w}_{\text{ap}} & = \left(\begin{array}{c} a\cos\left(\psi - \theta\right) - v \\ a\sin\left(\psi - \theta\right) \end{array}\right) \\ \psi_{\text{ap}} & = \text{angle } \mathbf{w}_{\text{ap}} \end{array}\right.$$

where (x, y, θ) corresponds to the posture of the boat, v is its forward speed, ω is its angular speed, f_s ("*s*" for *sail*) is the force of the wind on the sail, f_r ("*r*" for *rudder*) is the force of the water on the rudder, δ_s is the angle of the sail, a is the true wind speed, ψ is the true wind angle (see Figure 3.9) and $\mathbf{w}_{\text{ap}}$ is the apparent wind vector. The quantity σ is an indicator of the sheet tension. Thus, if $\sigma \leq 0$, the sheet is released and the sail is flapped. If $\sigma \geq 0$, the sheet is stretched and inflated by the wind. In these equations, the p_i are design parameters of the sailboat. We will take the following values, given in international units: $p_1 = 0.1$ (drift coefficient), $p_2 = 1$ (drag coefficient), $p_3 = 6\,000$ (angular friction of the hull against the water), $p_4 = 1\,000$ (sail lift), $p_5 = 2\,000$ (rudder lift), $p_6 = 1$ (position of the wind's center of thrust on the sail), $p_7 = 1$ (position of the mast), $p_8 = 2$ (position of the rudder), $p_9 = 300$ (mass of the sailboat) and $p_{10} = 10\,000$ (inertial momentum of the sailboat).

1) Simulate the boat in MATLAB.

2) Implement the controller proposed in section 3.3. We will use the following parameters $\zeta = \frac{\pi}{4}$ for the large-angle, $r = 10$ m for the radius of the air corridor, $\delta_r^{\max} = 1$ rad for the maximum angle of the rudder and $\beta = \frac{\pi}{4}$ for the angle of the sail in crosswind.

EXERCISE 3.6.– Flying drone

Consider a flying drone [BEA 12] such as the one represented in Figure 3.23a. This is a 1 kg fully autonomous plane. One possible model to describe its dynamics, very strongly inspired by those of *Faser Ultra Stick* [KLE 06] in Figure 3.23b, is given by:

$$\begin{pmatrix} \dot{\mathbf{p}} \\ \begin{pmatrix} \dot{\psi} \\ \dot{\theta} \\ \dot{\varphi} \end{pmatrix} \\ \dot{\mathbf{v}} \\ \dot{\omega} \end{pmatrix} = \begin{pmatrix} \mathbf{R}_{\text{euler}}(\varphi,\theta,\psi)\cdot\mathbf{v} \\ \begin{pmatrix} 0 & \frac{\sin\varphi}{\cos\theta} & \frac{\cos\varphi}{\cos\theta} \\ 0 & \cos\varphi & -\sin\varphi \\ 1 & \tan\theta\sin\varphi & \tan\theta\cos\varphi \end{pmatrix}\cdot\omega \\ 9.81\cdot\begin{pmatrix} -\sin\theta \\ \cos\theta\sin\varphi \\ \cos\theta\cos\varphi \end{pmatrix} + \mathbf{f}_a + \begin{pmatrix} u_1 \\ 0 \\ 0 \end{pmatrix} - \omega\wedge\mathbf{v} \\ \begin{pmatrix} -\omega_3\omega_2 - \frac{\|\mathbf{v}\|^2}{10}\left(\beta + 2u_3 + \frac{5\omega_1-\omega_3}{\|\mathbf{v}\|}\right) \\ \omega_3\omega_1 - \frac{\|\mathbf{v}\|^2}{100}\left(1 + 20\alpha - 2u_3 + 30u_2 + \frac{300\omega_2}{\|\mathbf{v}\|}\right) \\ \frac{\omega_1\omega_2}{10} + \frac{\|\mathbf{v}\|^2}{10}\left(\beta + \frac{u_3}{2} + \frac{\omega_1 - 2\omega_3}{2\|\mathbf{v}\|}\right) \end{pmatrix} \end{pmatrix}$$

with:

$$\alpha = \operatorname{atan}\left(\frac{v_3}{v_2}\right),\ \beta = \operatorname{asin}\left(\frac{v_2}{\|\mathbf{v}\|}\right)$$

$$\mathbf{f}_a = \frac{\|\mathbf{v}\|^2}{500}\begin{pmatrix} -\cos\alpha\cos\beta & \cos\alpha\sin\beta & \sin\alpha \\ \sin\beta & \cos\beta & 0 \\ -\sin\alpha\cos\beta & \sin\alpha\sin\beta & -\cos\alpha \end{pmatrix}$$

$$\cdot \begin{pmatrix} 4 + (-0.3 + 10\alpha + \frac{10\omega_2}{\|\mathbf{v}\|} + 2u_3 + 0.3u_2)^2 + |u_2| + 3\,|u_3| \\ -50\beta + \frac{10\omega_3 - 3\omega_1}{\|\mathbf{v}\|} \\ 10 + 500\alpha + \frac{400\omega_2}{\|\mathbf{v}\|} + 50u_3 + 10u_2 \end{pmatrix}$$

Figure 3.23. *a) μ-*STIC *plane made by the* ENSTA *Bretagne; b) Faser Ultra Stick plane from the University of Minnesota and; c) graphical representation used for the simulation*

In this model, all the quantities are given in international units. The vector $\mathbf{p} = (x, y, z)$ represents the position of the drone, with the z axis oriented toward the center of the Earth. The orientation of the drone is represented by the Euler angles (φ, θ, ψ) and the Euler matrix $\mathbf{R}_{\text{euler}}(\varphi, \theta, \psi)$ is given by formula [1.7] in section 1.2.1. The vector $\mathbf{v}$ represents the speed of the drone expressed in its own coordinate system. The rotation vector of the plane is denoted here by ω. It is linked to the derivatives of the Euler angles by formula [1.11] in section 1.3. Let us note that formula [1.11] gives the first three equations of our state model. The angles α and β correspond to the angle of attack and the sideslip angle. The vector $\mathbf{f}_a$ corresponds to the acceleration caused by the forces created by air. A simplified geometric view of the drone, given in Figure 3.23(c), shows a helix for propulsion and two ailerons for direction. The input vector $\mathbf{u} = (u_1, u_2, u_3)$ involved in our state model contained the propulsive acceleration $u_1 \in [0, 10]$ (in ms^{-2}), the sum $u_2 \in [-0.6, 0.6]$ (in radians) of the two aileron angles and $u_3 \in [-0.3, 0.3]$ (in radians) the differential between these two ailerons.

1) Simulate this drone in MATLAB. For the graphics, follow Figure 3.23(c).

2) Propose a heading, elevation and speed control law.

3) We would like the robot to be positioned on a circle of radius $\bar{r} = 100$ m, centered around 0 at an altitude of 50 m and a speed of $\bar{v} = 15$ ms^{-1}. Give the control law and illustrate the associated behavior of the robot in MATLAB.

3.5. Corrections

CORRECTION FOR EXERCISE 3.1.– (Robot tank on a line)

1) The complete simulation corresponding to this exercise is given in the program tank_line.m.

2) The heading controller is given by u=sawtooth(thetabar-x(3)); using the *sawtooth* function (see Figure 3.2) allows us to filter the jumps of angle $\pm 2\pi$.

3) As for following the line, we combine a heading controller with a vector field controller. The field used in order to make the line attractive is $\bar{\theta} = \varphi - \text{atan}(e)$ where e is the algebraic difference on the line. The program is:

```
a=[-40;-4]; b=[20;6]; % line
x=[-30;-10;pi;1]; % x,y,theta,v
dt=0.1;
for t=0:dt:50,
phi=angle(b-a); % angle of the line
m=[x(1);x(2)];
e=det([b-a,m-a])/norm(b-a); % distance to the line
thetabar=phi-atan(e); % desired heading
u=sawtooth(thetabar-x(3)); % heading control
x=x+f(x,u)*dt;
end;
```

CORRECTION FOR EXERCISE 3.2.– (Van der Pol car)

1) The simulation is performed with the following script:

```
x=[0;0.1;pi/6;2;0.6]; % x,y,theta,v,delta
dt=0.01; u=[0 0];
for t=0:dt:100,
x=x+f(x,u)*dt;
end;
```

The evolution function is the following:

```
function xdot=f(x,u)
xdot=[x(4)*cos(x(5))*cos(x(3));
x(4)*cos(x(5))*sin(x(3));
```

```
x(4)*sin(x(5))/3;
u(1);u(2) ];
```

2) We simply need to take:

$$\mathbf{u} = \rho(\mathbf{x}, \bar{\mathbf{u}}) = k\left(\bar{\mathbf{u}} - \begin{pmatrix} v\cos\delta \\ \frac{v\sin\delta}{L} \end{pmatrix}\right)$$

with large k. Thus, by following the same reasoning as for the operational amplifier, we obtain that if the feedback system is stable, we have

$$\begin{pmatrix} \bar{u}_1 \\ \bar{u}_2 \end{pmatrix} \simeq \begin{pmatrix} v\cos\delta \\ \frac{v\sin\delta}{L} \end{pmatrix}$$

and the feedback system becomes the tank model. For the simulation, we could take $k = 10$. Thus, the controller is expressed by:

```
u=10*(ubar-[x(4)*cos(x(5));x(4)*sin(x(5))/3]);
```

3) For $\bar{u}_1$, there is nothing to do. For $\bar{u}_2$, we could take a proportional heading controller. The controller corresponding to this second feedback is:

$$\bar{\mathbf{u}} = \sigma(\mathbf{x}, \mathbf{w}) = \begin{cases} \bar{u}_1 = a_1\, w_1 \\ \bar{u}_2 = a_2 \cdot \text{sawtooth}(w_2 - \theta) \end{cases}$$

with, for example, $a_1 = 1$ and $a_2 = 5$. The corresponding lines of code are:

```
ubar=[w(1);5*sawtooth(w(2)-x(3))];
```

4) We add a third feedback given by:

$$\mathbf{w} = \tau(\mathbf{x}) = \begin{pmatrix} v_0 \\ \text{angle}\begin{pmatrix} y \\ -\left(0.01\, x^2 - 1\right) y - x \end{pmatrix} \end{pmatrix}$$

where v_0 is the desired speed for the car and the `angle` function returns the angle of a vector (see `angle.m`). We will take, for example, $v_0 = 10\ \text{ms}^{-1}$.

The complete code (refer to the file vanderpol.m) corresponding to these three feedbacks is the following

```
vdp=[x(2);-(0.01*x(1)^2-1)*x(2)-x(1)]; % Van der Pol
dynamics
w=[10;angle(vdp)]; % heading and speed setpoint (here
10m/s)
ubar=[w(1);5*sawtooth(w(2)-x(3))];
u=10*(ubar-[x(4)*cos(x(5));x(4)*sin(x(5)/3]);
```

The code that draws the vector field in order to verify the correct behavior of the car is:

```
function draw(x)
Lx=50;Ly=30;
Mx = -Lx:2:Lx; My = -Ly:2:Ly;
[X1,X2] = meshgrid(Mx,My);
VX=X2; VY=-(0.01*X1.^2-1).*X2-X1;
VX=VX./sqrt(VX.^2+VY.^2); VY=VY./sqrt(VX.^2+VY.^2);
quiver(Mx,My,VX,VY);
```

CORRECTION FOR EXERCISE 3.3.– (Anchoring)

1) Since $\dot{x}^2 + \dot{y}^2 = 1$, the robot cannot be stopped.

2) The required state equations are the following:

$$\begin{cases} \text{(i)} \quad \dot{\varphi} = \frac{\sin\varphi}{d} + u \\ \text{(ii)} \quad \dot{d} \;= -\cos\varphi \\ \text{(iii)} \; \dot{\alpha} = -\frac{\sin\varphi}{d} \end{cases}$$

The proofs for (ii) and (iii) are direct. Let us now prove (i). Following Figure 3.24, we have $\varphi - \theta + \alpha = \pi$. In other words, after differentiating:

$$\dot{\varphi} = -\dot{\alpha} + \dot{\theta} = \frac{\sin\varphi}{d} + u$$

3) This representation allows us to switch from a three-dimensional (3D) system to a 2D system (since α is no longer necessary). A graphical representation of the vector field then becomes possible (see exercise 3.4).

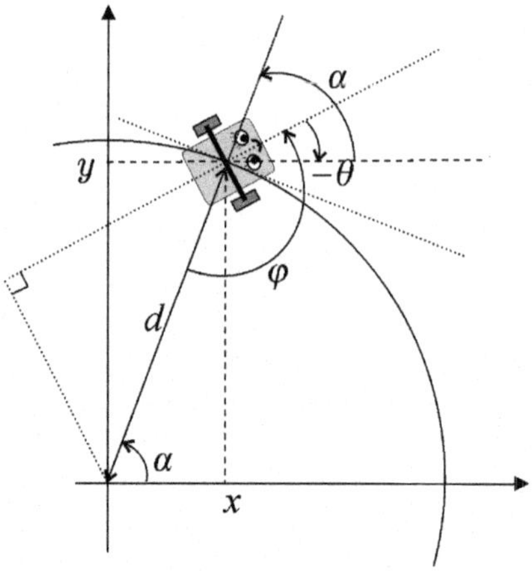

Figure 3.24. *Using cylindrical symmetry to switch from an order 3 system to an order 2 system*

4) The idea behind this control law is to use a proportional controller when the robot approximately points toward 0 (in other words, if $\cos\varphi > \frac{1}{\sqrt{2}}$) and turns to the left otherwise, in order to point toward 0 later.

5) The simulation program can be found in the file `anchor.m`.

CORRECTION FOR EXERCISE 3.4.– (Discretization of the state space)

The graph is represented in Figure 3.25 in two different forms. The state will be trapped under the bold line.

In matrix form, the graph is given by:

$$\mathbf{G} = \begin{pmatrix} 0\,1\,0\,1\,0 & 0 \\ 0\,0\,0\,0\,1 & 0 \\ 1\,1\,0\,0\,1 & 1 \\ 0\,0\,0\,0\,1 & 1 \\ 0\,0\,0\,0\,0 & 1 \\ 0\,0\,0\,1\,0 & [0,1] \end{pmatrix}$$

where the Boolean interval $[0,1]$ means either 0 or 1. The transitive closure is:

$$\mathbf{G}^+ = \mathbf{G} + \mathbf{G}^2 + \mathbf{G}^3 + \cdots = \begin{pmatrix} 0\,1\,0\,1\,1\,1 \\ 0\,0\,0\,1\,1\,1 \\ 1\,1\,0\,1\,1\,1 \\ 0\,0\,0\,1\,1\,1 \\ 0\,0\,0\,1\,1\,1 \\ 0\,0\,0\,1\,1\,1 \end{pmatrix}$$

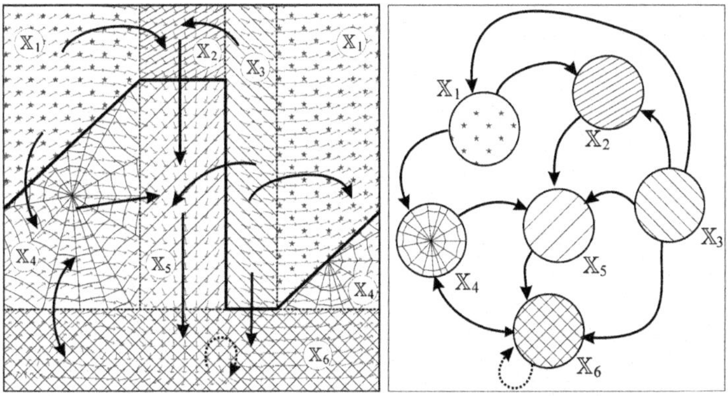

Figure 3.25. *Discretization of the system's dynamics by a partitioning of the state space in order to obtain a transition graph*

The diagonal gives us the attractor of the graph $\mathbb{X}_\infty = \mathbb{X}_4 \cup \mathbb{X}_5 \cup \mathbb{X}_6$. Thus, the attractor of the system $\mathbb{A}$ satisfies:

$$\mathbb{A} \subset \mathbb{X}_4 \cup \mathbb{X}_5 \cup \mathbb{X}_6$$

CORRECTION FOR EXERCISE 3.5.– (Sailing robot)

1) For the simulation, we can use an Euler method, by taking as evolution function:

```
function xdot = f(x,u)
theta=x(3); v=x(4); w=x(5); deltar=u(1); deltasmax=u(2);
w_ap=[awind*cos(psi-theta)-v;awind*sin(psi-theta)];
```

```
psi_ap=angle(w_ap); a_ap=norm(w_ap);
sigma=cos(psi_ap)+cos(deltasmax);
if (sigma < 0), deltas=pi+psi_ap;
else deltas=-sign(sin(psi_ap))*deltasmax;
end;
fr = p5*v*sin(deltar); fs = p4*a_ap*sin(deltas-psi_ap);
dx=v*cos(theta)+p1*awind*cos(psi);
dy=v*sin(theta)+p1*awind*sin(psi);
dtheta=w;
dv=(1/p9)*(sin(deltas)*fs-sin(deltar)*fr-p2*v^2);
dw=(1/p10)*((p6-p7*cos(deltas))*fs-p8*cos(deltar)*fr
-p3*w*v);
xdot=([dx;dy;dtheta;dv;dw]);
end
```

2) For line tracking, we write the controller as follows:

```
function [u,q] = control(x,q)
theta=x(3); r=10; zeta=pi/4; m=[x(1);x(2)];
e=det([(b-a)/norm(b-a),m-a]);
phi=angle(b-a);
if (abs(e) > r), q=sign(e); end;
thetabar=phi-atan(e/r);
if (cos(psi-thetabar)+cos(zeta) < 0)||((abs(e) <
r)&&(cos(psi-phi)+cos(zeta)<0))
thetabar=pi+psi-zeta*q;
end;
deltar=0.3*(sawtooth(theta-thetabar));
deltasmax=pi/4*(cos(psi-thetabar)+1);
u=[deltar;deltasmax];
end
```

CORRECTION FOR EXERCISE 3.6.– (Flying drone)

1) For the control law, we need to control the propulsion by u_1, the elevation by u_2 and the heading by u_3.

Propulsion. On our drone, the propulsion must be within the interval $[0, 10]$ (in ms^{-2}). We will take:

$$u_1 = 5\left(1 + \tanh\left(\bar{v} - \|\mathbf{v}\|\right)\right)$$

where $\mathbf{v}$ is the speed vector of the drone and $\bar{v}$ is the speed setpoint. If the drone is moving at the correct speed, we have an average propulsion of 5 N. Otherwise, due to the hyperbolic tangent saturation function $\tanh(\cdot)$ (refer to formula [3.1]), we will always have a propulsion within the interval $[0, 10]$.

Elevation. The elevation will be controlled using the sum of the aileron angles that we assume to be in the interval $[-0.6, 0.6]$ rad. We will take:

$$u_2 = -0.3 \cdot \left(\tanh\left(5\left(\bar{\theta} - \theta\right)\right) + |\sin\varphi|\right)$$

Thus, we have proportional control over the elevation saturated by the *tanh* function. The gain of 5 tells us that the reaction becomes significant when the error in elevation is greater than $\frac{1}{5}$rad$=$ 11 deg. For a strong enough bank φ, the drone loses altitude, which explains the term in $|\sin\varphi|$.

Heading. In our model of the drone, it is the bank φ that allows a change in the drone's heading, just like a motorcycle. If we wish to have a heading of $\bar{\psi}$, we may choose the following bank:

$$\bar{\varphi} = \tanh(5 \cdot \text{sawtooth}(\bar{\psi} - \psi))$$

Indeed, we define the error in heading as $\bar{\psi} - \psi$ that we filter using the `sawtooth` function (see Figure 3.2). We multiply this by the gain of the proportional controller, which is equal to 5. This gain of 5 tells us that the controller reacts significantly to correct the heading by the bank when $\text{sawtooth}(\bar{\psi} - \psi) = \frac{1}{5}$, in other words when $\bar{\psi} - \psi = \frac{1}{5}$rad $\simeq$ 11 deg, which seems reasonable. We can then propose the controller:

$$u_3 = -0.3 \cdot \tanh\left(\bar{\varphi} - \varphi\right)$$

where the *tanh* function is again used as the saturation function. Thus, the setpoint $\bar{\varphi}$ for φ will always be within the interval $\left[-\frac{\pi}{4}, \frac{\pi}{4}\right]$.

Vector field. In order to define the behavior of the plane so that it may return to its circle, we assign to each position in space $\mathbf{p} = (p_x, p_y, p_z)$ a direction $\bar{\psi}$ and an elevation $\bar{\theta}$ given by:

$$\bar{\psi} = \text{angle}(\mathbf{p}) + \frac{\pi}{2} + \tanh\frac{\sqrt{p_1^2+p_2^2}-\bar{r}}{50}$$

$$\bar{\theta} = -0.3 \cdot \tanh\frac{\bar{z}-z}{10}$$

The resulting vector field has a limit cycle that corresponds to our circle or radius $\bar{r}$. For the calculation of $\bar{\psi}$, the expression of the error indicates us a precision of 50 m. The component $\tanh\frac{1}{50}(\sqrt{p_1^2+p_2^2}-\bar{r})$ makes our circle attractive and the angle component $(\mathbf{p})+\frac{\pi}{2}$ creates the rotating field. For the expression of $\bar{\theta}$, we have a saturation function that ensures a precision of 10 m in altitude and a maximum elevation of $0.2\cdot\frac{\pi}{2}\simeq 0.31$ rad.

Summary. The controller for our drone will have the following inputs: the setpoints $\bar{z},\bar{r},\bar{v}$, the Euler angles of the plane, its position $\mathbf{p}$ and its speed vector. It will generate the control vector $\mathbf{u}$ for us as given by the algorithm:

Controller in: $\mathbf{p},\mathbf{v},\varphi,\theta,\psi,\bar{z},\bar{r},\bar{v}$; out: $\mathbf{u}$
1 $\bar{\psi}=\text{angle}(\mathbf{p})+\frac{\pi}{2}+\tanh\frac{\sqrt{p_1^2+p_2^2}-\bar{r}}{50}$
2 $\bar{\theta}=-0.3\cdot\tanh\frac{\bar{z}-z}{10}$
3 $\bar{\varphi}=\tanh(5\cdot\text{sawtooth}(\bar{\psi}-\psi))$
4 $\mathbf{u}=\begin{pmatrix}5\left(1+\tanh\left(\bar{v}-\Vert\mathbf{v}\Vert\right)\right)\\ -0.3\cdot\left(\tanh\left(5\left(\bar{\theta}-\theta\right)\right)+\vert\sin\varphi\vert\right)\\ -0.3\cdot\tanh\left(\bar{\varphi}-\varphi\right)\end{pmatrix}$

Figure 3.26 shows the result of the simulation. The corresponding program can be found in the file `plane.m`.

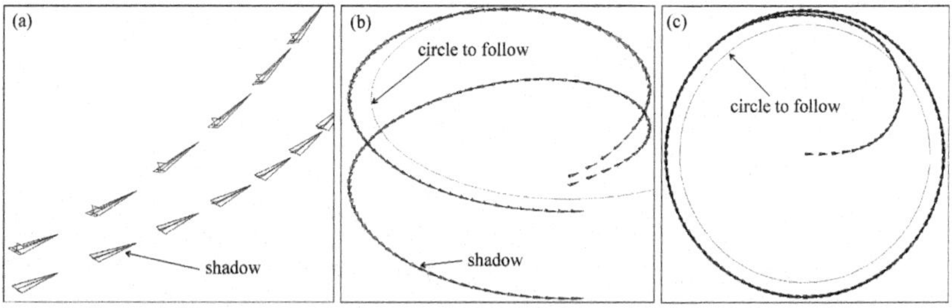

Figure 3.26. *a) Initial phase: the robot stands up and turns to its left; b) the robot returns to its circle; c) overhead view: we notice that there is a bias between the setpoint circle and the circle performed*

4

Guidance

In the previous chapters, we have studied how to build a control law for a robot described by nonlinear state equations (see Chapter 2) or when the robot's behavior is known (see Chapter 3). *Guidance* is performed on a higher level and focuses on the setpoint to give the controller in order for the robot to be able to accomplish its assigned mission. Therefore, it will have to take into account the knowledge of its surroundings, the presence of obstacles, the roundness of the Earth and so forth. Conventionally, guidance is applied in four different environments: terrestrial, marine, aerial and spatial. Given the fields of application covered in this book, we will not study the spatial environment.

4.1. Guidance on a sphere

For longer paths over the surface of the Earth, the Cartesian coordinate system, which assumes a flat Earth, can no longer be considered. We then have to rethink our control laws by navigating relative to a spherical coordinate system (also referred to as *geographical coordinates*), which rotates together with the Earth. Let us denote by ℓ_x the longitude and by ℓ_y the latitude of the point being considered. The transformation in the geographical coordinate system is written as:

$$\mathcal{T}: \begin{pmatrix} \ell_x \\ \ell_y \\ \rho \end{pmatrix} \to \begin{pmatrix} x \\ y \\ z \end{pmatrix} = \begin{pmatrix} \rho \cos \ell_y \cos \ell_x \\ \rho \cos \ell_y \sin \ell_x \\ \rho \sin \ell_y \end{pmatrix} \qquad [4.1]$$

When $\rho = 6\ 370$ km, we are on the surface of the Earth, which we will assume to be spherical (see Figure 4.1(a)).

Let us consider two points $\mathbf{a}$, $\mathbf{m}$ on the surface of the Earth, as shown in Figure 4.1(b), located by their geographical coordinates. Here, $\mathbf{a}$, for instance, represents a reference point to reach and $\mathbf{m}$ is the center of our robot. By assuming that the two points $\mathbf{a}$, $\mathbf{m}$ are not too far apart (by no more than 100 km), we may consider being in the plane and use a local map, as shown in Figure 4.2(a).

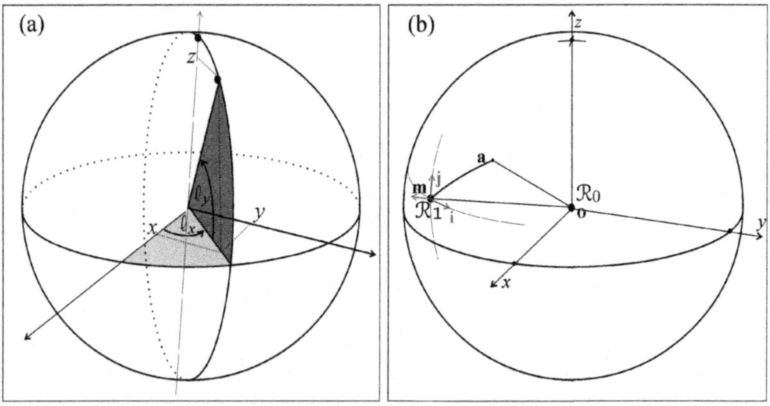

Figure 4.1. *a) Geographical coordinate system; b) we would like to express* $\mathbf{a}$ *in the local map* $\mathcal{R}_1$

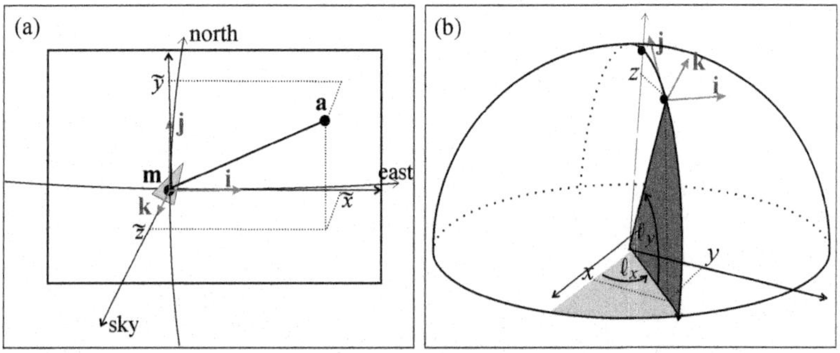

Figure 4.2. *a) The map gives a local Cartesian viewpoint around the robot; b) a slight shift* $d\ell_x, d\ell_y, d\rho$ *creates a displacement* dx, dy, dz *in the local map*

Let us differentiate relation [4.1]. We obtain:

$$\begin{pmatrix} dx \\ dy \\ dz \end{pmatrix} = \underbrace{\begin{pmatrix} -\rho\cos\ell_y\sin\ell_x & -\rho\sin\ell_y\cos\ell_x & \cos\ell_y\cos\ell_x \\ \rho\cos\ell_y\cos\ell_x & -\rho\sin\ell_y\sin\ell_x & \cos\ell_y\sin\ell_x \\ 0 & \rho\cos\ell_y & \sin\ell_y \end{pmatrix}}_{=\mathbf{J}} \cdot \begin{pmatrix} d\ell_x \\ d\ell_y \\ d\rho \end{pmatrix}$$

This formula can be used to find the geographical coordinates of the cardinal directions, which change depending on the location of the robot $\mathbf{m}$. For instance, the vector corresponding to *East* is found in the first column of the matrix $\mathbf{J}$, North in the second column and the altitude in the third column. Therefore, we will be able to build a coordinate system (East-North-Altitude) $\mathcal{R}_1$ centered on the robot (in gray in Figure 4.2(a)) that corresponds to the local map. The corresponding rotation matrix is obtained by normalizing each column of the Jacobian matrix $\mathbf{J}$. We obtain:

$$\mathbf{R} = \begin{pmatrix} -\sin\ell_x & -\sin\ell_y\cos\ell_x & \cos\ell_y\cos\ell_x \\ \cos\ell_x & -\sin\ell_y\sin\ell_x & \cos\ell_y\sin\ell_x \\ 0 & \cos\ell_y & \sin\ell_y \end{pmatrix} \quad [4.2]$$

The transformation that allows switching from the geographic coordinate system $\mathcal{R}_0$ to the local map $\mathcal{R}_1$ is:

$$\mathbf{v}_{|\mathcal{R}_1} = \mathbf{R}^{\mathrm{T}} \cdot \mathbf{v}_{|\mathcal{R}_0} \quad [4.3]$$

This coordinate system change relation can be applied in different contexts such as when establishing coherence in the data collected by two different robots.

EXAMPLE 4.1.– A robot situated at $\mathbf{m} : \left(\ell_x^{\mathbf{m}}, \ell_y^{\mathbf{m}}\right)$ is moving with a speed vector $\mathbf{v}^{\mathbf{m}}$, relative to a fixed ground. This vector is expressed in the local map of the robot. Find the speed with which this vector $\mathbf{v}^{\mathbf{m}}$ is perceived in the local map of an observer situated at $\mathbf{a} : \left(\ell_x^{\mathbf{a}}, \ell_y^{\mathbf{a}}\right)$. Following [4.3], we have:

$$\begin{cases} \mathbf{v}^{\mathbf{m}} = \mathbf{R}^{\mathrm{T}}\left(\ell_x^{\mathbf{m}}, \ell_y^{\mathbf{m}}\right) \cdot \mathbf{v}_{|\mathcal{R}_0} \\ \mathbf{v}^{\mathbf{a}} = \mathbf{R}^{\mathrm{T}}\left(\ell_x^{\mathbf{a}}, \ell_y^{\mathbf{a}}\right) \cdot \mathbf{v}_{|\mathcal{R}_0} \end{cases}$$

Therefore:

$$\mathbf{v}^{\mathbf{a}} = \mathbf{R}^{\mathrm{T}}\left(\ell_x^{\mathbf{a}}, \ell_y^{\mathbf{a}}\right) \cdot \mathbf{R}\left(\ell_x^{\mathbf{m}}, \ell_y^{\mathbf{m}}\right) \cdot \mathbf{v}^{\mathbf{m}}$$

This kind of calculation is useful when, for instance, two robots are trying to meet.

Switching to a local map. Let us take a local map $\mathcal{R}_{\mathbf{m}}$ centered on a point $\mathbf{m}$, (in other words, a coordinate system located on the surface of the Earth whose origin is $\mathbf{m}$ and whose directions are East-North-upward. Let us now try to express in $\mathcal{R}_{\mathbf{m}}$ the coordinates $(\widetilde{x}, \widetilde{y}, \widetilde{z})$ of a point $\mathbf{a}$ located by its GPS coordinates (ℓ_x, ℓ_y, ρ). We can switch from the geographical coordinates to the local coordinates with the following relation:

$$\begin{pmatrix} \widetilde{x} \\ \widetilde{y} \\ \widetilde{z} \end{pmatrix} + \begin{pmatrix} 0 \\ 0 \\ \rho^{\mathbf{m}} \end{pmatrix} \overset{[4.3]}{=} \underbrace{\begin{pmatrix} -\sin \ell_x^{\mathbf{m}} & \cos \ell_x^{\mathbf{m}} & 0 \\ -\cos \ell_x^{\mathbf{m}} \sin \ell_y^{\mathbf{m}} & -\sin \ell_x^{\mathbf{m}} \sin \ell_y^{\mathbf{m}} & \cos \ell_y^{\mathbf{m}} \\ \cos \ell_x^{\mathbf{m}} \cos \ell_y^{\mathbf{m}} & \cos \ell_y^{\mathbf{m}} \sin \ell_x^{\mathbf{m}} & \sin \ell_y^{\mathbf{m}} \end{pmatrix}}_{\mathbf{R}^{\mathrm{T}}(\ell_x^{\mathbf{m}}, \ell_y^{\mathbf{m}})} \cdot \underbrace{\begin{pmatrix} \rho \cos \ell_y \cos \ell_x \\ \rho \cos \ell_y \sin \ell_x \\ \rho \sin \ell_y \end{pmatrix}}_{\mathbf{a}_{|\mathcal{R}_0}}$$

When $\ell_x^{\mathbf{m}} \simeq \ell_x$ and $\ell_y^{\mathbf{m}} \simeq \ell_y$, a first-order approximation is directly obtained with the aid of Figure 4.2(b):

$$\begin{pmatrix} \widetilde{x} \\ \widetilde{y} \\ \widetilde{z} \end{pmatrix} \simeq \begin{pmatrix} \rho \cos \ell_y \cdot (\ell_x - \ell_x^{\mathbf{m}}) \\ \rho\left(\ell_y - \ell_y^{\mathbf{m}}\right) \\ \rho - \rho^{\mathbf{m}} \end{pmatrix} \qquad [4.4]$$

We could also have formally obtained these results using trigonometric relations, but it would have been more difficult. For example, we would obtain the last of the three equations by using the following line of reasoning:

$$\begin{aligned}\widetilde{z} &= \rho\left(\cos\ell_x^{\mathbf{m}}\cos\ell_y^{\mathbf{m}}\cos\ell_y\cos\ell_x + \cos\ell_y^{\mathbf{m}}\sin\ell_x^{\mathbf{m}}\cos\ell_y\sin\ell_x\right.\\ &\left.+\sin\ell_y^{\mathbf{m}}\sin\ell_y\right) - \rho^{\mathbf{m}}\\ &= \rho\left(\cos\ell_y^{\mathbf{m}}\cos\ell_y\left(\cos\ell_x^{\mathbf{m}}\cos\ell_x + \sin\ell_x^{\mathbf{m}}\sin\ell_x\right)\right.\\ &\left.+\sin\ell_y^{\mathbf{m}}\sin\ell_y\right) - \rho^{\mathbf{m}}\\ &= \rho(\cos\ell_y^{\mathbf{m}}\cos\ell_y\underbrace{\cos\left(\ell_x^{\mathbf{m}} - \ell_x\right)}_{\simeq 1} + \sin\ell_y^{\mathbf{m}}\sin\ell - \rho^{\mathbf{m}})\\ &\simeq \rho\underbrace{\cos\left(\ell_y^{\mathbf{m}} - \ell_y\right)}_{\simeq 1} - \rho^{\mathbf{m}} \simeq \rho - \rho^{\mathbf{m}}\end{aligned}$$

Let us note that when the robot is moving in a small-diameter area, we sometimes choose a reference point other than the center $\mathbf{m}$ of the robot, such as its launch position.

4.2. Path planning

When the robot is completely autonomous, the desired path must be planned out [LAV 06]. Very often, these paths are polynomials, for two reasons. First, the space of polynomials has a vector space structure and can, therefore, utilize the power of linear algebra. Second, they are easier to differentiate, which is useful for feedback linearization, since it requires the successive derivatives of the setpoints.

4.2.1. *Simple example*

Let us illustrate how such a planning is performed on the example of a robot tank. Assume that at the initial moment $t = 0$, the robot is located at the point (x_0, y_0) and that we would like to reach the point (x_1, y_1) at time t_1 with a speed equal to (v_x^1, v_y^1). We suggest a polynomial path of the form:

$$\begin{aligned}x_d &= a_x t^2 + b_x t + c_x\\ y_d &= a_y t^2 + b_y t + c_y\end{aligned}$$

We need to solve the system of equations:

$$\begin{array}{ll} c_x = x_0, & c_y = y_0 \\ a_x t_1^2 + b_x t_1 + c_x = x_1 & a_y t_1^2 + b_y t_1 + c_y = y_1, \\ 2a_x t_1 + b_x = v_x^1, & 2a_y t_1 + b_y = v_y^1 \end{array}$$

which is linear. We easily obtain:

$$\begin{pmatrix} a_x \\ a_y \\ b_x \\ b_y \\ c_x \\ c_y \end{pmatrix} = \begin{pmatrix} \frac{1}{t_1^2} x_0 - \frac{1}{t_1^2} x_1 + \frac{1}{t_1} v_x^1 \\ \frac{1}{t_1^2} y_0 - \frac{1}{t_1^2} y_1 + \frac{1}{t_1} v_y^1 \\ -v_x^1 - \frac{2}{t_1} x_0 + \frac{2}{t_1} x_1 \\ -v_y^1 - \frac{2}{t_1} y_0 + \frac{2}{t_1} y_1 \\ x_0 \\ y_0 \end{pmatrix}$$

Therefore, we have:

$$\begin{array}{ll} \dot{x}_d = 2a_x t + b_x & \dot{y}_d = 2a_y t + b_y \\ \ddot{x}_d = 2a_x, & \ddot{y}_d = 2a_y \end{array}$$

By inserting these quantities into a control law obtained using linearizing feedback (as is the case, for instance, with equation [2.9]), we obtain a controller that meets our objectives.

4.2.2. *Bézier polynomials*

Here, we will look at generalizing the approach presented in the previous section. Given the control points $\mathbf{p}_0, \mathbf{p}_1, \ldots, \mathbf{p}_n$, we can generate a polynomial $\mathbf{f}(t)$ such that $\mathbf{f}(0) = \mathbf{p}_0$, $\mathbf{f}(1) = \mathbf{p}_n$ and such that for $t \in [0, 1]$, the polynomial $\mathbf{f}(t)$ is successively attracted by the $\mathbf{p}_i$ with $i \in \{0, \ldots, n\}$. In order to correctly understand the method of building Bézier polynomials, let us examine various cases:

– case $n = 1$. We take the standard linear interpolation:

$$\mathbf{f}(t) = (1 - t)\, \mathbf{p}_0 + t\mathbf{p}_1$$

The point $\mathbf{f}(t)$ corresponds to a barycenter between the control points $\mathbf{p}_0$ and $\mathbf{p}_1$, and the weights assigned to these two points change with time:

– case $n = 2$. We now have three control points $\mathbf{p}_0, \mathbf{p}_1, \mathbf{p}_2$. We create an auxiliary control point $\mathbf{p}_{01}$ which moves on the segment $[\mathbf{p}_0, \mathbf{p}_1]$ and another $\mathbf{p}_{12}$ which is associated with the segment $[\mathbf{p}_1, \mathbf{p}_2]$. We take:

$$\begin{aligned}\mathbf{f}(t) &= (1-t)\,\mathbf{p}_{01} + t\mathbf{p}_{12} \\ &= (1-t)\underbrace{((1-t)\,\mathbf{p}_0 + t\mathbf{p}_1)}_{\mathbf{p}_{01}} + t\underbrace{((1-t)\,\mathbf{p}_1 + t\mathbf{p}_2)}_{\mathbf{p}_{12}} \\ &= (1-t)^2\,\mathbf{p}_0 + 2\,(1-t)\,t\mathbf{p}_1 + t^2\mathbf{p}_2\end{aligned}$$

Thus, we obtain a second-order polynomial;

– case $n = 3$. We apply the previous method for four control points. We obtain:

$$\begin{aligned}\mathbf{f}(t) &= (1-t)\,\mathbf{p}_{012} + t\mathbf{p}_{123} \\ &= (1-t)\underbrace{((1-t)\,\mathbf{p}_{01} + t\mathbf{p}_{12})}_{\mathbf{p}_{012}} + t\underbrace{((1-t)\,\mathbf{p}_{12} + t\mathbf{p}_{23})}_{\mathbf{p}_{123}} \\ &= (1-t)^3\,\mathbf{p}_0 + 3\,(1-t)^2\,t\mathbf{p}_1 + 3\,(1-t)\,t^2\mathbf{p}_2 + t^3\mathbf{p}_3\end{aligned}$$

– for a given n, we obtain:

$$\mathbf{f}(t) = \sum_{i=0}^{n} \underbrace{\frac{n!}{i!\,(n-i)!}\,(1-t)^{n-i}\,t^i}_{b_{i,n}(t)}\,\mathbf{p}_i$$

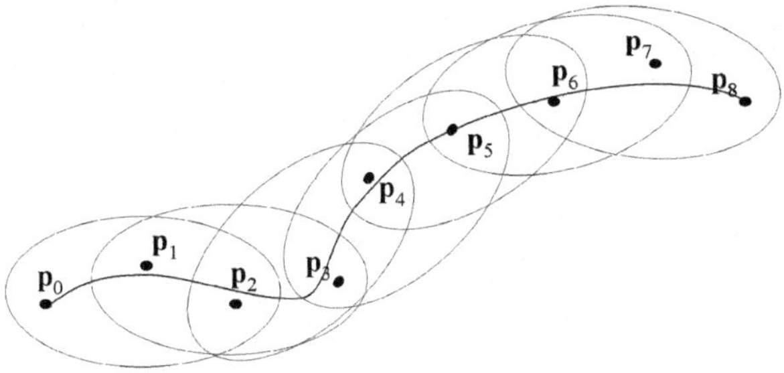

Figure 4.3. *Illustration of second-order B-splines*

The polynomials $b_{i,n}(t)$, called *Bernstein polynomials*, form a basis of the space of n-degree polynomials. When we increase the degree (in other words, the number of control points), numerical instability and oscillations appear. This is called *Runge's phenomenon*. For complex curves with hundreds of control points, it is preferable to use B-splines corresponding to a concatenation of Bézier curves of limited order. Figure 4.3 shows such a concatenation in which, for each group of three points, we can calculate a second-order Bézier polynomial.

4.3. Voronoi diagram

Let us consider n points $\mathbf{p}_1, \ldots, \mathbf{p}_n$. Contrarily to the previous sections, the $\mathbf{p}_i$ here do not correspond to control points, but to point obstacles that we try to avoid. To each of these points, we associate the set:

$$\mathbb{P}_i = \left\{ \mathbf{x} \in \mathbb{R}^d, \forall j, \|\mathbf{x} - \mathbf{p}_i\| \leq \|\mathbf{x} - \mathbf{p}_j\| \right\}$$

For all i, this set is a polygon. The collection of these $\mathbb{P}_i$ is called a *Voronoi diagram*. Figure 4.4 represents a set of points with the associated Voronoi diagram. If an environment contains obstacles, the robot will have to plan a path that remains on the borders of the $\mathbb{P}_i$.

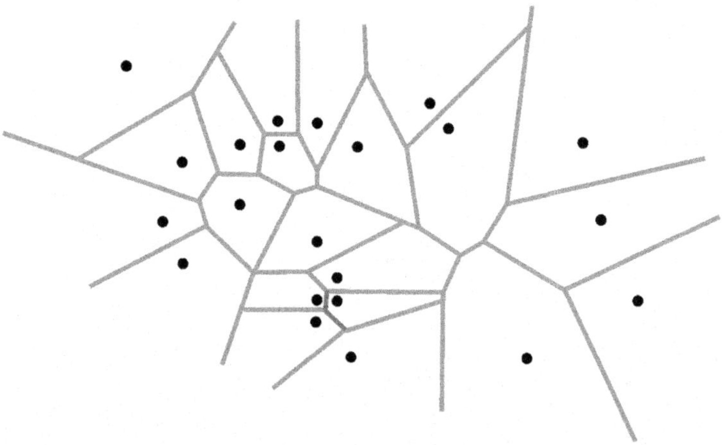

Figure 4.4. *Voronoi diagram*

Delaunay triangulation. Given n points in space, we can use the Voronoi diagrams to perform a triangulation of the space. This corresponding triangulation, referred to as *Delaunay triangulation*, allows maximizing the quantity of acute angles and thus avoid elongated triangles. It is obtained by connecting the neighboring points of the corresponding regions with an edge in the Voronoi diagram. In a Delaunay triangulation, none of the triangles contains another point within its circumscribed circle. Figure 4.5 represents the Delaunay triangulation associated with the Voronoi diagram in Figure 4.4. A Delaunay triangulation is often used in robotics to represent space such as the area already explored, restricted areas, lakes, etc. We often associate a color with each triangle following the characteristics of the space the triangle belongs to (water, land, road, etc.).

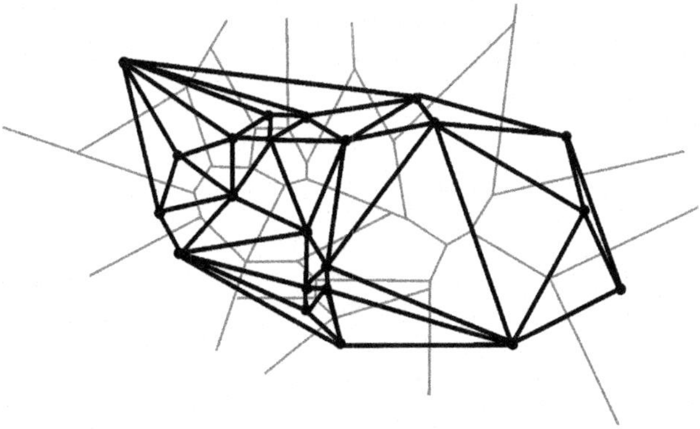

Figure 4.5. *Delaunay triangulation*

4.4. Artificial potential field method

A mobile robot has to move in a congested environment that contains mobile and stationary obstacles. The artificial potential field method [LAT 91] consists of imagining that the robot can behave like an electric particle that can be attracted or repelled by other objects following the sign of their electric charge. This is a reactive approach to guidance in which the path is not planned in advance. In physics, we have the following relation:

$$\mathbf{f} = -\text{grad}V(\mathbf{p})$$

where $\mathbf{p}$ is the position of point particle in space, V is the potential and $\mathbf{f}$ is the force applied on the particle. We will have the same relation in mobile robotics, but with $\mathbf{p}$ the position of the center of the robot, V a potential imagined by the robot and $\mathbf{f}$ the speed vector to follow. The potential fields will help us express a desired behavior for a robot. The obstacles will be represented by potentials exerting a repulsive force on the robot, while the objective to follow will exert an attractive force. In a situation where several robots need to remain grouped while avoiding collisions, we can use a near-field repulsive potential and a far-field attractive potential. More generally, the vector fields used might not derive from a potential, as is the case if we would like the robot to have a cyclic behavior. The following table gives several types of potential that can be used:

Potential	$V(\mathbf{p})$	$-\text{grad}(V(\mathbf{p}))$
attractive conical	$\|\mathbf{p}-\hat{\mathbf{p}}\|$	$-\frac{\mathbf{p}-\hat{\mathbf{p}}}{\|\mathbf{p}-\hat{\mathbf{p}}\|}$
attractive quadratic	$\|\mathbf{p}-\hat{\mathbf{p}}\|^2$	$-2(\mathbf{p}-\hat{\mathbf{p}})$
attractive plane or line	$(\mathbf{p}-\hat{\mathbf{p}})^{\mathrm{T}} \cdot \hat{\mathbf{n}}\,\hat{\mathbf{n}}^{\mathrm{T}} \cdot (\mathbf{p}-\hat{\mathbf{p}})$	$-2\,\hat{\mathbf{n}}\,\hat{\mathbf{n}}^{\mathrm{T}}(\mathbf{p}-\hat{\mathbf{p}})$
repulsive	$\frac{1}{\|\mathbf{p}-\hat{\mathbf{q}}\|}$	$\frac{(\mathbf{p}-\hat{\mathbf{q}})}{\|\mathbf{p}-\hat{\mathbf{q}}\|^3}$
uniform	$-\hat{\mathbf{v}}^{\mathrm{T}} \cdot \mathbf{p}$	$\hat{\mathbf{v}}$

In this table, $\hat{\mathbf{p}}$ represents an attractive point, $\hat{\mathbf{q}}$ represents a repulsive point and $\hat{\mathbf{v}}$ represents a desired speed for the robot. In the case of the attractive plane, $\hat{\mathbf{p}}$ is a point of the plane and $\hat{\mathbf{n}}$ is a vector orthonormal to the plane. By adding several potentials, we can ask the robot (which is supposed to follow the direction that tends to decrease the potential) to accomplish its objectives while moving away from the obstacles. Figure 4.6 represents three vector fields derived from artificial potentials. The one on the left corresponds to a uniform field, the one in the middle corresponds to a repulsive potential that is added to a uniform field and the one on the right corresponds to the sum of an attractive potential and a repulsive potential.

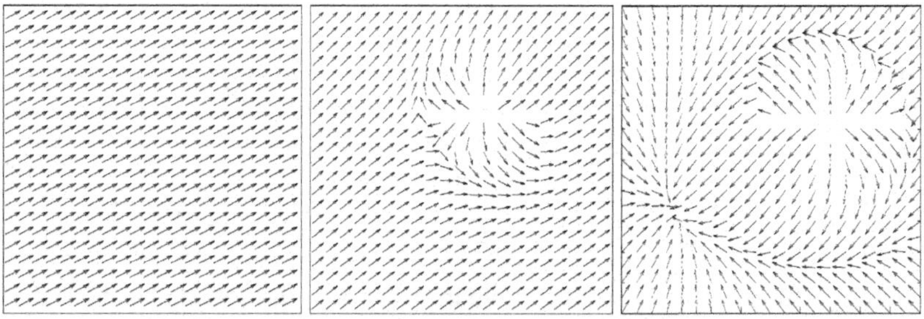

Figure 4.6. *Artificial potential fields*

4.5. Exercises

EXERCISE 4.1.– Pursuit on a sphere

Consider a robot $\mathcal{R}$ moving on the surface of a sphere similar to that of the Earth, with a radius of $\rho = 30$ m. This robot is located by its longitude ℓ_x, latitude ℓ_y and heading ψ, relative to the East. In a local coordinate system, the state equations of the robot are of the type:

$$\begin{cases} \dot{x} = \cos\psi \\ \dot{y} = \sin\psi \\ \dot{\psi} = u \end{cases}$$

1) Give the state equations in the case in which the state vector is (ℓ_x, ℓ_y, ψ).

2) Simulate this evolving system graphically in three-dimensional (3D), in MATLAB.

3) A second robot $\mathcal{R}_a$, described by the same equations, is moving randomly on the sphere (see Figure 4.7). Suggest a control law that allows the robot $\mathcal{R}$ to meet the robot $\mathcal{R}_a$.

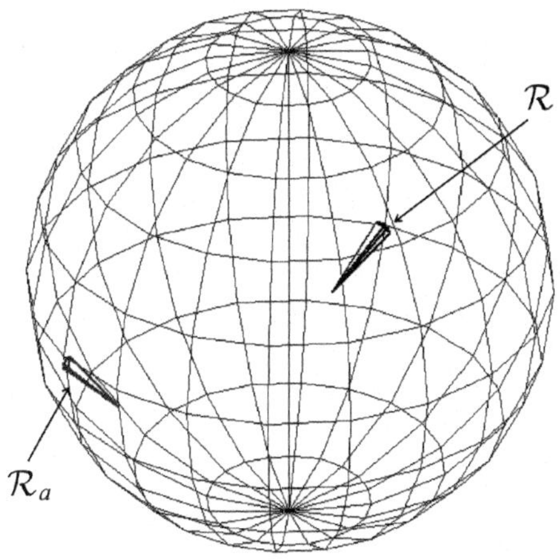

Figure 4.7. *On the sphere, the robot $\mathcal{R}$ follows the robot $\mathcal{R}_a$*

EXERCISE 4.2.– Planning a path

Let us consider a scene with two triangles as shown in Figure 4.8, and a robot described by the state equations:

$$\begin{cases} \dot{x} = v\cos\theta \\ \dot{y} = v\sin\theta \\ \dot{\theta} = u_1 \\ \dot{v} = u_2 \end{cases}$$

with initial state $(x, y, \theta, v) = (0, 0, 0, 1)$. This robot has to reach the point with coordinates $(8, 8)$.

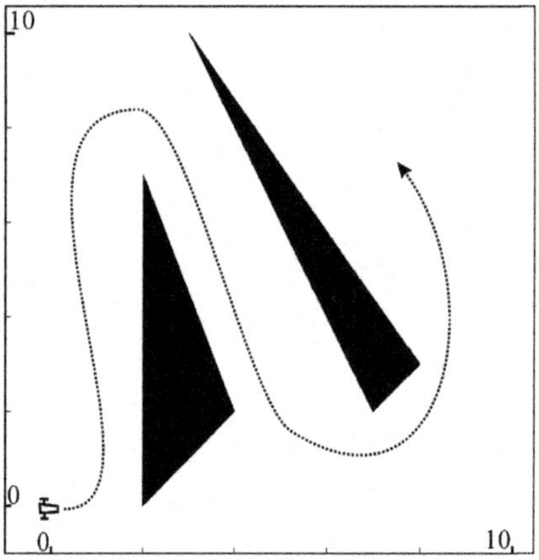

Figure 4.8. *The robot has to follow a path without hitting the obstacles*

1) In MATLAB, find the control points for a Bézier polynomial that connects the initial position to the desired position as shown in the figure.

2) Using feedback linearization, deduce the control law that allows us to reach the objective in 50 sec.

EXERCISE 4.3.– Drawing a Voronoi diagram

Let us consider the 10 points in Figure 4.9. Draw the associated Voronoi diagram on a piece of paper as well as the corresponding Delaunay triangulation.

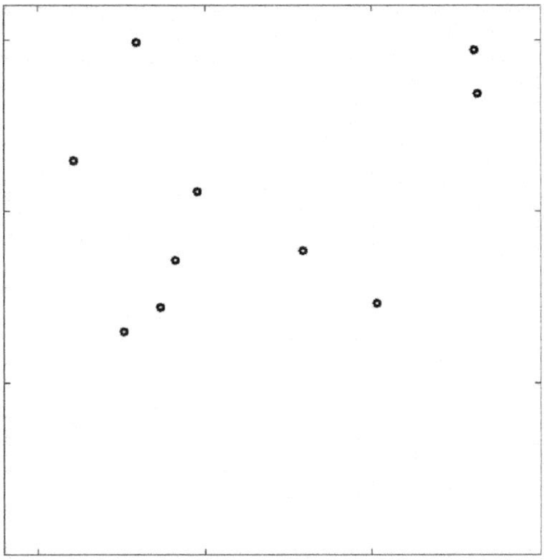

Figure 4.9. *Ten points for which we want to build a Voronoi diagram*

EXERCISE 4.4.– Calculating a Voronoi diagram

1) Show that if $\mathbf{x}$ and $\mathbf{y}$ are two vectors of $\mathbb{R}^n$, we have the so-called *polarization* equations:

$$\left\{ \|\mathbf{x}-\mathbf{y}\|^2 = \|\mathbf{x}\|^2 + \|\mathbf{y}\|^2 - 2\langle \mathbf{x}, \mathbf{y} \rangle \right.$$

For this, develop the expression of the scalar product $\langle \mathbf{x}-\mathbf{y}, \mathbf{x}-\mathbf{y} \rangle$.

2) Let us consider $n+1$ points $\mathbf{a}^1, \ldots, \mathbf{a}^{n+1}$ and their circumscribed sphere denoted by $\mathcal{S}$. With the aid of the previous question, give an expression, as a function of the $\mathbf{a}^i$, of the center $\mathbf{c}$ of $\mathcal{S}$ and its radius r.

3) Let us now consider three points in the plane $\mathbf{a}^1, \mathbf{a}^2, \mathbf{a}^3$. Which conditions must be verified for $\mathbf{m}$ to be within the circle circumscribed to the triangle $\left(\mathbf{a}^1, \mathbf{a}^2, \mathbf{a}^3\right)$?

4) Consider m points in the plane $\mathbf{p}^1, \ldots, \mathbf{p}^m$, a Delaunay triangulation is a partition of the space into triangles $\mathcal{T}(k) = \left(\mathbf{a}^1(k), \mathbf{a}^2(k), \mathbf{a}^3(k)\right)$, whose vertices are taken from the $\mathbf{p}^i$ such that each circle $\mathcal{C}(k)$ circumscribed to this triangle $\mathcal{T}(k)$ contains none of the $\mathbf{p}^i$. Write a MATLAB program that takes $m = 10$ random points of the plane and draws a Delaunay triangulation. What is the complexity of the algorithm?

5) Given the triangulation established in the previous question, build a Voronoi diagram associated with the points $\mathbf{a}^1, \ldots, \mathbf{a}^{n+1}$.

EXERCISE 4.5.– Heading control of a Dubins car

The results of this exercise will be used in exercise 4.6 for calculating Dubins paths. A Dubins car is described by the state equations:

$$\begin{cases} \dot{x} = \cos\theta \\ \dot{y} = \sin\theta \\ \dot{\theta} = u \end{cases}$$

θ is the robot's heading and (x, y) are the coordinates of its center. This robot has to be aligned with a heading setpoint $\bar{\theta}$.

1) We assume that the input $u \in [-1, 1]$. Give an analytic expression of the error angle $\delta \in [-\pi, \pi]$ as a function of θ and $\bar{\theta}$ that indicates the angle the robot has to turn in order to reach its setpoint as fast as possible. Take into account that the expression of δ has to be periodic relative to θ and $\bar{\theta}$. Indeed, the angles of $-\pi, \pi$ or 3π have to be considered as equivalent. Give the associated control law and simulate it in MATLAB.

2) The same as above, with the exception that the robot can only turn left (in the direct trigonometric sense), in other words $u \in [0, 1]$. Let us note that now, $\delta \in [0, 2\pi]$.

3) The same as above, but the robot can now only turn right, i.e. $u \in [-1, 0]$.

EXERCISE 4.6.– Dubins paths

As in the previous exercise, let us consider a robot moving on a plane, described by:

$$\begin{cases} \dot{x} = \cos\theta \\ \dot{y} = \sin\theta \\ \dot{\theta} = u \end{cases}$$

where θ is the robot's heading and (x, y) are the coordinates of its center. Its state vector is given by $\mathbf{x} = (x, y, \theta)$ and its input u must remain within the interval $[-u_{\max}, u_{\max}]$. This is the Dubins car [DUB 57] corresponding to the simplest possible non-holonomic mobile vehicle. Despite its simplicity, it illustrates many difficulties that may appear within the context of non-holonomic robots.

1) Calculate the maximum radius of curvature r that can be executed by the path of the robot.

2) Dubins showed that in order to switch from a configuration $\mathbf{a} = (x_a, y_a, \theta_a)$ to a configuration $\mathbf{b} = (x_b, y_b, \theta_b)$ that are not too close together (in other words, separated by a distance superior to $4r$), the minimum time strategy always consists of (1) turning to the maximum in one direction (in other words, $u = \pm u_{\max}$); (2) moving straight ahead and; (3) then turning to the maximum again. The path corresponding to such a maneuver is called a *Dubins path*, and is thus composed of a starting arc, segment and termination arc. There are four ways to construct a Dubins path: LSL, LSR, RSL and RSR, where L means left, R means right and S stands for straight ahead. Give a configuration for $\mathbf{a}$ and $\mathbf{b}$ such that none of the four paths corresponds to an optimal strategy (and therefore $\mathbf{a}$ and $\mathbf{b}$ are quite close together). A situation in which the optimal strategy is RLR will be chosen.

3) In the case of an RSL strategy as shown in Figure 4.10, calculate the length L of the Dubins path as a function of $\mathbf{a}$ and $\mathbf{b}$.

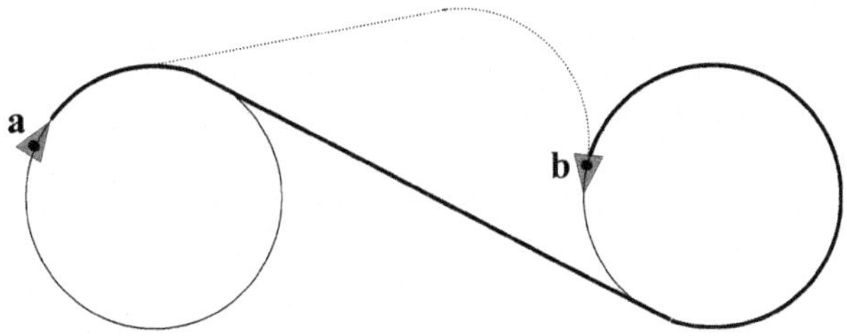

Figure 4.10. *In bold:* a right-straight-left *(RSL)-type Dubins path leading from* $\mathbf{a}$ *to* $\mathbf{b}$*; in dotted: an RLR-type Dubins path*

4) In the case of an LSL strategy, calculate the length L of the Dubins path as a function of $\mathbf{a}$ and $\mathbf{b}$.

5) By using the reflection symmetry of the problem, deduce L in the case of RSR and LSR strategies, then write a MATLAB function that calculates L in all situations. For this, the two Booleans ε_a ε_b will be used, which are equal to 1 if the corresponding arc is in the forward direction (i.e. to the left) and -1 otherwise.

6) Use the previous questions to write a MATLAB program that calculates the minimum length path for a Dubins car.

EXERCISE 4.7.– Artificial potentials

A robot situated at $\mathbf{p} = (x, y)$ must reach a target of unknown movement whose position $\hat{\mathbf{p}}$ and speed $\hat{\mathbf{v}}$ are known at the present time. This pair $(\hat{\mathbf{p}}, \hat{\mathbf{v}})$ might, for instance, correspond to a setpoint given by a human operator. A fixed obstacle located at position $\hat{\mathbf{q}}$ must be avoided. We model the desired behavior of our robot by the potential:

$$V(\mathbf{p}) = -\hat{\mathbf{v}}^{\mathrm{T}} \cdot \mathbf{p} + \|\mathbf{p} - \hat{\mathbf{p}}\|^2 + \frac{1}{\|\mathbf{p} - \hat{\mathbf{q}}\|}$$

where the potential $-\hat{\mathbf{v}}^{\mathrm{T}} \cdot \mathbf{p}$ represents the speed setpoint, the potential $\|\mathbf{p} - \hat{\mathbf{p}}\|^2$ makes the target position $\hat{\mathbf{p}}$ attractive and the potential $\frac{1}{\|\mathbf{p} - \hat{\mathbf{q}}\|}$ makes the obstacle $\hat{\mathbf{q}}$ repulsive.

1) Calculate the gradient of the potential $V(\mathbf{p})$ and deduce the speed vector setpoint $\mathbf{w}(\mathbf{p}, t)$ to apply to our robot so that it responds correctly to this potential.

2) We assume that our robot obeys the following state equations:

$$\begin{cases} \dot{x} = v \cos \theta \\ \dot{y} = v \sin \theta \\ \dot{v} = u_1 \\ \dot{\theta} = u_2 \end{cases}$$

Give the control law that corresponds to the desired potential field. We will use the same principle as the one shown in Figure 4.11. First, disassemble the robot (in gray in the figure) into a chain made up of two blocks. The first block forms the speeds from the actuators and the second block builds the position vector $\mathbf{p} = (x, y)$. Then, calculate the left inverse of the first block in order to end up with a system of the type:

$$\begin{cases} \dot{x} = \bar{v} \cos \bar{\theta} \\ \dot{y} = \bar{v} \sin \bar{\theta} \end{cases}$$

Use a simple proportional control to perform this approximative inversion. Then, generate the new input $(\bar{v}, \bar{\theta})$ using the potential to be satisfied. Illustrate the behavior of the robot in MATLAB with a target $\hat{\mathrm{p}} = (t, t)$ and a fixed obstacle placed at $\hat{\mathrm{q}} = (4, 5)$.

3) We would now like to follow the target $\hat{\mathrm{p}}$ with a mobile obstacle at $\hat{\mathrm{q}}$ with:

$$\hat{\mathrm{p}} = \begin{pmatrix} \cos \frac{t}{10} \\ 2 \sin \frac{t}{10} \end{pmatrix} \text{ and } \hat{\mathrm{q}} = \begin{pmatrix} 2 \cos \frac{t}{5} \\ 2 \sin \frac{t}{5} \end{pmatrix}$$

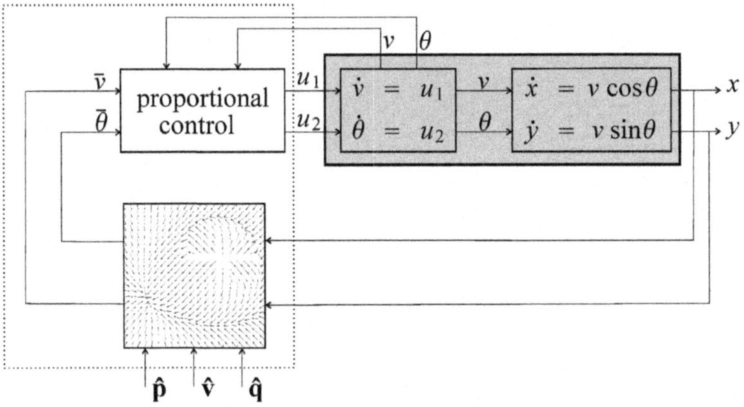

Figure 4.11. *Controller (dotted) obtained by the potential method*

Adjust the parameters of the potential in order to follow the target without hitting the obstacle. Illustrate the behavior of the controlled robot in MATLAB.

4.6. Corrections

CORRECTION FOR EXERCISE 4.1.– (Pursuit on a sphere)

1) Following [4.4], we have the following first-order approximation:

$$\begin{pmatrix} dx \\ dy \end{pmatrix} \simeq \begin{pmatrix} \rho \cos \ell_y \, d\ell_x \\ \rho d\ell_y \end{pmatrix}$$

Therefore, since $\dot{x} = \cos\psi$ and $\dot{y} = \sin\psi$, we have:

$$\begin{cases} \dot{\ell}_x = \dfrac{\cos\psi}{\rho\cos\ell_y} \\ \dot{\ell}_y = \dfrac{\sin\psi}{\rho} \\ \dot{\theta} \;= u \end{cases}$$

2) The meridians and parallels for representing the sphere are drawn in MATLAB as follows:

```
function draw_earth()
M=[]; a=pi/10;
Ly=-pi/2:a:pi/2; Lx=0:a:2*pi;
for ly=Ly, for lx=Lx, M=[M,T(lx,ly,rho)]; end; end;
for lx=Lx, for ly=Ly, M=[M,T(lx,ly,rho)]; end; end;
plot3(M(1,:),M(2,:),M(3,:));
end
```

The transformation function T used before is given by:

```
function p=T(lx,ly,rho)
p=rho*[cos(ly)*cos(lx);cos(ly)*sin(lx);sin(ly)];
end
```

In order to draw the robot, we first create a 3D pattern, then subject it to a coordinate system change which will lead it to the desired area (see equation [4.2]) and finally apply an Euler transformation in order to give it the correct local orientation. Note that in our case, the elevation θ and the bank φ are zero, since the robot is moving horizontally. We obtain the following instructions:

```
R1=[-sin(lx),-sin(ly)*cos(lx),cos(ly)*cos(lx);
cos(lx),-sin(ly)*sin(lx),cos(ly)*sin(lx);
0,cos(ly),sin(ly)];
E=eulermat(0,0,psi);
R=[R1*E,T(lx,ly,rho);0,0,0,1];
```

3) The robot $\mathcal{R}$ located at $\mathbf{m} : (\ell_x, \ell_y)$ with a heading of ψ relative to the East must reach robot $\mathcal{R}_a$ located at point $\mathbf{a} : \left(\ell_x^{\mathbf{a}}, \ell_y^{\mathbf{a}}\right)$. If $\mathbf{a}$ is in a close neighborhood of $\mathbf{m}$ (less than 100 km), we can calculate the coordinates of $\mathbf{a}$ within a map centered around $\mathbf{m}$. We obtain, following formula [4.4]:

$$\begin{cases} \widetilde{x}^{\mathbf{a}} = \rho\cos\ell_y^{\mathbf{a}} \cdot \left(\ell_x^{\mathbf{a}} - \ell_x\right) \\ \widetilde{y}^{\mathbf{a}} = \rho\left(\ell_y^{\mathbf{a}} - \ell_y\right) \end{cases}$$

In order to find out whether the robot $\mathcal{R}_a$ is on the left or right of $\mathcal{R}$, we look at the sine of the angle between the two vectors $(\cos\psi, \sin\psi)$ and $(\widetilde{x}^{\mathbf{a}}, \widetilde{y}^{\mathbf{a}})$. Therefore, we will take the proportional controller:

$$u = \det\begin{pmatrix} \cos\psi & \cos\ell_y \cdot (\ell_x^{\mathbf{a}} - \ell_x) \\ \sin\psi & \ell_y^{\mathbf{a}} - \ell_y \end{pmatrix}$$

This control law which is based on a local representation of the target $\mathcal{R}_a$ position works correctly, even when $\mathcal{R}_a$ is far away from $\mathcal{R}$. The program given in the file `earth.m` shows the correct behavior of this control law.

CORRECTION FOR EXERCISE 4.2.– (Path planning)

1) Given the control points, the associated setpoint function is given by:

```
function w=setpoint(t)
w=0;
for i=0:n, w=w+b(i,n,t)*P(:,i+1); end;
end
```

The coefficients are given by the function:

```
function y=b(i,n,t)
y=prod(1:n)/(prod(1:i)*(prod(1:n-i)))*(1-t)^(n-i)*t^i;
end
```

We can easily adjust the control points in order to obtain a desired path. For example, we will take the control points $\mathbf{p}_i$ given in the matrix:

$$\begin{pmatrix} 1 & 1 & 1 & 1 & 2 & 3 & 4 & 5 & 5 & 7 & 8 & 10 & 9 & 8 \\ 1 & 4 & 7 & 9 & 10 & 8 & 6 & 4 & 1 & 0 & 0 & 1 & 4 & 8 \end{pmatrix}$$

and which are represented in Figure 4.12(a).

2) We can take the controller found in section 2.4.1. To reach the objective in $t_{\max} = 50$ sec, we must take the setpoint:

$$\mathbf{w}(t) = \sum_{i=0}^{n} b_{i,n}\left(\tfrac{t}{50}\right)\mathbf{p}_i$$

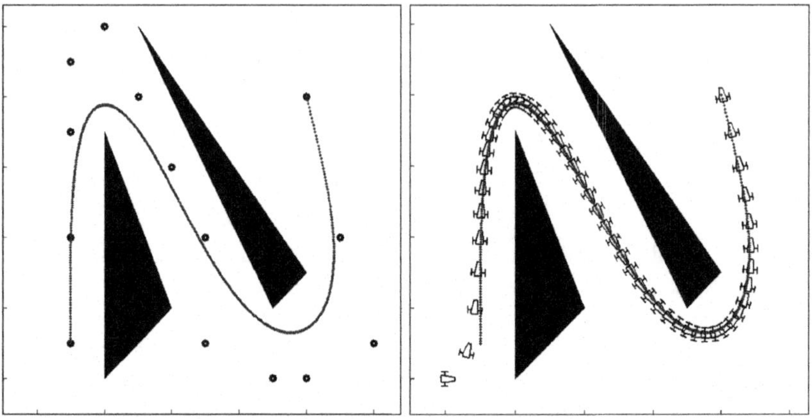

Figure 4.12. *a) Control points with the associated Bezier polynomial; b) path for the robot that follows the polynomial by avoiding the obstacles*

In order to apply our controller, we need the derivative $\dot{\mathbf{w}}(t)$ of the setpoint $\mathbf{w}(t)$. It is given by:

$$\dot{\mathbf{w}}(t) = \frac{1}{50}\sum_{i=0}^{n} \dot{b}_{i,n}\left(\frac{t}{50}\right)\mathbf{p}_i$$

with:

$$\dot{b}_{i,n}(t) = \frac{n!}{i!\,(n-i)!}\left(i\,(1-t)^{n-i}\,t^{i-1} - (n-i)\,(1-t)^{n-i-1}\,t^{i}\right)$$

We must make sure to take into account the particular case of $i = 0$ for which $\dot{b}_{i,n}(t) = -n(1-t)^{n-1}$ and the case of $i = n$ for which $\dot{b}_{n,n}(t) = nt^{n-1}$. The resulting path is represented in Figure 4.12(b). The MATLAB program that illustrates this control law can be found in the file `bezier.m`.

CORRECTION FOR EXERCISE 4.3.– (Drawing a Voronoi diagram)

In Figure 4.13, the Delaunay triangulation is represented by the bold lines. The Voronoi diagram is represented by the thin lines.

Figure 4.14 shows that the circle circumscribed to each triangle does not enclose any other point.

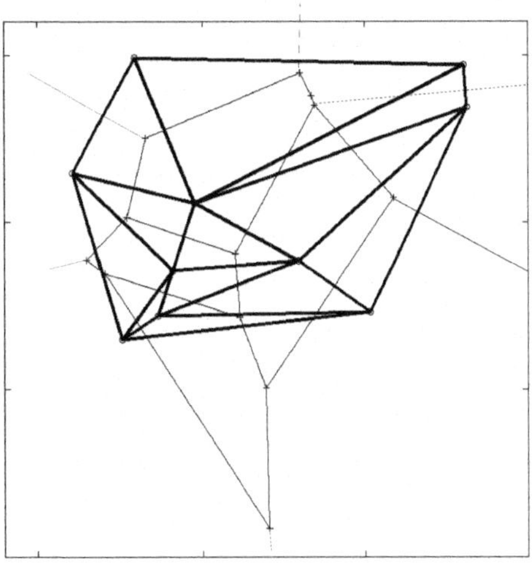

Figure 4.13. *Voronoi diagram*

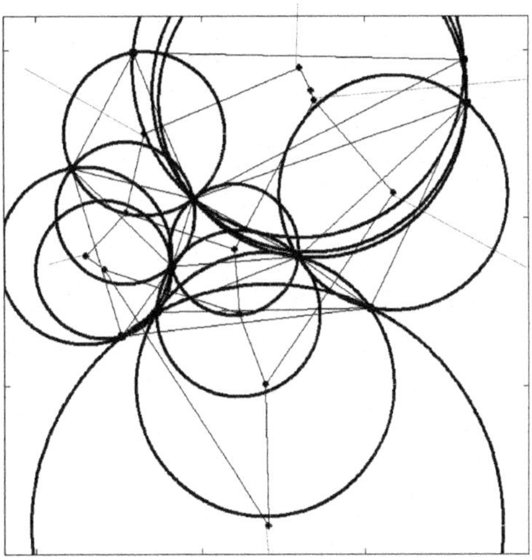

Figure 4.14. *Circles circumscribed to the Delaunay triangles*

CORRECTION FOR EXERCISE 4.4.– (Calculating a Voronoi diagram)

1) Recall that the scalar product is a bilinear form. Therefore, we have:

$$\begin{aligned}\|\mathbf{x}-\mathbf{y}\|^2 &= \langle \mathbf{x}-\mathbf{y}, \mathbf{x}-\mathbf{y}\rangle = \langle \mathbf{x},\mathbf{x}\rangle + \langle \mathbf{y},\mathbf{y}\rangle - \langle \mathbf{x},\mathbf{y}\rangle - \langle \mathbf{y},\mathbf{x}\rangle \\ &= \|\mathbf{x}\|^2 + \|\mathbf{y}\|^2 - 2\langle \mathbf{x},\mathbf{y}\rangle\end{aligned}$$

2) Let $\mathbf{c}$ be the center of the sphere $\mathcal{S}$. For $i \in \{1, \ldots, n+1\}$, we have:

$$\left\|\mathbf{a}^i - \mathbf{c}\right\|^2 = r^2$$

And, by using the polarization equation:

$$\left\|\mathbf{a}^i\right\|^2 + \|\mathbf{c}\|^2 - 2\left\langle \mathbf{a}^i, \mathbf{c}\right\rangle - r^2 = 0$$

Of course, we have $n+1$ unknowns which are r and $\mathbf{c}$. In order to obtain a linear system in these unknowns, let us take:

$$c_{n+1} = \|\mathbf{c}\|^2 - r^2$$

Thus:

$$2\left\langle \mathbf{a}^i, \mathbf{c}\right\rangle - c_{n+1} = \left\|\mathbf{a}^i\right\|^2$$

Therefore, we have:

$$\begin{pmatrix} 2a_1^1 & \cdots & 2a_n^1 & -1 \\ \vdots & & \vdots & \vdots \\ 2a_1^{n+1} & \cdots & 2a_n^{n+1} & -1 \end{pmatrix} \begin{pmatrix} c_1 \\ \vdots \\ c_n \\ c_{n+1} \end{pmatrix} = \begin{pmatrix} \left\|\mathbf{a}^1\right\|^2 \\ \vdots \\ \left\|\mathbf{a}^{n+1}\right\|^2 \end{pmatrix}$$

Solving the linear system yields the center of the circumscribed circle $\mathbf{c} = (c_1, \ldots, c_n)$ as well as its radius $r = \sqrt{\|\mathbf{c}\|^2 - c_{n+1}}$.

3) The point $\mathbf{m}$ is within the circle circumscribed to the triangle $\left(\mathbf{a}^1, \mathbf{a}^2, \mathbf{a}^3\right)$ if:

$$\|\mathbf{m} - \mathbf{c}\| \leq r$$

with:

$$\begin{pmatrix} c_1 \\ c_2 \\ c_3 \end{pmatrix} = \begin{pmatrix} 2a_1^1 \; 2a_2^1 \; -1 \\ 2a_1^2 \; 2a_2^2 \; -1 \\ 2a_1^2 \; 2a_2^3 \; -1 \end{pmatrix}^{-1} \begin{pmatrix} \left\|\mathbf{a}^1\right\|^2 \\ \left\|\mathbf{a}^2\right\|^2 \\ \left\|\mathbf{a}^3\right\|^2 \end{pmatrix}$$

$$\mathbf{c} = \begin{pmatrix} c_1 \\ c_2 \end{pmatrix} \text{ et } r = \sqrt{\|\mathbf{c}\|^2 - c_3}$$

4) The program is the following:

```
m=10;
P = 10*rand(2,m); % choose the points
C=[]; % list of circle centers
K=[]; % list of triangles
for i=1:1:m-2
for j=i+1:1:m-1
for k=j+1:1:m
A=P(:,[i,j,k]); c=[2*A',[-1;-1;-1]]\sum(A.^2,1)';
r=sqrt(norm(c(1:2))^2-c(3));
ok=true;
for q=1:m
ok=ok&&(q==i||q==j||q==k||norm(P(:,q)-c(1:2))>r);
end
if ok,
C=[C,c]; % save the centers
K=[K,[i;j;k]]; % save the triangles
end, end, end, end;
```

The complexity of the algorithm is in m^4, where m is the number of points being considered. There are algorithms in $m \log m$ such as the *sweep line algorithm* and *Fortune's algorithm*.

5) The following program draws the Voronoi diagram. The principle consists of connecting the centers of the circles circumscribed to the neighboring triangles:

```
for i=1:size(K,2), for j=1:size(K,2), % take all the
triangles two-by-two
if sum(ismember(K(:,i),K(:,j)))==2, % if they are neighbors
plot(C(1,[i j]),C(2,[i j])); % connect them
end, end, end
```

CORRECTION FOR EXERCISE 4.5.– (Heading control of a Dubins car)

1) Let us recall the expression of the sawtooth function (see formula [3.2]):

$$\text{sawtooth}\,(\theta) = \text{mod}(\theta + \pi, 2\pi) - \pi$$

Let us take $\widetilde{\theta} = \bar{\theta} - \theta$. In order to have $\delta \in [-\pi, \pi]$, we will take $\delta = \text{sawtooth}(\widetilde{\theta})$. Note that if $\widetilde{\theta} \in [-\pi, \pi]$, then $\delta = \widetilde{\theta}$. In order to have $u \in [-1, 1]$, we can take a proportional control law of the type:

$$u = \frac{\delta}{\pi} = \frac{1}{\pi}\text{sawtooth}(\widetilde{\theta})$$

2) We write the directional sawtooth function in MATLAB as follows:

```
function y = sawtooth(x,d)
y=d*pi+mod(x+pi-d*pi,2*pi)-pi;
end
```

This function returns the distance to the left ($d = 1$), to the right ($d = -1$) or the shortest distance ($d = 0$) as shown in Figure 4.15. The shortest distance case corresponds to the usual sawtooth function. In our case, $d = 1$ and therefore $\delta = \text{sawtooth}(\widetilde{\theta}, 1)$. In order to have $u \in [0, 1]$, we need to take:

$$u = \frac{\delta}{2\pi} = \frac{\text{sawtooth}(\widetilde{\theta}, 1)}{2\pi}$$

3) Now, $d = -1$. We take $\delta = \text{sawtooth}(\widetilde{\theta}, -1) - \pi$. To have $u \in [-1, 0]$, we take:

$$u = \frac{\delta}{2\pi} = \frac{\text{sawtooth}(\widetilde{\theta}, -1)}{2\pi}$$

The program associated with this exercise is given in `dubins.m`.

CORRECTION FOR EXERCISE 4.6.– (Dubins path)

1) For u constant, we have $\theta = ut + \theta_0$. And, therefore $\dot{x}\,(t) = \cos\,(ut + \theta_0)$ and $\dot{y}\,(t) = \sin\,(ut + \theta_0)$. And, after integrating: $x\,(t) = x\,(0) + \frac{1}{u}\sin\,(ut + \theta_0)$ and $y\,(t) = y\,(0) - \frac{1}{u}\cos\,(ut + \theta_0)$. The associated curvature radius is $r = \frac{1}{u}$. Since $u \in [-u_{\max}, u_{\max}]$, we have $r \in \frac{1}{[-u_{\max}, u_{\max}]} = [-\infty, -\frac{1}{u_{\max}}] \cup [\frac{1}{u_{\max}}, \infty]$.

2) We can take $\mathbf{b}$ at a distance of $3.9r$ to the right of $\mathbf{a}$. The optimal strategy is RLR.

3) As shown in Figure 4.17, the vectors $\mathbf{d}_a - \mathbf{c}_a$ are orthogonal. Thus, by using the Pythagorean theorem, the half-length ℓ of the Dubins path segment is given by:

$$\ell = \sqrt{\left(\frac{\|\mathbf{c}_b - \mathbf{c}_a\|}{2}\right)^2 - r^2}$$

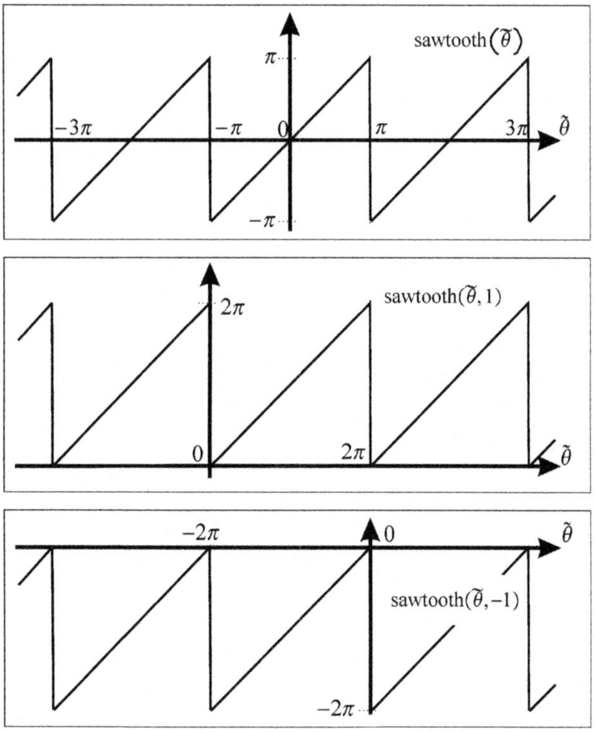

Figure 4.15. *Sawtooth functions for performing the angular difference and rejoining the desired interval*

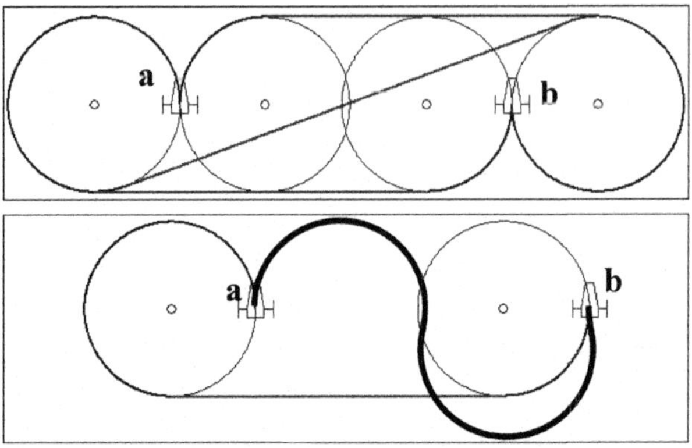

Figure 4.16. *Top: the four superimposed Dubins paths of type RSR, LSR, RSL and LSL. Bottom: the best of the four paths together with the RLR strategy (in bold) which proves to be the best among all strategies*

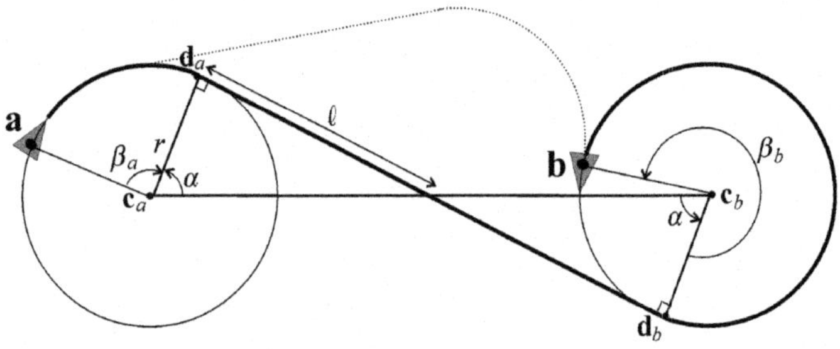

Figure 4.17. *Calculating the length of the RSL Dubins path based on the fact that* $d_a - c_a \perp d_b - d_a \perp d_b - c_b$

If this square root is not defined (i.e. if $\|\mathbf{c}_b - \mathbf{c}_a\| \leq 2r$), then this path is not valid and will not correspond to a Dubins path. As a matter of fact, it is in this situation that an RLR strategy could be optimal.

In the case where $\|\mathbf{c}_b - \mathbf{c}_a\| \geq 2r$, calculating the path length is done as follows:

$$\mathbf{c}_a = \mathbf{a} - r\begin{pmatrix} -\sin\theta_a \\ \cos\theta_a \end{pmatrix}; \ \mathbf{c}_b = \mathbf{b} + r\begin{pmatrix} -\sin\theta_b \\ \cos\theta_b \end{pmatrix}$$

$$\ell = \sqrt{\tfrac{1}{4}\left\|\mathbf{c}_b - \mathbf{c}_a\right\|^2 - r^2}. \text{ If } \ell \text{ is not real, return } L = \infty \text{ (fail)}$$

$$\alpha = \text{atan2}\left(\ell, r\right)$$

$$\mathbf{d}_a = \mathbf{c}_a + \frac{r}{\left\|\mathbf{c}_b - \mathbf{c}_a\right\|}\begin{pmatrix} \cos\alpha & -\sin\alpha \\ \sin\alpha & \cos\alpha \end{pmatrix}\left(\mathbf{c}_b - \mathbf{c}_a\right)$$

$$\mathbf{d}_b = \mathbf{c}_b + \mathbf{c}_a - \mathbf{d}_a$$

$$\beta_a = \text{sawtooth}(\text{angle}\left(\mathbf{a} - \mathbf{c}_a, \mathbf{d}_a - \mathbf{c}_a\right), -1)$$

$$\beta_b = \text{sawtooth}(\text{angle}\left(\mathbf{d}_b - \mathbf{c}_b, \mathbf{b} - \mathbf{c}_b\right), 1)$$

$$L = r\left|\beta_a\right| + r\left|\beta_b\right| + 2\ell$$

The *angle* function can be found in the file `angle.m`. It is the following recursive function:

```
function theta = angle(u,v)
if (exist('v')==0), theta=atan2(u(2),u(1));
else theta=sawtooth(angle(v)-angle(u)); end;
end
```

4) Calculating the length of the path is done as follows:

$$\mathbf{c}_a = \mathbf{a} + r\begin{pmatrix} -\sin\theta_a \\ \cos\theta_a \end{pmatrix}; \ \mathbf{c}_b = \mathbf{b} + r\begin{pmatrix} -\sin\theta_b \\ \cos\theta_b \end{pmatrix}$$

$$\ell = \frac{1}{2}\left\|\mathbf{c}_b - \mathbf{c}_a\right\|; \ \alpha = -\frac{\pi}{2}$$

$$\mathbf{d}_a = \mathbf{c}_a + \frac{r}{\left\|\mathbf{c}_b - \mathbf{c}_a\right\|}\begin{pmatrix} \cos\alpha & -\sin\alpha \\ \sin\alpha & \cos\alpha \end{pmatrix}\left(\mathbf{c}_b - \mathbf{c}_a\right)$$

$$\mathbf{d}_b = \mathbf{c}_b - \mathbf{c}_a + \mathbf{d}_a$$

$$\beta_a = \text{sawtooth}(\text{angle}\left(\mathbf{a} - \mathbf{c}_a, \mathbf{d}_a - \mathbf{c}_a\right), 1)$$

$$\beta_b = \text{sawtooth}(\text{angle}\left(\mathbf{d}_b - \mathbf{c}_b, \mathbf{b} - \mathbf{c}_b\right), 1)$$

$$L = r\left|\beta_a\right| + r\left|\beta_b\right| + 2\ell$$

5) The following MATLAB function computes L in all situations:

```
function L =path(a,b,r,epsa,epsb)
ca=a(1:2)+epsa*r*[-sin(a(3));cos(a(3))];
```

```
cb=b(1:2)+epsb*r*[-sin(b(3));cos(b(3))];
if (epsa*epsb==-1),
ell2=0.25*norm(cb-ca)^2-r^2;
if ell2<0, L=inf; return; end;
ell=sqrt(ell2);
alpha=-epsa*atan2(ell,r);
else ell=0.5*norm(cb-ca); alpha=-epsa*pi/2;
end
R=[cos(alpha),-sin(alpha);sin(alpha),cos(alpha)]
da=ca+(r/norm(cb-ca))*R*(cb-ca);
db=cb+epsa*epsb*(da-ca);
betaa=sawtooth(angle(a(1:2)-ca,da-ca),epsa);
betab=sawtooth(angle(db-cb,b(1:2)-cb),epsb);
L=r*(abs(betaa)+abs(betab))+2*ell;
end
```

6) The following MATLAB function computes the minimum time Dubins function:

```
L1=path(a,b,r,-1,-1); %RSR
L2=path(a,b,r,-1, 1); % RSL
L3=path(a,b,r, 1,-1); % LSR
L4=path(a,b,r, 1, 1); %LSL
L=min([L1,L2,L3,L4]);
if (L==L1) path(a,b,r,-1,-1); end;
if (L==L2) path(a,b,r,-1, 1); end;
if (L==L3) path(a,b,r, 1,-1); end;
if (L==L4) path(a,b,r, 1, 1); end;
```

This program tests all four possible paths and takes the best one. For $r = 10, \mathbf{a} = (-20, 10, -3)$ and $\mathbf{b} = (20, -10, 2)$, we obtain the results of Figure 4.18. For $\mathbf{a} = (-3, 1, 0.9)$ and $\mathbf{b} = (2, -1, 1)$ we obtain those of Figure 4.19. The program is given in the file `dubinspath.m`.

CORRECTION FOR EXERCISE 4.7.– (Artificial potentials)

1) We have:

$$\frac{dV}{d\mathbf{p}}(\mathbf{p}) = -\hat{\mathbf{v}}^{\mathrm{T}} + 2(\mathbf{p} - \hat{\mathbf{p}})^{\mathrm{T}} - \frac{(\mathbf{p} - \hat{\mathbf{q}})^{\mathrm{T}}}{\|\mathbf{p} - \hat{\mathbf{q}}\|^3}$$

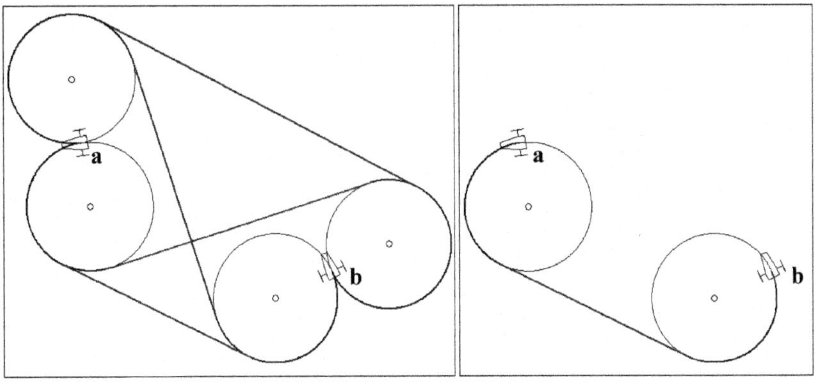

Figure 4.18. *a) The four Dubins paths possible; b) the best of the four paths*

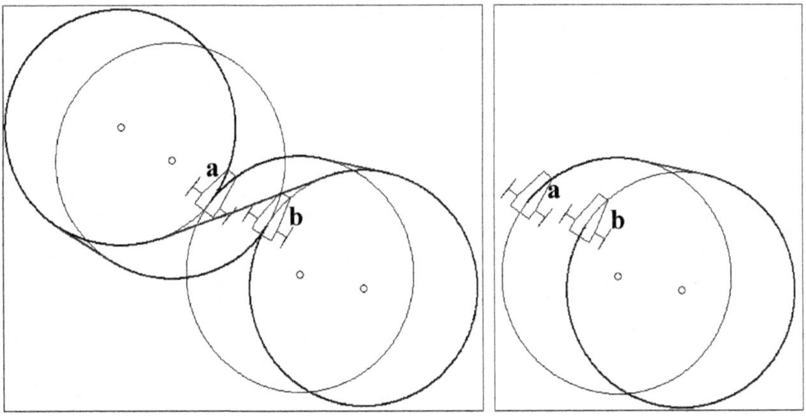

Figure 4.19. *Another situation in which the robot has to perform quite a complex maneuver in order to move very little*

Given that $\mathbf{w} = -\text{grad}(V(\mathbf{p}))$, we can deduce that:

$$\mathbf{w}(\mathbf{p}, t) = -\text{grad}\,\mathbf{V}(\mathbf{p}) = -\left(\frac{dV}{d\mathbf{p}}(\mathbf{p})\right)^{\mathrm{T}} = \hat{\mathbf{v}} - 2(\mathbf{p} - \hat{\mathbf{p}}) + \frac{(\mathbf{p} - \hat{\mathbf{q}})}{\|\mathbf{p} - \hat{\mathbf{q}}\|^3}$$

2) For the control, we can take a proportional speed and heading control, which give us:

$$\mathbf{u} = \begin{pmatrix} \|\mathbf{w}(\mathbf{p},t)\| - v \\ \text{sawtooth}\left(\text{angle}\left(\mathbf{w}(\mathbf{p},t)\right) - \theta\right) \end{pmatrix}$$

where the *sawtooth* function avoids the jumps of 2π. Given that the target is $\hat{\mathbf{p}} = (t,t)$, the target speed has to be taken as $\hat{\mathbf{v}} = (1,1)$.

3) Let us take the vector field:

$$\mathbf{w} = \bar{\mathbf{v}} - \frac{\mathbf{p} - \hat{\mathbf{p}}}{2} + \frac{\mathbf{p} - \hat{\mathbf{q}}}{\|\mathbf{p} - \hat{\mathbf{q}}\|^3} \text{ with } \bar{\mathbf{v}} = \frac{d\hat{\mathbf{p}}}{dt} = \frac{1}{10}\begin{pmatrix} -\sin\frac{t}{10} \\ 2\cos\frac{t}{10} \end{pmatrix}$$

which gives an acceptable behavior. We have simply reduced the proportional term (which corresponds to the attractiveness of the target) in order not to be too sensitive to the variations of the vector field around the target, which would cause violent and undesirable changes in heading. The program is given in the file `potential.m`.

5

Instantaneous Localization

Localization consists of finding the position of the robot (in other words, the coordinates of its center as well as its orientation), or more generally all its degrees of freedom. This problem is encountered in navigation, where we need to approximate the position, orientation and speed of the robot. The problem of localization is often considered to be a particular case of state estimation, which will be presented in the following chapters. However, in the case where an accurate state model is not available for our robot, an instantaneous localization often remains possible and may be sufficient for making a decision. Let us take, for instance, the situation in which we are aboard a ship and have just detected a lighthouse whose absolute position and height are known. By measuring the perceived height of the lighthouse and its angle relative to the ship, we may deduce the position of the ship using a compass, without using a state model for the ship. Instantaneous, or *model-free* localization, is an approach to localization that does not utilize the evolution equation of the robot, in other words it does not seek to make the measures coherent through time. This localization mainly consists of solving equations of geometric nature which are often nonlinear. The variables involved may be position variables or kinematic variables such as speed and accelerations. Since these localization methods are specific and quite far from state estimation methods, we will devote an entire chapter to them. After introducing the main sensors used for localization, we will present goniometric localization (in which the robot uses the angles of perception of landmarks) followed by multilateration which uses distances between the robot and landmarks.

5.1. Sensors

The robots are equipped with numerous sensors that are used for their localization. We will now present some of these:

Odometers: robots with wheels are generally equipped with odometers that measure the angular movements of the wheels. Given only the odometers, it is possible to calculate an estimation of the position of the robot. The precision of such a localization is very low given the systematic integration of the estimation error. We say that the estimation is drifting.

Doppler log: this type of sensor, mainly used in underwater robotics, allows us to calculate the speed of the robot. A Doppler log emits ultrasounds that are reflected by the ocean bed. Since the ocean bed is immobile, the sensor is able to estimate the speed of the robot by using the Doppler effect with a very high precision (around $0.1\ m/s$).

Accelerometers: these sensors provide measurements of the instantaneous forward acceleration. The principle of the axis-based accelerometer is shown in Figure 5.1. Generally, three accelerometers are used by the robot. Due to gravity, the value measured according to the vertical axis must be compensated.

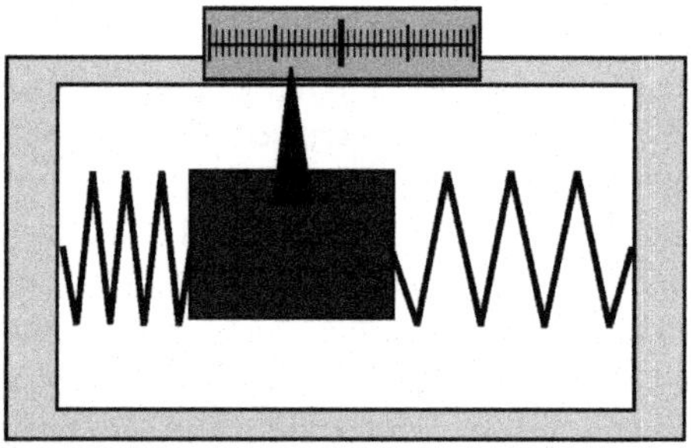

Figure 5.1. *Operating principle of an accelerometer*

Gyro or gyroscope: these sensors provide measurements of the instantaneous rotation speed. There are three types of gyros: the Coriolis vibratory gyro, mechanical gyro and optical gyro. The principle of the Coriolis vibratory gyro is shown in Figure 5.2(a). A vertical rod placed on a horizontal disk vibrates from left to right. As a result of the Coriolis force, if the disk rotates there is an angular vibration following the axis of the rod whose amplitude allows us to get the rotation speed of the disk. If the disk is not rotating, there is a forward rotation, but it is not angular. Piezoelectric gyros, very widely used for low-cost robotics, form a subclass of Coriolis vibratory gyroscopes. These gyros exploit the variation of the amplitude of a piezoelectric oscillator induced by the Coriolis force, due to the rotation applied to the sensor. Mechanical gyros make use of the fact that a rotating body tends to preserve its rotational axis if no torque is subjected to it. A well-known example is the gimbal gyroscope invented by Foucault, represented in Figure 5.2(b). A flywheel at the center rotates with high speed. If the base of the gyroscope moves, the two gimbal angles ψ, θ will change, but the rotation axis of the flywheel will not. From the values of $\psi, \theta, \dot{\psi}, \dot{\theta}$, we can find the rotation speed of the base (which is fixed on the robot). If the rotation axis of the flywheel is initialized correctly, and in a perfect situation in which no torque is exerted on this flywheel, such a system would theoretically give us the orientation of the robot. Unfortunately, there is always a small drift and only the rotation speed can be given in a reliable and drift-free way.

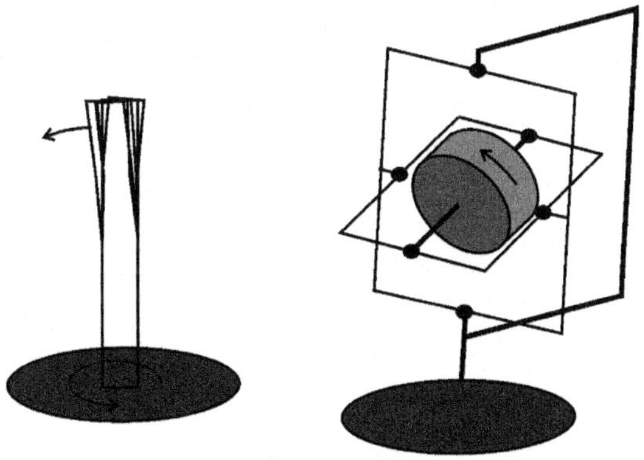

Figure 5.2. *Coriolis vibratory gyroscope and gimbal gyroscope*

More recently, optical gyroscopes can be as precise as mechanical ones. They make use of the Sagnac effect (for a circular optical path, the time taken by light to make a complete lap depends on the direction of the path) and have a precision of around 0.001 deg/s. Their principle is shown in Figure 5.3. In Figure 5.3(a), the laser leaves the light source represented by the black disk. In Figure 5.3(b), the beam splitter creates two other beams which travel in opposite directions in the optical loop. After rebounding several times on the three mirrors represented in gray, the two beams meet. Since the beams intersect on the left, the gyro rotates in the opposite trigonometric direction,

Figure 5.3(c). The beams are separated again in Figure 5.3(d). The two beams that arrive at the receiver are not in phase. Their phase offset allows us to find the rotation speed of the gyro, which is fixed on the robot.

Inertial unit: an *inertial measurement unit* associates a gyro and an accelerometer in order to increase the precision of the estimation. More recent ones merge other types of information such as the estimated speed and even take into account the Earth's rotation. For instance, the Octans III unit of the IXBLUE company uses the Sagnac effect together with the Earth's rotation in order to deduce the direction of the Earth's North-South axis in the robot's coordinate system. Knowing this direction gives us two equations involving the Euler angles of the robot (see section 1.2) which are the bank ϕ, elevation θ and heading ψ of the robot, expressed in a local coordinate system. Due to the accelerometer included in the unit, it is possible to deduce the gravity vector from the above and thus to generate an additional equation which will allow us to calculate the three Euler angles. Let us note that the accelerometers also give us the accelerations in all directions (the *surge* in the direction of the robot, the *heave* (in the vertical direction) and the *sway* for lateral accelerations). The knowledge of the gravity vector and the axis of rotation theoretically allows, using a simple scalar product, to find the latitude of the robot. However, the obtained precision is too low to be taken into account in localization.

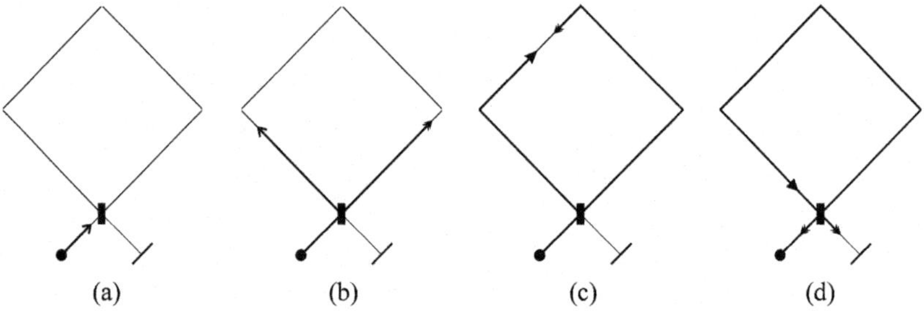

Figure 5.3. *Principle of the Sagnac effect for optical gyroscopes*

Barometer: this measures pressure. In the case of underwater robots, it allows us to deduce the depth of the robot with a precision of 1 cm. For indoor flying robots, the barometer is used to measure the altitude with a precision of around 1 m.

Global positioning system (GPS): the *global navigation satellite system* (GNSS) is a satellite navigation system that provides a geolocalization service covering the entire world. Nowadays, the American *NAVigation System by Timing And Ranging* (NAVSTAR) system and the Russian Globanaya Navigazionnaya Sputnikovaya Sistema (GLONASS) system are operational. Two other systems are being developed: the Chinese *Compass* system and the European *Galileo* system. In practice, our mobile robots will use the American system, operational since 1995, which we will refer to as GPS. Originally designed for exclusive military use, the precision of civil applications was limited to several hundreds of meters by a deliberate degradation of civil signals. The deactivation of this degradation in 2000 allowed the precision to increase to about 10 m. Given that electromagnetic waves (here around 1.2 MHz) do not propagate underwater or across walls, the GPS does not function within buildings or in water. Thus, during a diving experiment, an underwater robot can only be localized by GPS when it begins its dive or when it resurfaces. When a georeferenced station is near the robot and advises it about the errors in distance calculated for each satellite, a localization with a precision of ± 1 m is possible. This operating mode forms the so-called differential GPS or DGPS. Finally, by using the phase, it is possible to achieve even a centimeter precision. This is the principle of the *kinematic GPS*. A detailed and educational presentation of the GPS can be found in the thesis by Vincent Drevelle [DRE 11]. In practice, a GPS gives us a longitude ℓ_x and a latitude ℓ_y and it is often comfortable to convert it into Cartesian coordinates in a local coordinate system $(\mathbf{o}, \mathbf{i}, \mathbf{j}, \mathbf{k})$ fixed within the area in which the robot is evolving. Let us denote by ℓ_x^0 and ℓ_y^0 the longitude and latitude expressed in radians at the origin $\mathbf{o}$ of this coordinate system. We will assume that the vector $\mathbf{i}$ indicates the North, $\mathbf{j}$ indicates the East and $\mathbf{k}$ is oriented toward the center of the Earth. Let $\mathbf{p} = (p_x, p_y, p_z)$ be the coordinates of the robot expressed in the coordinate system $(\mathbf{o}, \mathbf{i}, \mathbf{j}, \mathbf{k})$. From the latitude and longitude given by the GPS, we can deduce the first two coordinates of the robot, expressed in meters in the local coordinate system, by using the following relation (see equation [4.4] in section 4.1):

$$\begin{pmatrix} p_x \\ p_y \end{pmatrix} = \rho \begin{pmatrix} 0 & 1 \\ \cos(\ell_y) & 0 \end{pmatrix} \begin{pmatrix} \ell_x - \ell_x^0 \\ \ell_y - \ell_y^0 \end{pmatrix} = \begin{pmatrix} \rho \left(\ell_y - \ell_y^0\right) \\ \rho \cos(\ell_y) \left(\ell_x - \ell_x^0\right) \end{pmatrix}$$

where ρ corresponds to the distance between $\mathbf{o}$ and the center of the Earth ($\rho \simeq$ 6 371 km, if $\mathbf{o}$ is not too far from sea level). This formula is valid everywhere on the Earth, if we assume that the Earth is spherical and if the robot is in the neighborhood of the origin $\mathbf{o}$ (let us say a distance less than 100 km). In order to understand this formula, we must note that $\rho \cos(\ell_y)$ corresponds to the distance between $\mathbf{o}$ and the rotational axis of the Earth. Thus, if a robot is moving on a latitude parallel ℓ_y, by modifying its longitude by an angle $\alpha > 0$, it will have traveled $\alpha\rho\cos(\ell_y)$ meters. Similarly, if this robot is moving on a meridian with an angle β in latitude, it will have traveled $\beta\rho$ meters.

Radar or sonar: the robot emits electromagnetic or ultrasound waves. It recovers their echoes and builds an image that it interprets in order to map its surroundings. The radar is mainly used by surface or flying robots. The sonar is used as a low-cost rangefinder by robots with wheels as well as in underwater robotics.

Cameras: cameras are low-cost sensors used for the recognition of objects. In localization, they are used as goniometers, in other words they allow us to find the angles relative to landmarks that will then be used for localization. Cameras are also used for recognizing objects in an underwater context [BAZ 12].

5.2. Goniometric localization

5.2.1. *Formulation of the problem*

The problem consists of using angles measured between the robot and landmarks, whose position as a function of time is known, for localization. Let us consider the robot in Figure 5.4 moving on a plane. We call *bearing* the angle α_i between the axis of the robot and the vector pointing toward the landmark. These angles could have been obtained, for instance, using a camera.

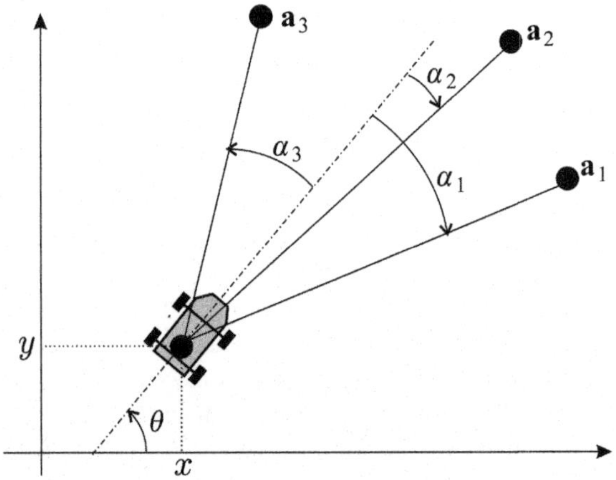

Figure 5.4. *A robot moving on a plane, measures the angles in order to locate itself*

Recall that two vectors $\mathbf{u}, \mathbf{v}$ of $\mathbb{R}^2$ are collinear if their determinant is zero, in other words if $\det(\mathbf{u}, \mathbf{v}) = 0$. Thus, for each landmark, we have the relation:

$$\det\left(\begin{pmatrix} x_i - x \\ y_i - y \end{pmatrix}, \begin{pmatrix} \cos(\theta + \alpha_i) \\ \sin(\theta + \alpha_i) \end{pmatrix}\right) = 0$$

in other words:

$$(x_i - x)\sin(\theta + \alpha_i) - (y_i - y)\cos(\theta + \alpha_i) = 0 \qquad [5.1]$$

where (x_i, y_i) are the coordinates of the landmark $\mathbf{a}_i$ and θ is the robot's heading.

5.2.2. *Inscribed angles*

THEOREM 5.1.– Inscribed Angle Theorem: consider a triangle $\mathbf{abm}$ as represented in Figure 5.5. Let us denote by $\mathbf{c}$ the center of the circle circumscribed to this triangle (in other words that $\mathbf{c}$ is at the intersection of the three perpendicular bisectors). Let $\alpha = \widehat{\mathbf{amc}}$, $\beta = \widehat{\mathbf{cmb}}$, $\gamma = \widehat{\mathbf{acb}}$. We have the angular relation:

$$\gamma = 2(\alpha + \beta)$$

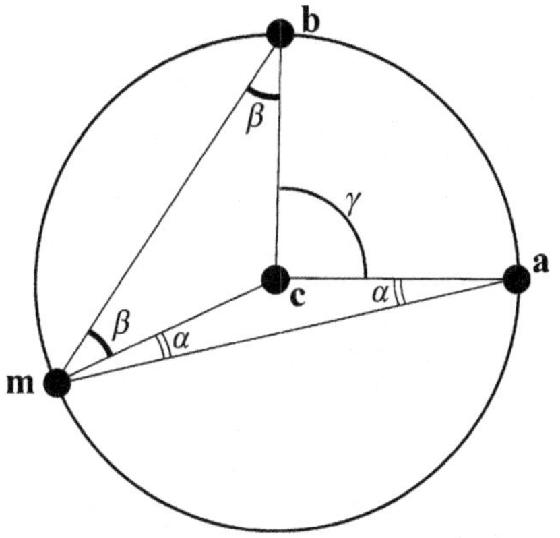

Figure 5.5. *Illustration of the inscribed angle theorem*

PROOF.– First, the two triangles $\mathbf{amc}$ and $\mathbf{cmb}$ are isosceles. Thus, we find the angles α and β as shown in the figure. By going around the point $\mathbf{c}$, we obtain:

$$\gamma + (\pi - 2\beta) + (\pi - 2\alpha) = 2\pi$$

In other words, $\gamma = 2\alpha + 2\beta$. ■

A consequence of this theorem is that if $\mathbf{m}$ moves on the circle, the angle $\alpha + \beta$ will not move.

Inscribed arcs. Let us consider two points $\mathbf{a}_1$ and $\mathbf{a}_2$. The set of points $\mathbf{m}$ such that the angle $\widehat{\mathbf{a}_1\mathbf{m}\mathbf{a}_2}$ is equal to α is a circle arc, referred to as an *inscribed arc*. We can show this from relations [5.1] or from the inscribed angle theorem. Goniometric localization often breaks down to intersecting arcs. Figure 5.6 shows the concept of an inscribed arc.

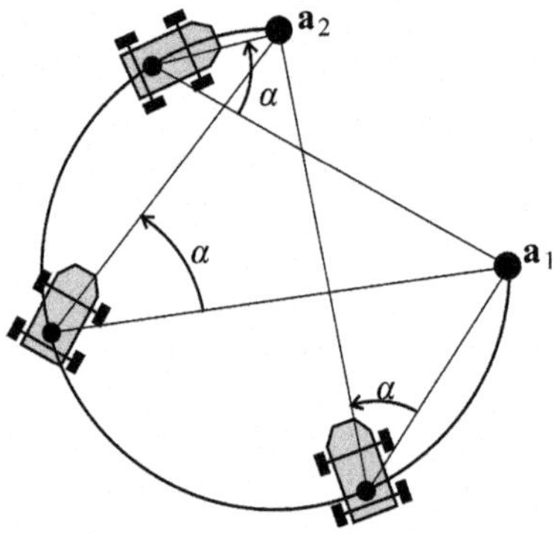

Figure 5.6. *The three cars perceive the landmarks with the same angle*

5.2.3. *Static triangulation of a plane robot*

5.2.3.1. *Two landmarks and a compass*

In the case where we have two landmarks and a compass, we have, following [5.1], the two relations:

$$\begin{cases} (x_1 - x)\sin(\theta + \alpha_1) - (y_1 - y)\cos(\theta + \alpha_1) = 0 \\ (x_2 - x)\sin(\theta + \alpha_2) - (y_2 - y)\cos(\theta + \alpha_2) = 0 \end{cases}$$

in other words:

$$\underbrace{\begin{pmatrix} \sin(\theta+\alpha_1) & -\cos(\theta+\alpha_1) \\ \sin(\theta+\alpha_2) & -\cos(\theta+\alpha_2) \end{pmatrix}}_{\mathbf{A}(\theta,\alpha_1,\alpha_2)} \begin{pmatrix} x \\ y \end{pmatrix} = \underbrace{\begin{pmatrix} x_1 \sin(\theta+\alpha_1) - y_1 \cos(\theta+\alpha_1) \\ x_2 \sin(\theta+\alpha_2) - y_2 \cos(\theta+\alpha_2) \end{pmatrix}}_{\mathbf{b}(\theta,\alpha_1,\alpha_2,x_1,y_1,x_2,y_2)}$$

i.e.:

$$\begin{pmatrix} x \\ y \end{pmatrix} = \mathbf{A}^{-1}(\theta,\alpha_1,\alpha_2) \cdot \mathbf{b}(\theta,\alpha_1,\alpha_2,x_1,y_1,x_2,y_2)$$

The problem of localization is, therefore, a linear one, which can be solved analytically. We have an identifiability problem if the matrix to invert has zero determinant, in other words:

$$\begin{aligned} & \sin(\theta+\alpha_1)\cos(\theta+\alpha_2) = \cos(\theta+\alpha_1)\sin(\theta+\alpha_2) \\ \Leftrightarrow\ & \tan(\theta+\alpha_2) = \tan(\theta+\alpha_1) \\ \Leftrightarrow\ & \theta+\alpha_2 = \theta+\alpha_1+k\pi,\ k\in\mathbb{N} \\ \Leftrightarrow\ & \alpha_2 = \alpha_1+k\pi,\ k\in\mathbb{N} \end{aligned}$$

This corresponds to a situation in which the two landmarks and robot are aligned.

5.2.3.2. *Three landmarks*

If we no longer have a compass, we need at least three landmarks. We then need to solve the system of three equations and three unknowns:

$$(x_i - x)\sin(\theta+\alpha_i) - (y_i - y)\cos(\theta+\alpha_i) = 0,\ i\in\{1,2,3\}$$

It can be shown that this system always has a unique solution, except when the robot is located on the circle that passes through all three landmarks. Indeed, in such a case, the inscribed angles are superimposed.

5.2.4. *Dynamic triangulation*

5.2.4.1. *One landmark, a compass and several odometers*

In the case of dynamic state observation, we are looking for the relation that connects the position of the robot to the derivatives of the values measured. For localization, we will assume that a single landmark is available to us. We will use the equations:

$$\begin{cases} \dot{x} = v\cos\theta \\ \dot{y} = v\sin\theta \end{cases} \qquad [5.2]$$

where v represents the speed of the robot measured by the odometer and θ represents its heading measured by the compass. These equations, which are kinematic in nature, are not supposed to describe the behavior of a particular robot with the aim of controlling it. The inputs v and θ are not necessarily the real inputs of the system that we can act on. These equations have to be understood as a simple differential relation between the variables of a plane robot. By differentiating relation [5.1], we obtain:

$$\begin{aligned}&(\dot{x}_i - \dot{x})\sin(\theta+\alpha_i) + (x_i - x)\left(\dot{\theta}+\dot{\alpha}_i\right)\cos(\theta+\alpha_i)\\ &-(\dot{y}_i - \dot{y})\cos(\theta+\alpha_i) + (y_i - y)\left(\dot{\theta}+\dot{\alpha}_i\right)\sin(\theta+\alpha_i) = 0\end{aligned} \qquad [5.3]$$

Let us take relations [5.1] and [5.3] for $i = 1$. By isolating x and y, we obtain:

$$\begin{aligned}\begin{pmatrix} x \\ y \end{pmatrix} = &\begin{pmatrix} \sin(\theta+\alpha_1) & \cos(\theta+\alpha_1) \\ -\cos(\theta+\alpha_1) & \sin(\theta+\alpha_1) \end{pmatrix}\\ &\cdot\begin{pmatrix} -y_1 & x_1 \\ x_1 - \frac{\dot{y}_1 - v\sin\theta}{\dot{\theta}+\dot{\alpha}_1} & \frac{\dot{x}_1 - v\cos\theta}{\dot{\theta}+\dot{\alpha}_1} + y_1 \end{pmatrix}\begin{pmatrix} \cos(\theta+\alpha_1) \\ \sin(\theta+\alpha_1) \end{pmatrix}\end{aligned} \qquad [5.4]$$

This relation can allow us to be located using a single mobile or fixed landmark and other proprioceptive sensors. For instance, in the case where we have a compass and several odometers (for a robot on wheels), we are able to measure the heading θ using the compass, the speed v and $\dot{\theta}$ using the odometers. Relation [5.4] then allows us to calculate the positions x and y at the given moment in time.

5.2.4.2. *One landmark and no compass*

In the situation where the compass is not present, we are missing an equation. We either need to add a second landmark, or differentiate again. Let us remain with a single landmark and differentiate relation [5.3], we obtain:

$$\begin{aligned}&(\ddot{x}_1 - \ddot{x})\sin(\theta+\alpha_1) - (\ddot{y}_1 - \ddot{y})\cos(\theta+\alpha_1)\\ &+(x_1 - x)\left(\ddot{\theta}+\ddot{\alpha}_1\right)\cos(\theta+\alpha_1) + (y_1 - y)\left(\ddot{\theta}+\ddot{\alpha}_1\right)\sin(\theta+\alpha_1)\\ &+2(\dot{x}_1 - \dot{x})\left(\dot{\theta}+\dot{\alpha}_1\right)\cos(\theta+\alpha_1) + 2(\dot{y}_1 - \dot{y})\left(\dot{\theta}+\dot{\alpha}_1\right)\sin(\theta+\alpha_1)\\ &-(x_1 - x)\left(\dot{\theta}+\dot{\alpha}_1\right)^2\sin(\theta+\alpha_1) + (y_1 - y)\left(\dot{\theta}+\dot{\alpha}_1\right)^2\cos(\theta+\alpha_1) = 0\end{aligned}$$

Moreover:

$$\begin{cases} \ddot{x} = \dot{v}\cos\theta - v\dot{\theta}\sin\theta \\ \ddot{y} = \dot{v}\sin\theta + v\dot{\theta}\cos\theta \end{cases}$$

Thus, we obtain a system of three equations with three unknowns x, y, θ:

$$\begin{aligned}
&(x_1 - x)\sin(\theta + \alpha_1) - (y_1 - y)\cos(\theta + \alpha_1) = 0 \\
&(\dot{x}_1 - v\cos\theta)\sin(\theta + \alpha_1) + (x_1 - x)\left(\dot{\theta} + \dot{\alpha}_1\right)\cos(\theta + \alpha_1) \\
&\quad - (\dot{y}_1 - v\sin\theta)\cos(\theta + \alpha_1) + (y_1 - y)\left(\dot{\theta} + \dot{\alpha}_1\right)\sin(\theta + \alpha_1) = 0 \\
&\left(\ddot{x}_1 - \dot{v}\cos\theta + v\dot{\theta}\sin\theta\right)\sin(\theta + \alpha_1) - \left(\ddot{y}_1 - \dot{v}\sin\theta - v\dot{\theta}\cos\theta\right)\cos(\theta + \alpha_1) \\
&\quad + (x_1 - x)\left(\ddot{\theta} + \ddot{\alpha}_1\right)\cos(\theta + \alpha_1) + (y_1 - y)\left(\ddot{\theta} + \ddot{\alpha}_1\right)\sin(\theta + \alpha_1) \\
&\quad + 2(\dot{x}_1 - v\cos\theta)\left(\dot{\theta} + \dot{\alpha}_1\right)\cos(\theta + \alpha_1) + 2(\dot{y}_1 - v\sin\theta)\left(\dot{\theta} + \dot{\alpha}_1\right)\sin(\theta + \alpha_1) \\
&\quad - (x_1 - x)\left(\dot{\theta} + \dot{\alpha}_1\right)^2\sin(\theta + \alpha_1) + (y_1 - y)\left(\dot{\theta} + \dot{\alpha}_1\right)^2\cos(\theta + \alpha_1) = 0
\end{aligned}$$

The quantities $x_1, y_1, \dot{x}_1, \dot{y}_1, \ddot{x}_1, \ddot{y}_1$ are calculated from the path of landmark 1, for which we know an analytic expression. The quantities α_1, $\dot{\alpha}_1, \ddot{\alpha}_1$ are assumed to be measured. The quantities $\dot{\theta}$, $\ddot{\theta}$ can be obtained using a gyro. The speed v can be measured using odometers. It is clear that this system is not easy to solve analytically and does not always admit a unique solution. For instance, if the landmark is fixed, by rotational symmetry we can see that we will not be able to find the angle θ. In such a case, we need at least two landmarks for localization.

5.3. Multilateration

Multilateration is a localization technique based on measuring the difference of the distances between the robot and landmarks. Indeed, in a number of situations (such as in GPS localization), the clocks between the landmarks and robot are not synchronized and therefore we cannot directly measure the absolute distance between the landmarks and robot (by the propagation time of airwaves or soundwaves), but we can measure the difference between these distances. We will now give the principles of this technique.

Four landmarks emit a brief signal at the same time t_0 which propagates with a speed c. Each emitted signal contains the identifier of the landmark, its position and the emission time t_0. The robot (which does not have an accurate clock, only an accurate chronometer) receives the four signals at times t_i. From this, it easily deduces the offsets between the reception times

$\tau_2 = t_2 - t_1$, $\tau_3 = t_3 - t_1$, $\tau_4 = t_4 - t_1$ (see Figure 5.7). Thus, we obtain the four equations:

$$\begin{aligned}
\sqrt{(x-x_1)^2 + (y-y_1)^2 + (z-z_1)^2} &= c(t_1 - t_0) \\
\sqrt{(x-x_2)^2 + (y-y_2)^2 + (z-z_2)^2} &= c(\tau_2 + t_1 - t_0) \\
\sqrt{(x-x_3)^2 + (y-y_3)^2 + (z-z_3)^2} &= c(\tau_3 + t_1 - t_0) \\
\sqrt{(x-x_4)^2 + (y-y_4)^2 + (z-z_4)^2} &= c(\tau_4 + t_1 - t_0)
\end{aligned}$$

where the parameters, whose values are known with high precision, are c, t_0, $x_1, y_1, z_1, \ldots, x_4, y_4, z_4, \tau_2, \tau_3, \tau_4$. The four unknowns are x, y, z, t_1. Solving this system allows it to be localized and also to readjust its clock (through t_1). In the case of the GPS, the landmarks are mobile. They use a similar principle to be localized and synchronized, from fixed landmarks on the ground.

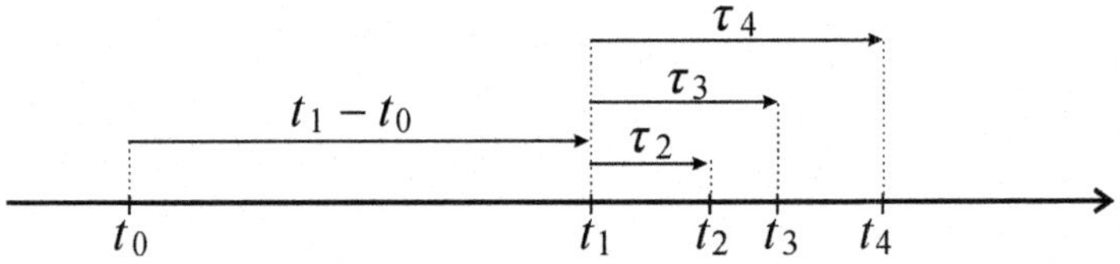

Figure 5.7. *The emission time t_0 and the offsets between the arrival times τ_2, τ_3, τ_4 are known*

5.4. Exercises

EXERCISE 5.1.– Instantaneous state estimation

Localization consists of finding the position and orientation of the robot. This problem can sometimes be reduced to a state estimation problem, if the state model for our robot is available. In this exercise, we will give a method that is sometimes used for the state estimation of nonlinear systems. Let us consider the tricycle given in section 2.5, described by the state equations:

$$\begin{pmatrix} \dot{x} \\ \dot{y} \\ \dot{\theta} \\ \dot{v} \\ \dot{\delta} \end{pmatrix} = \begin{pmatrix} v\cos\delta\cos\theta \\ v\cos\delta\sin\theta \\ v\sin\delta \\ u_1 \\ u_2 \end{pmatrix}$$

We measure the positions x and y with such high precision that we may assume that $\dot{x}, \dot{y}, \ddot{x}, \ddot{y}$ are known. Express the other state variables θ, v, δ as a function of x, y, $\dot{x}$, $\dot{y}$, $\ddot{x}$, $\ddot{y}$.

EXERCISE 5.2.– Localization by lidar

Here, we are interested in developing a fast localization method for a robot using a rotating laser rangefinder, or lidar (*light radar*) of type Hokuyo, in a rectangular room whose length and width are unknown.

1) Let $\mathbf{a}_1, \ldots, \mathbf{a}_{n_p}$ be points of $\mathbb{R}^2$ located on the same line. Find this line using a least squares method. Represent this line in normal form:

$$x \cos \alpha + y \sin \alpha = d, \text{ with } d \geq 0$$

where α, d are the parameters of the line.

2) Consider 100 measurements θ_i of the same angle θ of a robot by 100 compasses placed on the robots, of which 30 are defective. Estimate θ using the median method. For this, we define the *disambiguation* function:

$$\mathbf{f} : \begin{cases} \mathbb{R} \to \mathbb{R}^2 \\ \theta \to \begin{pmatrix} \cos \theta \\ \sin \theta \end{pmatrix} \end{cases}$$

Such a function, which must be continuous, aims to associate a single image with two equivalent angles (i.e. congruent to 2π), in other words:

$$\theta_1 \sim \theta_2 \Leftrightarrow f(\theta_1) = f(\theta_2)$$

3) The robot's lidar, which has an aperture angle of 180 °, gives us 512 points that belong to the rectangle representing our room. These points can be found in the file `lidar_data.mat`. We will take them in groups of 10 (i.e. 51 groups) and try to find the line that passes the best through each group (using the least squares method). We will only keep the groups with a small residue. Thus, we obtain m lines, represented by n points of the form (α_i, d_i), $i \in \{1, \ldots, m\}$ in the so-called *Hough space*. A pair (α_i, d_i) is called an *alignment*. By using a median estimator, find the four possible directions for our room (knowing that it is rectangular). Why is this median estimator considered robust?

4) How are the angles α_i filtered from the alignments?

5) How are the d_i filtered?

6) Deduce from the above a method for localizing the robot.

EXERCISE 5.3.– Instantaneous goniometric localization

Consider a robot boat $\mathcal{R}$ described by the state equations:

$$\begin{cases} \dot{x} = v\cos\theta \\ \dot{y} = v\sin\theta \\ \dot{\theta} = u_1 \\ \dot{v} = u_2 \end{cases}$$

where v is the speed of the robot, θ is its orientation and (x, y) are the coordinates of its center. Its state vector is given by $\mathbf{x} = (x, y, \theta, v)$. In the surroundings of the robot, there is a point landmark (for instance, a lighthouse) $\mathbf{m} = (x_m, y_m)$ whose position is known (see Figure 5.8). The robot $\mathcal{R}$ is equipped with five sensors: an omnidirectional camera (allowing it to measure the bearing angle α of the landmark's direction), odometers for measuring its speed, a compass that gives its heading θ, an accelerometer for u_2 and a gyro for u_1.

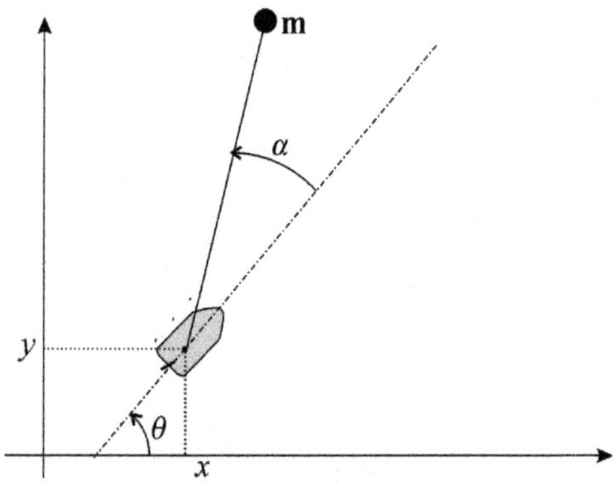

Figure 5.8. *Goniometric localization*

1) Simulate this system in a situation where the robot is moving around $\mathbf{m}$ in order to generate the signals $\alpha, \dot{\alpha}, \theta, v, u_1, u_2$. For generating α, use the two-argument arctangent function *atan2*. As we have seen in equation [1.8] in

section 1.2.1, the atan2(b, a) function returns the argument of the coordinate vector (a, b). For generating $\dot{\alpha}$, we could use the fact that:

$$\frac{\partial \text{atan2}\,(b,a)}{\partial a} = -\frac{b}{a^2+b^2} \quad \text{and} \quad \frac{\partial \text{atan2}\,(b,a)}{\partial b} = \frac{a}{a^2+b^2}$$

2) Design an instantaneous localization system for $\mathcal{R}$. Verify this localization system using a simulation. Add a small white Gaussian noise in your measurements to test its robustness.

3) The robot $\mathcal{R}$ is no longer equipped with odometers. How would you adjust the previous approach to allow localization?

4) Assuming once again that there are no odometers and that $\dot{\alpha}$ is not available, use a Kalman filter for localization.

5) Let us add a second robot $\mathcal{R}_b$ to the scene which is of the same type as the first (which we will now refer to as $\mathcal{R}_a$), with the exception that it has no odometers, contrarily to $\mathcal{R}_a$. Robot $\mathcal{R}_b$ can see $\mathcal{R}_a$ (see Figure 5.9) which gives it an angle β_b. Both robots can communicate via WiFi. Design a localization system for the robot $\mathcal{R}_b$ which also allows it to find its speed.

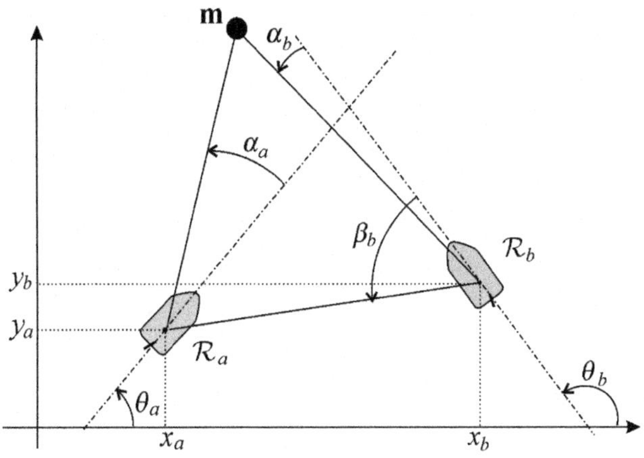

Figure 5.9. *The robots $\mathcal{R}_a$ and $\mathcal{R}_a$ can see each other, which will aid them in the localization*

6) Robot $\mathcal{R}_a$ can see $\mathcal{R}_b$, which gives it a new angular measurement β_a. Deduce a localization method from this for $\mathcal{R}_a$ and $\mathcal{R}_b$ that is more reliable

(in other words, one that presents less singularities) than that developed in the first question.

EXERCISE 5.4.– Localization by distance measuring

We consider a robot described by the state equations:

$$\begin{cases} \dot{x} = v\cos\theta \\ \dot{y} = v\sin\theta \\ \dot{\theta} = u_1 \\ \dot{v} = u_2 \end{cases}$$

where v is the speed of the robot, θ is its orientation and (x, y) are the coordinates of its center $\mathbf{c}$. Its state vector is given by $\mathbf{x} = (x, y, \theta, v)$.

In the surroundings of the robot, there is a fixed point landmark $\mathbf{m}$ whose position is known (see Figure 5.10). Each instant, the robot measures the distance d between its center $\mathbf{c}$ and the landmark. If the landmark and robot both have synchronized clocks, such a system for measuring the distance can be done by measuring the propagation time of a sound wave between the landmark and robot. Moreover, by using the Doppler effect, the robot is also capable of measuring $\dot{d}$ with very high precision. In addition to the microphone that allows it to measure d and $\dot{d}$, the robot is equipped with odometers to measure its speed v, a compass that gives its heading θ. In this exercise, we are looking to build an instantaneous localization system in order to determine its position (x, y) from $d, \dot{d}, \theta, v$.

1) For a given vector $\left(d, \dot{d}, \theta, v\right)$, there may be several configurations possible for the robot. Draw, on the picture, all the configurations for the robot that are compatible with the one that is represented.

2) Consider the coordinate system $\mathcal{R}_1$ centered at $\mathbf{m}$ as represented in the figure. Give the expression of the coordinates (x_1, y_1) of the center $\mathbf{c}$ of the robot as a function of x, y, θ, x_m, y_m. This is in fact a coordinate system change equation.

3) Express x_1 and y_1 as a function of $d, \dot{d}, v$.

4) From this, deduce the position(s) of the robot (x, y) as a function of $(d, \dot{d}, \theta, v, x_m, y_m)$. What are the singularities of this localization system? In which case do we have a single solution?

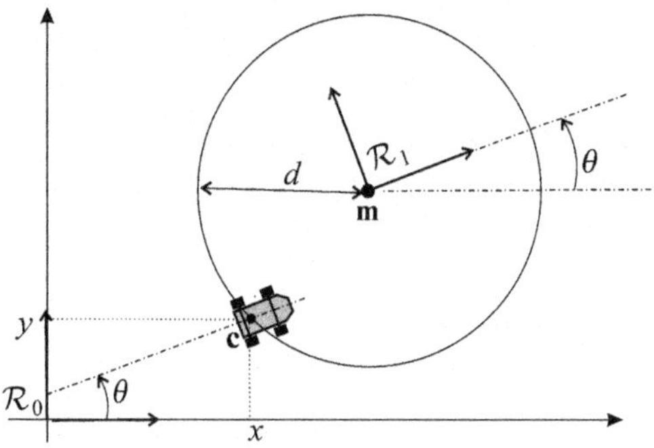

Figure 5.10. *The robot measures the distance to the landmark*

5.5. Corrections

CORRECTION FOR EXERCISE 5.1.– (Instantaneous state estimation)

We have:

$$\theta = \arctan\left(\frac{\dot{y}}{\dot{x}}\right)$$

By differentiating this equation, we obtain:

$$\dot{\theta} = \frac{1}{\left(\frac{\dot{y}}{\dot{x}}\right)^2 + 1}\left(\frac{\ddot{y}\dot{x} - \ddot{x}\dot{y}}{\dot{x}^2}\right)$$

Moreover, following the state equations of the tricycle:

$$\begin{cases} v\cos\delta = \dfrac{\dot{x}}{\cos\theta} \\ v\sin\delta = \dot{\theta} \end{cases}$$

By isolating v and δ, we obtain:

$$\begin{cases} v = \sqrt{(v\cos\delta)^2 + (v\sin\delta)^2} = \sqrt{\dfrac{\dot{x}^2}{\cos^2\theta} + \dot{\theta}^2} \\ \delta = \arctan\left(\dfrac{\dot{\theta}}{\frac{\dot{x}}{\cos(\theta)}}\right) \end{cases}$$

Thus, the state vector is expressed as an analytic function of the output (x, y) and its derivatives:

$$\begin{pmatrix} x \\ y \\ \theta \\ v \\ \delta \end{pmatrix} = \begin{pmatrix} x \\ y \\ \arctan\left(\frac{\dot{y}}{\dot{x}}\right) \\ \sqrt{\frac{\dot{x}^2}{\cos^2\left(\arctan\left(\frac{\dot{y}}{\dot{x}}\right)\right)} + \frac{1}{\left(\left(\frac{\dot{y}}{\dot{x}}\right)^2 + 1\right)^2}\left(\frac{\ddot{y}\dot{x} - \ddot{x}\dot{y}}{\dot{x}^2}\right)^2} \\ \arctan\left(\frac{\frac{1}{\left(\frac{\dot{y}}{\dot{x}}\right)^2 + 1}\left(\frac{\ddot{y}\dot{x} - \ddot{x}\dot{y}}{\dot{x}^2}\right)}{\frac{\dot{x}}{\cos\left(\arctan\left(\frac{\dot{y}}{\dot{x}}\right)\right)}}\right) \end{pmatrix}$$

Thus, we have a state observer that we may describe as *quasi-static* since, contrarily to the standard observers, this does not require the integration of differential equations (they are not dynamic). Obtaining such an observer can be done for a large class of nonlinear systems encompassing the class of flat systems [FLI 95].

CORRECTION FOR EXERCISE 5.2.– (Localization by lidar)

1) For the line equations, there is the explicit form $y = ax + b$, the standard form $ax + by + c = 0$, the normal form $x \cos \alpha + y \sin \alpha = d$, as well as other forms which we will not discuss here. The explicit form can represent vertical lines and therefore this form contains singularities. The standard form does not have any singularities, however, the associated model is non-identifiable (since different sets of parameters may lead to the same line). The normal form is identifiable, without singularities but it is nonlinear. Let us take the form $p_1 x + p_2 y = 1$ (singular if the line passes through 0, which will not be the case for our robot). We have:

$$p_1 x_i + p_2 y_i = 1$$

in other words, by considering that a small amount of noise is possible within the measurements:

$$\underbrace{\begin{pmatrix} x_1 & y_1 \\ \vdots & \vdots \\ x_n & y_n \end{pmatrix}}_{\mathbf{A}} \begin{pmatrix} p_1 \\ p_2 \end{pmatrix} \simeq \begin{pmatrix} 1 \\ \vdots \\ 1 \end{pmatrix}$$

Thus, we take as estimator:

$$\hat{\mathbf{p}} = \underbrace{\left(\mathbf{A}^{\mathrm{T}}\mathbf{A}\right)^{-1} \mathbf{A}^{\mathrm{T}}}_{\mathbf{K}} \begin{pmatrix} 1 \\ \vdots \\ 1 \end{pmatrix}$$

However, we need an equation of the type:

$$x\frac{\cos\hat{\alpha}}{\hat{d}} + y\frac{\sin\hat{\alpha}}{\hat{d}} = 1, \text{ with } d \geq 0$$

Therefore, we will take:

$$\hat{d} = \frac{1}{\|\hat{\mathbf{p}}\|} \text{ and } \hat{\alpha} = \text{angle}\left(\hat{\mathbf{p}}\right)$$

In order to verify our calculations, let us generate 1,000 points in MATLAB that form an approximatively aligned point cloud.

```
N=1000; xi=5*randn(N,1); yi=-2*xi+1+randn(N,1);
plot(xi,yi,'+');
```

The estimation of the line is performed as follows:

```
A=[xi,yi]; y=ones(N,1); K=inv(A'*A)*A'; phat=K*y;
dhat=1/norm(phat);
alphahat=angle(phat);yhat=A*phat;residu=norm(yhat-y);
```

To draw the estimated line, we execute the following instructions:

```
x=-15:0.1:15; y=(-phat(1)*x+1)/phat(2); plot(x,y,'.');
```

2) We take the median $\hat{x}$ of the x_i and $\hat{y}$ of the y_i. The angle estimator is $\hat{\theta} = \arg(\hat{x}, \hat{y})$. In order to illustrate the robustness of the estimator, let us generate a data set in MATLAB:

```
theta=1; N=100; b=0.1*randn(N,1)+round(randn(N,1));
thetai=theta+b;
```

The estimator is the following:

```
xi=cos(thetai); yi=sin(thetai); xhat=median(xi);
yhat=median(yi);
theta_hat=atan2(yhat,xhat); hold on; plot(xi,yi,'+');
plot(xhat,yhat,'*');
```

3) Two angles α_1 and α_2 are equivalent if:

$$\begin{array}{ll} \alpha_2 = \alpha_1 + \frac{k\pi}{2}, & k \in \mathbb{Z} \\ \Leftrightarrow 4\alpha_2 = 4\alpha_1 + 2k\pi, & \\ \Leftrightarrow \begin{cases} \cos 4\alpha_2 = \cos 4\alpha_1 \\ \sin 4\alpha_2 = \sin 4\alpha_1 \end{cases} & \end{array}$$

with $k \in \mathbb{Z}$. At each alignment (α_i, d_i), we generate the m points of the unit circle:

$$\begin{pmatrix} x_i \\ y_i \end{pmatrix} = \begin{pmatrix} \cos 4\alpha_i \\ \sin 4\alpha_i \end{pmatrix}$$

which represents the disambiguation function. Let us take the median $\hat{x}$ of the x_i and the median $\hat{y}$ of the y_i. The argument of the vector $(\hat{x}, \hat{y})$ gives us the angle $4\hat{\alpha}$ of the room (in the coordinate system of the robot) with a quarter turn margin. Therefore, we will take:

$$\hat{\alpha} = \frac{1}{4} \operatorname{atan2}(\hat{y}, \hat{x}) \in \left[-\frac{\pi}{4}, \frac{\pi}{4}\right]$$

It is of course important at this point to remove the outliers in the orientation, in other words the alignments (α_i, d_i) such that the quantity $|\hat{x} - x_i| + |\hat{y} - y_i|$ is not negligible. In order to illustrate this principle, we generate N angles α_i as follows:

```
alpha=1; N=100;
b=0.01*randn(N,1)+pi/2*round(10*randn(N,1));
alphai=alpha+b;
```

The obtained estimator is:

```
xi=cos(4*alphai); yi=sin(4*alphai); xhat=median(xi);
yhat=median(yi);
alpha_hat=atan2(yhat,xhat)/4; plot(xi,yi,'+');
plot(cos(alpha_hat),sin(alpha_hat),'*');
```

4) For each α_i of the alignment (α_i, d_i), we calculate the wall number:

$$k_i = \text{mod}\left(\text{round}\left(\frac{\alpha_i - \hat{\alpha}}{\pi/4}\right), 4\right) \in \{0, 1, 2, 3\}$$

The filtered angle is then given by:

$$\alpha_i = \hat{\alpha} + k_i \frac{\pi}{4}$$

5) For each $k \in \{0, 1, 2, 3\}$, we calculate (using the median estimator) the distance:

$$\hat{\delta}_k = \text{median}\{d_i \text{ with } i \text{ such that } k_i = k\}$$

6) We define η_k as the number of alignments compatible with wall k, in other words $\eta_k = \text{card}(\{i \mid k_i = k\})$. Therefore, we have knowledge of two or three distances $\hat{\delta}_k$ corresponding to the orientation $\hat{\alpha} + k_i \frac{\pi}{4}$ and whose confidence index is given by η_k. In order to break the symmetry, we need additional information (such as the lengths and widths ℓ_x and ℓ_y of the rectangle and information about the heading). Figure 5.11 gives an illustration of the robot's localization principle from the telemetric measurements.

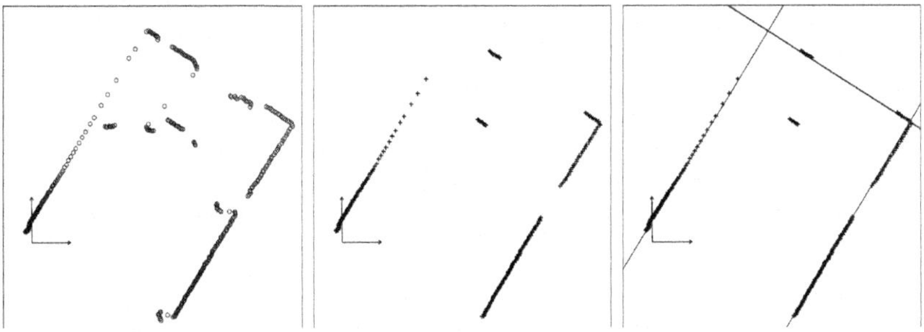

Figure 5.11. *Left: raw lidar data; middle: data associated with the small-residue alignments; right: recreated walls*

CORRECTION FOR EXERCISE 5.3.– (Instantaneous goniometric localization)

1) For the simulation, we need α and $\dot{\alpha}$. First, we have:

$$\alpha = -\theta + \text{atan}(y_m - y, x_m - x)$$

By differentiating this relation, we obtain:

$$\begin{aligned}
\dot{\alpha} &= -\dot{\theta} + \frac{d}{dt}\text{atan}(y_m - y, x_m - x) \\
&= -\dot{\theta} + \left(\frac{-(y_m - y)}{(x_m - x)^2 + (y_m - y)^2} \quad \frac{(x_m - x)}{(x_m - x)^2 + (y_m - y)^2} \right) \\
&\quad \times \begin{pmatrix} \dot{x}_m - \dot{x} \\ \dot{y}_m - \dot{y} \end{pmatrix}
\end{aligned}$$

And therefore, since $\dot{x} = v\cos\theta$, $\dot{y} = v\sin\theta$ and $\dot{\theta} = u_1$, we have:

$$\dot{\alpha} = -u_1 + \frac{(x_m - x)(\dot{y}_m - v\sin\theta) - (y_m - y)(\dot{x}_m - v\cos\theta)}{(x_m - x)^2 + (y_m - y)^2}$$

The MATLAB code of the evolution function is:

```
function y = g(x,u,m,mdot)
st=sin(x(3));ct=cos(x(3));
v=m-[x(1:2)]; alpha=-x(3)+angle(v)
alphadot=-u(1)+(1/norm(v)^2)*(v(1)*(mdot(2)-x(4)*st)-v(2)
*(mdot(1)-x(4)*ct));
y=[alpha;alphadot;x(3);x(4)];
end
```

2) We have the relation:

$$\det\left(\begin{pmatrix} x_m - x \\ y_m - y \end{pmatrix}, \begin{pmatrix} \cos(\theta + \alpha) \\ \sin(\theta + \alpha) \end{pmatrix} \right) = 0$$

i.e.:

$$(x_m - x)\sin(\theta + \alpha) - (y_m - y)\cos(\theta + \alpha) = 0$$

By differentiating this relation, we obtain:

$$\begin{aligned}&(\dot{x}_m - \dot{x})\sin(\theta+\alpha) + (x_m - x)\left(\dot{\theta}+\dot{\alpha}\right)\cos(\theta+\alpha)\\&-(\dot{y}_m - \dot{y})\cos(\theta+\alpha) + (y_m - y)\left(\dot{\theta}+\dot{\alpha}\right)\sin(\theta+\alpha) = 0\end{aligned}$$

Let us take the above relations. By isolating x and y, we obtain:

$$\begin{pmatrix} x \\ y \end{pmatrix} = \begin{pmatrix} \sin(\theta+\alpha) & \cos(\theta+\alpha) \\ -\cos(\theta+\alpha) & \sin(\theta+\alpha) \end{pmatrix} \begin{pmatrix} -y_m & x_m \\ x_m - \frac{\dot{y}_m - v\sin\theta}{\dot{\theta}+\dot{\alpha}} & y_m + \frac{\dot{x}_m - v\cos\theta}{\dot{\theta}+\dot{\alpha}} \end{pmatrix} \cdot \begin{pmatrix} \cos(\theta+\alpha) \\ \sin(\theta+\alpha) \end{pmatrix}$$

This relation can allow us to be located using a single landmark, whether it is fixed or mobile, and other proprioceptive sensors. For instance, if we have a compass and odometers (for a robot in motion), we are capable of measuring the heading θ_a using the compass, the speed v_a and $\dot{\theta}_a$ using the odometers. The code is given in the file `locboat.m`. The corresponding MATLAB script is:

```
function phat = loc(u,y,m,mdot)
alpha_m=y(1); dalpha_m=y(2); theta=y(3); v=y(4);
beta=theta+alpha_m; dbeta=(u(1)+dalpha_m);
A1=[sin(beta),cos(beta);-cos(beta),sin(beta)];
A2=[-m(2),m(1);m(1)-(mdot(2)-v*sin(theta))/dbeta,m(2)+(mdot(1)
-v*cos(theta))/dbeta];
phat=A1*A2*[cos(beta);sin(beta)];
end
```

3) The expression of $\ddot{\alpha}$ would have to be calculated to have the missing equation.

4) The idea is to write the problem in the form of a linear state estimation problem. For the observation equation, we have:

$$\begin{aligned}&(x_m - x)\sin(\theta+\alpha) - (y_m - y)\cos(\theta+\alpha) = 0\\ \Leftrightarrow\ &\underbrace{x_m\sin(\theta+\alpha) - y_m\cos(\theta+\alpha)}_{=z:\ \text{measured}} = \begin{pmatrix}\sin(\theta+\alpha) & -\cos(\theta+\alpha)\end{pmatrix}\begin{pmatrix} x \\ y \end{pmatrix}\end{aligned}$$

And therefore, the state equation that describes our system is:

$$\begin{cases} \begin{pmatrix} \dot{x} \\ \dot{y} \\ \dot{v} \end{pmatrix} = \begin{pmatrix} 0 & 0 & \cos\theta \\ 0 & 0 & \sin\theta \\ 0 & 0 & 0 \end{pmatrix} \begin{pmatrix} x \\ y \\ v \end{pmatrix} + \begin{pmatrix} 0 & 0 \\ 0 & 0 \\ 0 & 1 \end{pmatrix} \mathbf{u} \\ z = \begin{pmatrix} \sin(\theta+\alpha) & -\cos(\theta+\alpha) \end{pmatrix} \begin{pmatrix} x \\ y \\ v \end{pmatrix} \end{cases}$$

Therefore, we can apply a Kalman filter (see Chapter 7).

5) We have two more equations available for $\mathcal{R}_b$:

$$\begin{cases} (x_m - x_b)\sin(\theta_b + \alpha_b) - (y_m - y_b)\cos(\theta_b + \alpha_b) = 0 \\ (x_a - x_b)\sin(\theta_b + \beta_b) - (y_a - y_b)\cos(\theta_b + \beta_b) = 0 \end{cases}$$

which allow us to find (x_b, y_b):

$$\begin{pmatrix} x_b \\ y_b \end{pmatrix} = \begin{pmatrix} -\sin(\theta_b + \alpha_b) & \cos(\theta_b + \alpha_b) \\ -\sin(\theta_b + \beta_b) & \cos(\theta_b + \beta_b) \end{pmatrix}^{-1}$$

$$\begin{pmatrix} y_m \cos(\theta_b + \alpha_b) - x_m \sin(\theta_b + \alpha_b) \\ y_a \cos(\theta_b + \beta_b) - x_a \sin(\theta_b + \beta_b) \end{pmatrix}$$

6) We have a new equation:

$$(x_a - x_b)\sin(\theta_a + \beta_a) - (y_a - y_b)\cos(\theta_a + \beta_a) = 0$$

where β_a is the angle with which $\mathcal{R}_a$ sees $\mathcal{R}_b$. Thus, we have five equations with four unknowns:

$$\begin{array}{l} (x_m - x_a)\sin(\theta_a + \alpha_a) - (y_m - y_a)\cos(\theta_a + \alpha_a) = 0 \\ (\dot{x}_m - v_a\cos\theta_a)\sin(\theta_a + \alpha_a) + (x_m - x_a)\left(\dot{\theta}_a + \dot{\alpha}_a\right)\cos(\theta_a + \alpha_a) \\ -(\dot{y}_m - v_a\sin\theta_a)\cos(\theta_a + \alpha_a) + (y_m - y_a)\left(\dot{\theta}_a + \dot{\alpha}_a\right)\sin(\theta_a + \alpha_a) = 0 \\ (x_m - x_b)\sin(\theta_b + \alpha_b) - (y_m - y_b)\cos(\theta_b + \alpha_b) = 0 \\ (x_a - x_b)\sin(\theta_b + \beta_b) - (y_a - y_b)\cos(\theta_b + \beta_b) = 0 \\ (x_a - x_b)\sin(\theta_a + \beta_a) - (y_a - y_b)\cos(\theta_a + \beta_a) = 0 \end{array}$$

These equations are linear in (x_a, y_a, x_b, y_b) and therefore we can solve them by the least squares formula.

CORRECTION FOR EXERCISE 5.4.– (Localization by distance measurement)

1) There are two solutions: the one already represented in Figure 5.10 and its symmetry with respect to the line that passes through $\mathbf{m}$ and with direction vector $\mathbf{u} = (\cos\theta, \sin\theta)$.

2) We have:

$$\begin{pmatrix} x_1 \\ y_1 \end{pmatrix} = \begin{pmatrix} \cos\theta & \sin\theta \\ -\sin\theta & \cos\theta \end{pmatrix} \begin{pmatrix} x - x_m \\ y - y_m \end{pmatrix}$$

3) We have:

$$(x - x_m)^2 + (y - y_m)^2 = d^2$$

By differentiating, we obtain $2\dot{x}(x - x_m) + 2\dot{y}(y - y_m) = 2d\dot{d}$, in other words:

$$(v\cos\theta)(x - x_m) + (v\sin\theta)(y - y_m) = d\dot{d}$$

In the coordinate system $\mathcal{R}_1$, these two equations are written as:

$$\begin{cases} x_1^2 + y_1^2 = d^2 \\ \quad vx_1 \quad = d\dot{d} \end{cases}$$

The resolution of this system is simple. We obtain:

$$\begin{cases} x_1 = \frac{d\dot{d}}{v} \\ y_1 = \varepsilon d\sqrt{1 - \frac{\dot{d}^2}{v^2}} \end{cases}$$

with $\varepsilon = \pm 1$. These are the two symmetric equations.

4) The localization equation is:

$$\begin{pmatrix} x \\ y \end{pmatrix} = \begin{pmatrix} \cos\theta & -\sin\theta \\ \sin\theta & \cos\theta \end{pmatrix} \begin{pmatrix} \frac{d\dot{d}}{v} \\ \varepsilon d\sqrt{1 - \frac{\dot{d}^2}{v^2}} \end{pmatrix} + \begin{pmatrix} x_m \\ y_m \end{pmatrix}$$

If $v = 0$, we will not be able to be localized. Thus, we have a singularity. We have a unique solution if $1 - \frac{\dot{d}^2}{v^2} = 0$, in other words if $\dot{d} = \pm v$, which means that the robot is either moving in the direction of the landmark or in the opposite direction. Finally, we will never have $1 - \frac{\dot{d}^2}{v^2} < 0$ (except if there is a problem with the sensors) since this would mean the absence of a solution. This condition is logical since we cannot approach the landmark faster than our own speed.

6

Identification

The aim of identification is to estimate unmeasured quantities from other measured values, with high precision. In the particular case in which the quantity to estimate is the state vector of an invariant linear system, state observers using pole placement (or Luenberger observers) can be considered efficient tools for identification. In this chapter, we will present several basic concepts of estimation, with the aim of introducing Kalman filtering in Chapter 7. In summary, this filtering can be seen as state observation for dynamic linear systems with time-variable coefficients. However, in contrast to more standard observers using a pole placement method, Kalman filtering uses the probabilistic properties of signals. Here, we will consider the static (as opposed to the dynamic) case. The unknowns to estimate are all stored in a vector of parameters $\mathbf{p}$, while the measurements are stored in a vector of measurements $\mathbf{y}$. In order to perform this estimation, we will mainly look at the so-called *least squares* approach which seeks to find the vector $\mathbf{p}$ that minimizes the sum of the squares of the errors.

6.1. Quadratic functions

In the case in which the dependency between the vectors $\mathbf{p}$ and $\mathbf{y}$ is linear, the least squares method is used to minimize a quadratic function. This section recalls several concepts attached to these functions, which are of a particular nature.

6.1.1. *Definition*

A quadratic function $f : \mathbb{R}^n \to \mathbb{R}$ is a function of the form:

$$f(\mathbf{x}) = \mathbf{x}^{\mathrm{T}} \cdot \mathbf{Q} \cdot \mathbf{x} + \mathbf{L}\mathbf{x} + c$$

where $\mathbf{Q}$ is a symmetric matrix. This definition is equivalent to stating that $f(\mathbf{x})$ is a linear combination of a constant c, of the x_i, of their squares x_i^2 and of the cross products $x_i x_j$ where $i \neq j$. For instance, the function $f(x_1, x_2) = 2x_1^2 - 6x_1x_2 + x_2^2 - 2x_1 + x_2 + 1$ is a quadratic function. We have:

$$f(\mathbf{x}) = (x_1 \;\; x_2) \begin{pmatrix} 2 & -3 \\ -3 & 1 \end{pmatrix} \begin{pmatrix} x_1 \\ x_2 \end{pmatrix} + (-2 \;\; 1) \begin{pmatrix} x_1 \\ x_2 \end{pmatrix} + 1 \qquad [6.1]$$

We will show below that the derivative of f at point $\mathbf{x}$ is an affine function. In our example, the derivative of f at point $\mathbf{x}$ is given by:

$$\frac{df}{d\mathbf{x}}(\mathbf{x}) = \left(\frac{\partial f}{\partial x_1}(\mathbf{x}) \;\; \frac{\partial f}{\partial x_2}(\mathbf{x}) \right)$$

with $\frac{\partial f}{\partial x_1}(\mathbf{x}) = 4x_1 - 6x_2 - 2$ and $\frac{\partial f}{\partial x_2}(\mathbf{x}) = -6x_1 + 2x_2 + 1$, in other words:

$$\frac{df}{d\mathbf{x}}(\mathbf{x}) = (4x_1 - 6x_2 - 2 \;\; ; \;\; -6x_1 + 2x_2 + 1)$$

This is an affine function in $\mathbf{x}$. The function $\mathbf{x} \mapsto \mathbf{x}^{\mathrm{T}}\mathbf{Q}\mathbf{x}$ which composes $f(\mathbf{x})$ has terms only in $x_i x_j$ and x_i^2. Such a function is called a *quadratic form*.

6.1.2. *Derivative of a quadratic form*

Let us consider the following quadratic form:

$$f(\mathbf{x}) = \mathbf{x}^{\mathrm{T}}\mathbf{Q}\,\mathbf{x}$$

The first-order Taylor development of f at point $\mathbf{x}$ in the neighborhood of $\mathbf{x}$ yields:

$$f(\mathbf{x} + \delta\mathbf{x}) = f(\mathbf{x}) + \frac{df}{d\mathbf{x}}(\mathbf{x}) \cdot \delta\mathbf{x} + o(||\delta\mathbf{x}||)$$

where $o(||\delta\mathbf{x}||)$ means *negligible compared to* $||\delta\mathbf{x}||$, when $\delta\mathbf{x}$ is infinitely small. Of course, here $\frac{df}{d\mathbf{x}}(\mathbf{x})$ will be represented by a $1 \times n$ matrix since, just like the function we are linearizing, it goes from $\mathbb{R}^n$ to $\mathbb{R}$. However:

$$\begin{aligned} f(\mathbf{x} + \delta\mathbf{x}) &= (\mathbf{x} + \delta\mathbf{x})^{\mathrm{T}} \cdot \mathbf{Q} \cdot (\mathbf{x} + \delta\mathbf{x}) \\ &= \mathbf{x}^{\mathrm{T}} \cdot \mathbf{Q} \cdot \mathbf{x} + \mathbf{x}^{\mathrm{T}} \cdot \mathbf{Q} \cdot \delta\mathbf{x} + \delta\mathbf{x}^{\mathrm{T}} \cdot \mathbf{Q} \cdot \mathbf{x} + \delta\mathbf{x}^{\mathrm{T}} \cdot \mathbf{Q} \cdot \delta\mathbf{x} \\ &= \mathbf{x}^{\mathrm{T}} \cdot \mathbf{Q} \cdot \mathbf{x} + 2\mathbf{x}^{\mathrm{T}} \cdot \mathbf{Q} \cdot \delta\mathbf{x} + o(||\delta\mathbf{x}||) \end{aligned}$$

since $\mathbf{Q}$ is symmetric and $\delta\mathbf{x}^{\mathrm{T}} \cdot \mathbf{Q} \cdot \delta\mathbf{x} = o(||\delta\mathbf{x}||)$. By uniqueness of the Taylor development and given the expressions for $f(\mathbf{x} + \delta\mathbf{x})$, we have:

$$\frac{df}{d\mathbf{x}}(\mathbf{x}) = \left(\frac{\partial f}{\partial x_1}(\mathbf{x}), \ldots, \frac{\partial f}{\partial x_n}(\mathbf{x})\right) = 2\mathbf{x}^{\mathrm{T}}\mathbf{Q}$$

For instance, the derivative of the quadratic function [6.1] is:

$$2\,(x_1\ \ x_2)\begin{pmatrix} 2 & -3 \\ -3 & 1 \end{pmatrix} + (-2\ \ 1) = \begin{pmatrix} 4x_1 - 6x_2 - 2 & -6x_1 + 2x_2 + 1 \end{pmatrix}$$

6.1.3. *Eigenvalues of a quadratic function*

These are the eigenvalues of $\mathbf{Q}$. The eigenvalues are all real and the eigenvectors are all orthogonal two-by-two. The contour lines of a quadratic function $f(\mathbf{x}) = \alpha$ are of the form:

$$\mathbf{x}^{\mathrm{T}}\mathbf{Q}\mathbf{x} + \mathbf{L}\mathbf{x} = \alpha - c$$

and are called *quadrics*. These are ellipsoid if all the eigenvalues have the same sign or hyperboloid if they have different signs. If all the eigenvalues of $\mathbf{Q}$ are positive, we say that the quadratic form $\mathbf{x}^{\mathrm{T}}\mathbf{Q}\mathbf{x}$ is positive. If they are all non-zero, we say that the quadratic form is definite. If they are all strictly positive, we say that the quadratic form is positive definite. The quadratic function f has one and only one minimum if and only if its associated quadratic form is positive definite.

6.1.4. *Minimizing a quadratic function*

If $f(\mathbf{x}) = \mathbf{x}^{\mathrm{T}}\mathbf{Q}\mathbf{x} + \mathbf{L}\mathbf{x} + c$ has one and only one minimizer $\mathbf{x}^*$, $\mathbf{Q}$ is positive definite. In this case:

$$\frac{df}{d\mathbf{x}}(\mathbf{x}^*) = \mathbf{0}$$

Therefore, $2\mathbf{x}^{*\mathrm{T}}\mathbf{Q} + \mathbf{L} = \mathbf{0}$, in other words:

$$\mathbf{x}^* = -\frac{1}{2}\mathbf{Q}^{-1}\mathbf{L}^{\mathrm{T}}$$

Its minimum is given by:

$$\begin{aligned} f(\mathbf{x}^*) &= \left(-\frac{1}{2}\mathbf{Q}^{-1}\mathbf{L}^{\mathrm{T}}\right)^{\mathrm{T}} \mathbf{Q} \left(-\frac{1}{2}\mathbf{Q}^{-1}\mathbf{L}^{\mathrm{T}}\right) + \mathbf{L}\left(-\frac{1}{2}\mathbf{Q}^{-1}\mathbf{L}^{\mathrm{T}}\right) + c \\ &= \frac{1}{4}\mathbf{L}\mathbf{Q}^{-1}\mathbf{L}^{\mathrm{T}} - \frac{1}{2}\mathbf{L}\mathbf{Q}^{-1}\mathbf{L}^{\mathrm{T}} + c \\ &= -\frac{1}{4}\mathbf{L}\mathbf{Q}^{-1}\mathbf{L}^{\mathrm{T}} + c \end{aligned}$$

EXAMPLE 6.1.– The quadratic function:

$$f(\mathbf{x}) = (x_1 \ \ x_2) \begin{pmatrix} 2 & -1 \\ -1 & 1 \end{pmatrix} \begin{pmatrix} x_1 \\ x_2 \end{pmatrix} + (3\ 4) \begin{pmatrix} x_1 \\ x_2 \end{pmatrix} + 5$$

has a minimum since the matrix of its quadratic form $\mathbf{Q}$ is positive definite (its eigenvalues $\frac{3}{2} \pm \frac{1}{2}\sqrt{5}$ are both positive). The function has the following vector as minimizer:

$$\mathbf{x}^* = -\frac{1}{2} \begin{pmatrix} 2 & -1 \\ -1 & 1 \end{pmatrix}^{-1} \begin{pmatrix} 3 \\ 4 \end{pmatrix} = \begin{pmatrix} -\frac{7}{2} \\ -\frac{11}{2} \end{pmatrix}$$

Its minimum is:

$$f(\mathbf{x}^*) = -\frac{1}{4}(3\ 4) \begin{pmatrix} 2 & -1 \\ -1 & 1 \end{pmatrix}^{-1} \begin{pmatrix} 3 \\ 4 \end{pmatrix} + 5 = -\frac{45}{4}$$

EXAMPLE 6.2.– The function $f(x) = 3x^2 + 6x + 7$ has a minimum since the matrix of its quadratic form (which here corresponds to the scalar 3) is positive definite (since $3 > 0$). Its minimizer is the scalar:

$$x^* = -\frac{1}{2} \cdot \frac{1}{3} \cdot 6 = -1$$

and its minimum is $f(x^*) = 3 - 6 + 7 = 4$.

6.2. The least squares method

Estimating means obtaining an order of magnitude for certain quantities of a system from measurements of other quantities of the same system. The estimation problem we will consider in this chapter is the following. Consider a system for which we have made various measurements $\mathbf{y} = (y_1, \ldots, y_p)$ and a model $\mathcal{M}(\mathbf{p})$ depending on a vector of parameters $\mathbf{p}$. We need to estimate $\mathbf{p}$ such that the outputs $\mathbf{f}(\mathbf{p})$ generated by $\mathcal{M}(\mathbf{p})$ resemble $\mathbf{y}$ as much as possible.

6.2.1. *Linear case*

Let us assume that the vector of the outputs can be written in the form:

$$\mathbf{f}(\mathbf{p}) = \mathbf{M}\mathbf{p}$$

The model is then referred to as *linear with respect to the parameters*. We would like to have:

$$\mathbf{f}(\mathbf{p}) = \mathbf{y}$$

but this is generally not possible due to the presence of noise and the fact that the number of measurements is generally higher than the number of parameters (in other words, $\dim(\mathbf{y}) > \dim(\mathbf{p})$). Therefore, we will try to find the best $\mathbf{p}$, i.e. the one that minimizes the so-called *least squares* criterion:

$$j(\mathbf{p}) = ||\mathbf{f}(\mathbf{p}) - \mathbf{y}||^2$$

We have:

$$\begin{aligned} j(\mathbf{p}) &= ||\mathbf{f}(\mathbf{p}) - \mathbf{y}||^2 = ||\mathbf{M}\mathbf{p} - \mathbf{y}||^2 \\ &= (\mathbf{M}\mathbf{p} - \mathbf{y})^{\mathrm{T}} (\mathbf{M}\mathbf{p} - \mathbf{y}) = \left(\mathbf{p}^{\mathrm{T}}\mathbf{M}^{\mathrm{T}} - \mathbf{y}^{\mathrm{T}}\right)(\mathbf{M}\mathbf{p} - \mathbf{y}) \\ &= \mathbf{p}^{\mathrm{T}}\mathbf{M}^{\mathrm{T}}\mathbf{M}\mathbf{p} - \mathbf{p}^{\mathrm{T}}\mathbf{M}^{\mathrm{T}}\mathbf{y} - \mathbf{y}^{\mathrm{T}}\mathbf{M}\mathbf{p} + \mathbf{y}^{\mathrm{T}}\mathbf{y} \\ &= \mathbf{p}^{\mathrm{T}}\mathbf{M}^{\mathrm{T}}\mathbf{M}\mathbf{p} - 2\mathbf{y}^{\mathrm{T}}\mathbf{M}\mathbf{p} + \mathbf{y}^{\mathrm{T}}\mathbf{y} \end{aligned}$$

However, $\mathbf{M}^{\mathrm{T}}\mathbf{M}$ is symmetric (since $\left(\mathbf{M}^{\mathrm{T}}\mathbf{M}\right)^{\mathrm{T}} = \mathbf{M}^{\mathrm{T}}\mathbf{M}$). Therefore, we have a quadratic function. Moreover, all the eigenvalues of $\mathbf{M}^{\mathrm{T}}\mathbf{M}$ are positive or zero. The minimizer $\hat{\mathbf{p}}$ is obtained as follows:

$$\begin{aligned} \tfrac{dj}{d\mathbf{p}}(\hat{\mathbf{p}}) = \mathbf{0} &\Leftrightarrow 2\hat{\mathbf{p}}^{\mathrm{T}}\mathbf{M}^{\mathrm{T}}\mathbf{M} - 2\mathbf{y}^{\mathrm{T}}\mathbf{M} = \mathbf{0} \Leftrightarrow \hat{\mathbf{p}}^{\mathrm{T}}\mathbf{M}^{\mathrm{T}}\mathbf{M} = \mathbf{y}^{\mathrm{T}}\mathbf{M} \\ &\Leftrightarrow \mathbf{M}^{\mathrm{T}}\mathbf{M}\hat{\mathbf{p}} = \mathbf{M}^{\mathrm{T}}\mathbf{y} \qquad \Leftrightarrow \hat{\mathbf{p}} = \left(\mathbf{M}^{\mathrm{T}}\mathbf{M}\right)^{-1}\mathbf{M}^{\mathrm{T}}\mathbf{y} \end{aligned}$$

The matrix:

$$\mathbf{K} = \left(\mathbf{M}^{\mathrm{T}}\mathbf{M}\right)^{-1}\mathbf{M}^{\mathrm{T}}$$

is called the *generalized inverse* of the rectangular matrix $\mathbf{M}$. The vector $\hat{\mathbf{p}}$ is called the least squares *estimate*. The function:

$$\mathbf{y} \mapsto \mathbf{Ky}$$

is called the *estimator*. Note that this estimator is linear since the model function $\mathbf{f}$ is also linear. The vector:

$$\hat{\mathbf{y}} = \mathbf{M}\hat{\mathbf{p}} = \mathbf{MKy}$$

is the vector of the *filtered measurements* and the quantity:

$$\mathbf{r} = \hat{\mathbf{y}} - \mathbf{y} = (\mathbf{MK} - \mathbf{I})\,\mathbf{y}$$

is called the *vector of residuals*. The norm of this vector represents the distance between $\mathbf{y}$ and the hyperplane $\mathbf{f}\left(\mathbb{R}^n\right)$. If this norm is large, it often means that there is an error in the model or inaccuracies in the data.

6.2.2. *Nonlinear case*

If $\mathbf{y}$ is the vector of measurements and if $\mathbf{f}\left(\mathbf{p}\right)$ is the output generated by the model, then the least squares estimate is defined by:

$$\hat{\mathbf{p}} = \arg\min_{\mathbf{p}\in\mathbb{R}^n} ||\mathbf{f}(\mathbf{p}) - \mathbf{y}||^2$$

When $\mathbf{f}(\mathbf{p})$ is linear with respect to $\mathbf{p}$, in other words $\mathbf{f}(\mathbf{p}) = \mathbf{Mp}$, then the vector of parameters $\hat{\mathbf{p}}$ estimated using the least squares method is $\hat{\mathbf{p}} = \left(\mathbf{M}^{\mathrm{T}}\mathbf{M}\right)^{-1}\mathbf{My}$ and the vector of the filtered measurements is $\hat{\mathbf{y}} = \mathbf{M}\left(\mathbf{M}^{\mathrm{T}}\mathbf{M}\right)^{-1}\mathbf{My}$. In general, and even when $\mathbf{f}(\mathbf{p})$ is nonlinear, we can have the following geometric interpretation (see Figure 6.1):

– the vector of the filtered measurements $\hat{\mathbf{y}}$ represents the projection of $\mathbf{y}$ on the set $\mathbf{f}\left(\mathbb{R}^n\right)$;

– the vector estimated using the least squares method $\hat{\mathbf{p}}$ represents the inverse image of the vector of the filtered measurements $\mathbf{y}$ by $\mathbf{f}\left(.\right)$.

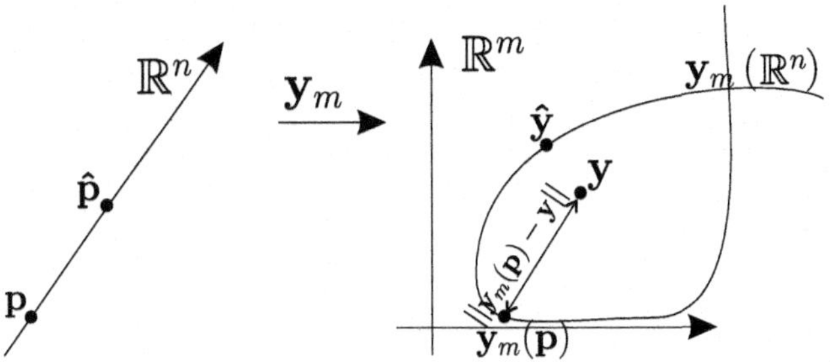

Figure 6.1. *Illustration of the least squares method in the nonlinear case*

When $\mathbf{f}(\mathbf{p})$ is nonlinear, we can use a local optimization algorithm to try to obtain $\hat{\mathbf{p}}$. The given algorithm proposes a simple version of such an optimization method.

Algorithm MINIMIZE(input: $\mathbf{p}$)
1 $j^+ = j(\mathbf{p})$
2 take a random movement δ
3 $\mathbf{q} = \mathbf{p} + \delta$
4 if $j(\mathbf{q}) < j^+$ then $\{\mathbf{p} = \mathbf{q}; j^+ = j(\mathbf{q})\}$;
5 go to 2

This algorithm converges toward a local optimum of the criterion $j(\mathbf{p}) = ||\mathbf{f}(\mathbf{p}) - \mathbf{y}||^2$. The quantity δ is the step that represents a small vector taken randomly from $\mathbb{R}^n$. In the case of the *simulated annealing* method, this step decreases with the iterations as a function of a parameter called *temperature* which decreases with time.

6.3. Exercises

EXERCISE 6.1.– Representation of a quadratic function

Consider the quadratic function $f(x, y) = x \cdot y$.

1) Find the gradient of f at point (x_0, y_0).

2) Put f in the form $(x\ \ y) \cdot \mathbf{Q} \cdot (x\ \ y)^{\mathrm{T}} + \mathbf{L}(x\ \ y)^{\mathrm{T}} + c$, where $\mathbf{Q}$ is a symmetric matrix. Verify that the gradient found in question 1 is given by

$2\,(x\ \ y)\,\mathbf{Q}$. Draw the vector field associated with this gradient in MATLAB using the `quiver` instruction. Discuss.

3) Using the `contour` instruction in MATLAB, draw the contour lines of f then draw the graph of f. Does f have a minimum?

4) Restart this exercise with the function $g(x, y) = 2x^2 + xy + 4y^2 + y - x + 3$.

EXERCISE 6.2.– Identification of a parabola

We would like to find a parabola $p_1 t^2 + p_2 t + p_3$ that passes through n points given by:

t	-3	-1	0	2	3	6
y	17	3	1	5	11	46

1) Give a least squares estimation of the parameters p_1, p_2, p_3.

2) What are the corresponding filtered measurements? Give the vector of residuals.

EXERCISE 6.3.– Identifying the parameters of a Direct Current (DC) motor

The angular speed Ω of a DC motor in permanent regime depends linearly on the supply voltage U and the resistive torque T_r:

$$\Omega = p_1 U + p_2 T_r$$

We perform a series of experiments on a particular motor. We measure:

U(V)	4	10	10	13	15
T_r(Nm)	0	1	5	5	3
Ω(rad/ sec)	5	10	8	14	17

1) Give a least squares estimation of the parameters p_1, p_2. Give the filtered measurements and the corresponding vector of residuals.

2) Deduce from the above an estimation of the angular speed of the motor $U = 20$ V and $T_r = 10$ Nm.

EXERCISE 6.4.– Estimation of a transfer function

Consider the system described by the recurrence equations:

$$y(k) + a_1 y(k-1) + a_0 y(k-2) = b_1 u(k-1) + b_0 u(k-2)$$

We perform noisy measurements on the input $u(k)$ and output $y(k)$ of this system for k varying from 0 to 7. We obtain:

k	0	1	2	3	4	5	6	7
$u(k)$	1	-1	1	-1	1	-1	1	-1
$y(k)$	0	-1	-2	3	7	11	16	36

Estimate the vector of parameters $\mathbf{p} = (a_1, a_0, b_1, b_0)$ by the least squares method. Discuss.

EXERCISE 6.5.– Monte Carlo method

Consider the discrete-time system given by its state representation:

$$\begin{cases} \mathbf{x}(k+1) = \begin{pmatrix} 1 & 0 \\ a & 0.9 \end{pmatrix} \mathbf{x}(k) + \begin{pmatrix} b \\ 1-b \end{pmatrix} u(k) \\ \quad y(k) \quad = \begin{pmatrix} 1 & 1 \end{pmatrix} \mathbf{x}(k) \end{cases}$$

where a, b are the two parameters to be estimated. The initial state is given by $\mathbf{x}(0) = (0, 0)$ and $u(k) = 1$. We collect six measurements:

$$(y(1), \cdots, y(6)) = \begin{pmatrix} 0 \; 1 \; 2.65 \; 4.885 \; 7.646 \; 10.882 \end{pmatrix}$$

Let us note that these values were obtained for the values $a^* = 0.9$ and $b^* = 0.75$, but we are not supposed to know them. We will only assume that $a \in [0, 2]$ and $b \in [0, 2]$.

1) Propose a MATLAB program that estimates the parameters a and b using a Monte Carlo method. For this, generate a cloud of vectors $\mathbf{p} = (a, b)$ using a uniform random draw. Then, by simulating the state equations, calculate for all the $\mathbf{p}$ the corresponding outputs $y_m(\mathbf{p},k)$. Draw on the screen the vectors $\mathbf{p}$ such that for each $k \in \{1, \ldots, 6\}$, $|y_m(k) - y(k)| < \varepsilon$, where ε is a small positive number.

2) Calculate the transfer function of the system as a function of a and b.

3) Let us assume that the real values $a^* = 0.9$ a,d $b^* = 0.75$ for a and b are known. Calculate the set of all pairs (a, b) that generate the same transfer function as the pair (a^*, b^*). Deduce from this an interpretation of the results obtained in question 1.

EXERCISE 6.6.– Localization by simulated annealing

The localization problem that we will now consider is inspired from [JAU 02]. The robot, represented in Figure 6.2, is equipped with eight laser telemeters capable of measuring its distance from the walls for angles equal to $\frac{k\pi}{4}$, $k \in \{0, \ldots, 7\}$. We assume that the obstacles are composed of n segments $[\mathbf{a}_i\mathbf{b}_i]$, $i = 1, \ldots, n$, where the coordinates of $\mathbf{a}_i$ and $\mathbf{b}_i$ are known. The eight distances are stored in the vector $\mathbf{y}$ and the localization problem amounts to estimating the position and orientation of the robot from $\mathbf{y}$.

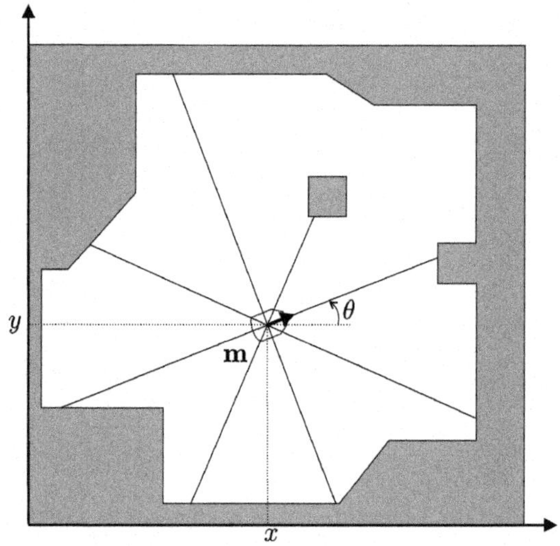

Figure 6.2. *Robot equipped with eight telemeters trying to localize itself*

1) Let $\mathbf{m}$, $\mathbf{a}$, $\mathbf{b}$ be three points of $\mathbb{R}^2$ and $\vec{\mathbf{u}}$ a unit vector. Show that the ray $\mathcal{E}(\mathbf{m}, \vec{\mathbf{u}})$ intersects the segment $[\mathbf{ab}]$ if and only if:

$$\begin{cases} \det(\mathbf{a} - \mathbf{m}, \vec{\mathbf{u}}) \cdot \det(\mathbf{b} - \mathbf{m}, \vec{\mathbf{u}}) \leq 0 \\ \det(\mathbf{a} - \mathbf{m}, \mathbf{b} - \mathbf{a}) \cdot \det(\vec{\mathbf{u}}, \mathbf{b} - \mathbf{a}) \geq 0 \end{cases}$$

If this condition is verified, show that the distance from $\mathbf{m}$ to $[\mathbf{ab}]$ following $\overrightarrow{\mathbf{u}}$ is:

$$d = \frac{\det(\mathbf{a} - \mathbf{m}, \mathbf{b} - \mathbf{a})}{\det(\overrightarrow{\mathbf{u}}, \mathbf{b} - \mathbf{a})}$$

2) Design a simulator $\mathbf{f}(\mathbf{p})$ that calculates the directional distances between the pose $\mathbf{p} = (x, y, \theta)$ and the walls.

3) Using a global simulated annealing-type optimization method, design a MATLAB program that gives a least squares estimation $\hat{\mathbf{p}}$ of the pose $\mathbf{p}$ from $\mathbf{y}$. For the segments $[\mathbf{a}_i, \mathbf{b}_i]$ of the room and for the vector of the measured distances, take the following quantities:

```
A=[0 7 7 9 9 7 7 4 2 0 5 6 6 5; 0 0 2 2 4 4 7 7 5 5 2 2 3 3
];
B=[7 7 9 9 7 7 4 2 0 0 6 6 5 5; 0 2 2 4 4 7 7 5 5 0 2 3 3
2];
y=[6.4;3.6;2.3;2.1;1.7;1.6;3.0;3.1];
```

6.4. Corrections

CORRECTION FOR EXERCISE 6.1.– (Representation of a quadratic function)

1) The gradient of $f(x, y) = x \cdot y$ at point (x_0, y_0) is:

$$\frac{df}{d(x,y)}(x_0, y_0) = \begin{pmatrix} y_0 & x_0 \end{pmatrix}$$

2) We have:

$$f(x, y) = (x \;\; y) \begin{pmatrix} 0 & \frac{1}{2} \\ \frac{1}{2} & 0 \end{pmatrix} \begin{pmatrix} x \\ y \end{pmatrix}$$

The gradient is given by:

$$\frac{df}{d(x,y)}(x, y) = 2(x \;\; y)\mathbf{Q} = 2(x \;\; y) \begin{pmatrix} 0 & \frac{1}{2} \\ \frac{1}{2} & 0 \end{pmatrix} = (y \;\; x)$$

This is the same result as in 1). The vector field associated with this gradient is obtained by writing:

```
Mx = -1:0.1:1; My = -1:0.1:1;
[X,Y] = meshgrid(Mx,My); GX=Y; GY=X;
quiver(Mx,My,GX,GY);
```

We obtain the left of Figure 6.3.

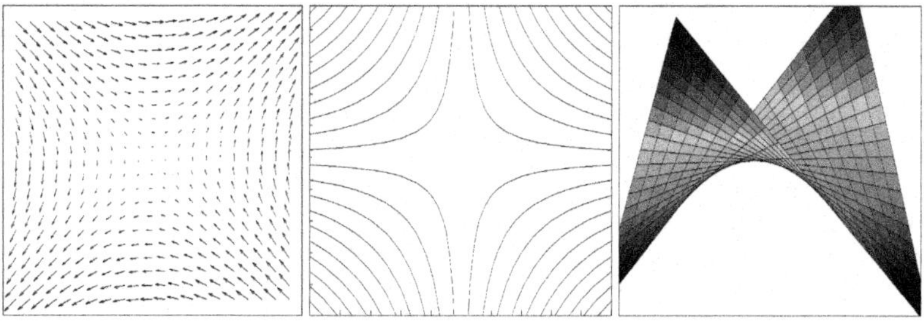

Figure 6.3. *Representation of the function $x \cdot y$ on the paving $[-1,1]^2$ left: vector field of the gradient; middle: level sets of the function; right: 3D view of the graph*

3) We form the function f by writing `Z = X.*Y`. The contour is obtained by writing `contour3(X,Y,Z,20)` and the graph of f by `surface(X,Y,Z)`. We, respectively, obtain the middle and right figures. The function does not have a minimum. The instructions associated with this question can be found in the script `quadra.m`.

4) For the function $g(x,y) = 2x^2 + xy + 4y^2 + y - x + 3$, we obtain:

$$g(x,y) = \begin{pmatrix} x & y \end{pmatrix} \begin{pmatrix} 2 & \frac{1}{2} \\ \frac{1}{2} & 4 \end{pmatrix} \begin{pmatrix} x \\ y \end{pmatrix} + \begin{pmatrix} -1 & 1 \end{pmatrix} \begin{pmatrix} x \\ y \end{pmatrix} + 3$$

Since the eigenvalues $3 + \frac{1}{2}\sqrt{5}, 3 - \frac{1}{2}\sqrt{5}$ of the matrix are strictly positive, the quadratic form associated with g is positive definite and therefore f has a minimum. To calculate it, we need to solve:

$$\frac{dg}{d(x,y)} = 2 \begin{pmatrix} x & y \end{pmatrix} \begin{pmatrix} 2 & \frac{1}{2} \\ \frac{1}{2} & 4 \end{pmatrix} + \begin{pmatrix} -1 & 1 \end{pmatrix} = \begin{pmatrix} 0 & 0 \end{pmatrix}$$

We obtain:

$$\begin{pmatrix} x \\ y \end{pmatrix} = -\frac{1}{2}\begin{pmatrix} 2 & \frac{1}{2} \\ \frac{1}{2} & 4 \end{pmatrix}^{-1}\begin{pmatrix} -1 \\ 1 \end{pmatrix} = \frac{1}{31}\cdot\begin{pmatrix} 9 \\ -5 \end{pmatrix}$$

By drawing the graphs in the same manner as for f (see Figure 6.4), we can verify that $g(x, y)$ has a minimum and that the contour lines are ellipses.

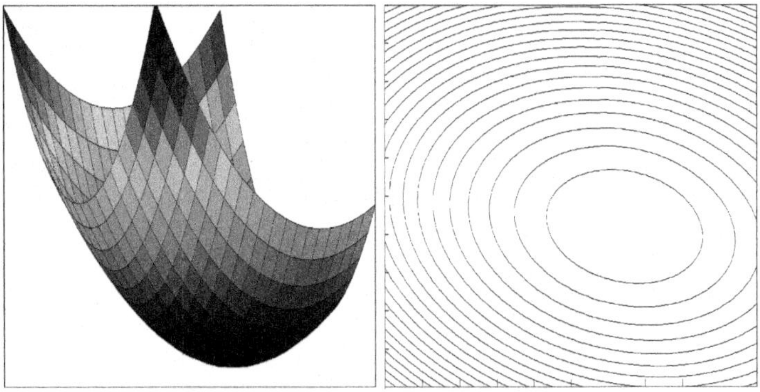

Figure 6.4. *a) Graph of the function $g(x, y)$; b) contour lines. The search space for x and y corresponds to the interval $[-1, 1]$*

CORRECTION FOR EXERCISE 6.2.– (Identification of a parabola)

NOTE 6.1.– First, let us note that to obtain these measurements, we have taken $p_1^* = \sqrt{2}, p_2^* = -1, p_3^* = 1$, in order to deduce the noiseless measurements:

$$\mathbf{y}^* = (16.72,\ 3.41,\ 1,\ 4.65,\ 10.73,\ 45.91)$$

We then truncated to the closest integer. The corresponding instruction in MATLAB is given by:

```
t=[-3;-1;0;2;3;6],y1=sqrt(2)*t.^2-t+1,y=round(y1)
```

The vector of measurements is, therefore, $\mathbf{y} = (17, 3, 1, 5, 11, 46)$, as given in the question and the vector of parameters to be estimated is $\mathbf{p} = (p_1, p_2, p_3)$. Of course, this process of data generation is not known and we are not supposed to use it in the solution of the exercise.

1) The output of the model is:

$$\mathbf{f}(\mathbf{p})=\begin{pmatrix} f_1(\mathbf{p}) \\ f_2(\mathbf{p}) \\ f_3(\mathbf{p}) \\ f_4(\mathbf{p}) \\ f_5(\mathbf{p}) \\ f_6(\mathbf{p}) \end{pmatrix}=\begin{pmatrix} 9p_1-3p_2+p_3 \\ p_1-p_2+p_3 \\ 0p_1-0p_2+p_3 \\ 4p_1+2p_2+p_3 \\ 9p_1+3p_2+p_3 \\ 36p_1+6p_2+p_3 \end{pmatrix}=\begin{pmatrix} 9 & -3 & 1 \\ 1 & -1 & 1 \\ 0 & 0 & 1 \\ 4 & 2 & 1 \\ 9 & 3 & 1 \\ 36 & 6 & 1 \end{pmatrix}\begin{pmatrix} p_1 \\ p_2 \\ p_3 \end{pmatrix}$$

The least squares estimated vector is, therefore:

$$\hat{\mathbf{p}}=\left(\mathbf{M}^{\mathrm{T}}\mathbf{M}\right)^{-1}\mathbf{M}^{\mathrm{T}}\mathbf{y}=\begin{pmatrix} 1.41 \\ -0.98 \\ 1.06 \end{pmatrix}$$

2) The filtered measurements are:

$$\hat{\mathbf{y}}=\mathbf{f}(\hat{\mathbf{p}})=\mathbf{M}\hat{\mathbf{p}}=(16.76,\ 3.46,\ 1.06,\ 4.76,\ 10.84,\ 46.11)$$

and the vector of residuals is:

$$\mathbf{r}=\hat{\mathbf{y}}-\mathbf{y}=(-0.24,\ 0.46,\ 0.06,\ -0.24,\ -0.15,\ 0.11)$$

The MATLAB script corresponding to this exercise can be found in the file `parab.m`.

CORRECTION FOR EXERCISE 6.3.– (Identifying the parameters of a DC motor)

1) We have:

$$\mathbf{f}(\mathbf{p})=\mathbf{M}\cdot\mathbf{p}$$

with:

$$\mathbf{M}=\begin{pmatrix} 4 & 0 \\ 10 & 1 \\ 10 & 5 \\ 13 & 5 \\ 15 & 3 \end{pmatrix},\ \mathbf{p}=\begin{pmatrix} p_1 \\ p_2 \end{pmatrix}\ \text{and}\ \mathbf{y}=\begin{pmatrix} 5 \\ 10 \\ 8 \\ 14 \\ 17 \end{pmatrix}$$

Therefore:

$$\hat{\mathbf{p}} = \left(\mathbf{M}^{\mathrm{T}}\mathbf{M}\right)^{-1}\mathbf{M}^{\mathrm{T}}\mathbf{y} = \begin{pmatrix} 1.188 \\ -0.516 \end{pmatrix}$$

The vector of filtered measurements is:

$$\hat{\mathbf{y}} = \mathbf{M} \cdot \hat{\mathbf{p}} = (4.75,\ 11.36,\ 9.3,\ 12.86,\ 16.27)$$

and the vector of residuals is:

$$\mathbf{r} = \hat{\mathbf{y}} - \mathbf{y} = (-0.25,\ 1.36,\ 1.3,\ -1.14,\ -0.73)$$

2) For $U = 20$ V and $T_r = 10$ Nm, we have:

$$\hat{\Omega} = \begin{pmatrix} U \ T_r \end{pmatrix} \cdot \hat{\mathbf{p}} = \begin{pmatrix} 20 \ 10 \end{pmatrix} \begin{pmatrix} 1.188 \\ -0.516 \end{pmatrix} = 18.6 \text{ rad/ sec}$$

CORRECTION FOR EXERCISE 6.4.– (Estimation of a transfer function)

The recurrence equations for $k = 2$ to 7 are:

$$y(2) = -a_1 y(1) - a_0 y(0) + b_1 u(1) + b_0 u(0)$$
$$y(3) = -a_1 y(2) - a_0 y(1) + b_1 u(2) + b_0 u(1)$$
$$\vdots$$
$$y(7) = -a_1 y(6) - a_0 y(5) + b_1 u(6) + b_0 u(5)$$

Therefore, the matrix $\mathbf{M}$ is given by:

$$\mathbf{M} = \begin{pmatrix} -y(1) & -y(0) & u(1) & u(0) \\ -y(2) & -y(1) & u(2) & u(1) \\ -y(3) & -y(2) & u(3) & u(2) \\ -y(4) & -y(3) & u(4) & u(3) \\ -y(5) & -y(4) & u(5) & u(4) \\ -y(6) & -y(5) & u(6) & u(5) \end{pmatrix} = \begin{pmatrix} 1 & 0 & -1 & 1 \\ 2 & 1 & 1 & -1 \\ -3 & 2 & -1 & 1 \\ -7 & -3 & 1 & -1 \\ -11 & -7 & -1 & 1 \\ -16 & -11 & 1 & -1 \end{pmatrix} = \begin{pmatrix} \mathbf{m}_2^{\mathrm{T}} \\ \mathbf{m}_3^{\mathrm{T}} \\ \mathbf{m}_4^{\mathrm{T}} \\ \mathbf{m}_5^{\mathrm{T}} \\ \mathbf{m}_6^{\mathrm{T}} \\ \mathbf{m}_7^{\mathrm{T}} \end{pmatrix}$$

The vector:

$$\mathbf{m}^{\mathrm{T}}(k) = (-y(k-1), -y(k-2), u(k-1), u(k-2))$$

contains the information connecting the kth measurement to the unknown $\mathbf{p}$. It is called the *regressor*. Since these k equations are not entirely satisfied given that the $u(i)$ and $y(i)$ are only approximatively known, we need to find the vector $\hat{\mathbf{p}}$ that minimizes the criterion:

$$j(\mathbf{p}) = ||\mathbf{M}\mathbf{p} - \mathbf{y}||^2$$

with:

$$\mathbf{y} = \begin{pmatrix} y(2) \\ \vdots \\ y(7) \end{pmatrix} = \begin{pmatrix} -2 \\ 3 \\ 7 \\ 11 \\ 16 \\ 36 \end{pmatrix} \text{ and } \mathbf{p} = \begin{pmatrix} a_1 \\ a_0 \\ b_1 \\ b_0 \end{pmatrix}$$

We should have $\hat{\mathbf{p}} = \left(\mathbf{M}^{\mathrm{T}}\mathbf{M}\right)^{-1} \mathbf{M}^{\mathrm{T}}\mathbf{y}$. Here, however, the matrix $\mathbf{M}$ is of rank 3. Indeed, given the particular form of the signal $u(k)$, the last two columns of $\mathbf{M}$ are dependent. In our case, this means that we cannot identify $\mathbf{p}$. This situation remains, however, unusual and we would need to choose another input to make the problem identifiable.

CORRECTION FOR EXERCISE 6.5.– (Monte Carlo method)

1) For the paving $[0, 2] \times [0, 2]$ and for $\varepsilon = 0.3$, the following MATLAB algorithm (which can also be found in `montecarlo.m`) allows us to characterize the required likelihood set:

```
y=[0;1;2.5;4.1;5.8;7.5];
for i=1:10000,
a=2*rand(1); b=2*rand(1); x=[0;0]; ym=0*y;
A=[1,0;a,0.3];B=[b;1-b];C=[1 1];
for k=1:6, x1=A*x+B;ym(k)=C*x;x=x1; end
if norm(ym-y,inf)<0.3, plot(a,b,'+black'); else
plot(a,b,'.blue'); end;
end
```

We then obtain the set of solutions represented in Figure 6.5. Note that there is a continuum of likely vectors.

2) The transfer function of the system is:

$$\left(1\ 1\right)\left(s\mathbf{I}-\begin{pmatrix}1 & 0\\ a & 0.9\end{pmatrix}\right)^{-1}\begin{pmatrix}b\\ 1-b\end{pmatrix}=\frac{10s+b\left(1+10a\right)-10}{\left(10s-9\right)\left(s-1\right)}$$

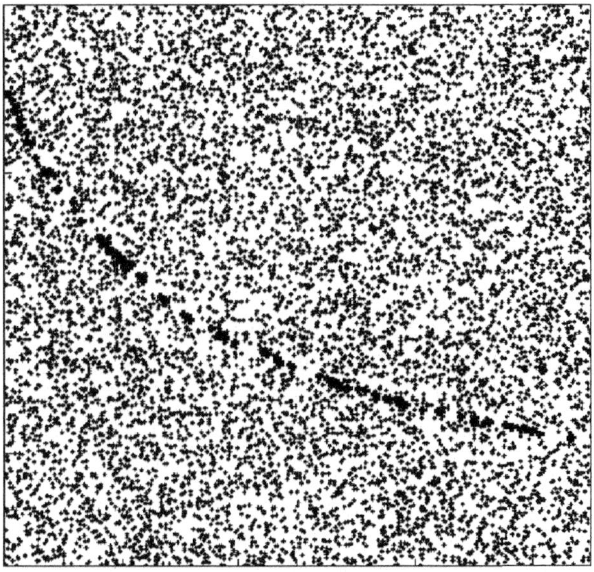

Figure 6.5. *Monte Carlo method for estimating the parameters a and b*

3) If we know that $a^* = 0.9$ and $b^* = 0.75$ are the real parameters, then, for each pair (a, b) such that:

$$\frac{10s+b\left(1+10a\right)-10}{\left(10s-9\right)\left(s-1\right)}=\frac{10s+b^*\left(1+10a^*\right)-10}{\left(10s-9\right)\left(s-1\right)}$$

we will have the same transfer function. This condition is translated by:

$$b\left(1+10a\right)=0.75\left(1+9\right)=7.5$$

i.e.:

$$b=\frac{7.5}{1+10a}$$

In such situations, when different values for the vector of parameters produce the same behavior, we say that the model is *non-identifiable*.

CORRECTION FOR EXERCISE 6.6.– (Localization by simulated annealing)

1) In order to understand the following proof, recall that (1) $\det(\vec{u}, \vec{v}) > 0$ if $\vec{v}$ is on the left of $\vec{u}$, (2) $\det(\vec{u}, \vec{v}) < 0$ if $\vec{v}$ is on the right of $\vec{u}$ and (3) $\det(\vec{u}, \vec{v}) = 0$ if $\vec{u}$ and $\vec{v}$ are collinear. For instance, in Figure 6.6, $\det(\mathbf{a} - \mathbf{m}, \vec{u}) > 0$ and $\det(\mathbf{b} - \mathbf{m}, \vec{u}) < 0$. Recall as well that the determinant is a multilinear form, in other words:

$$\begin{aligned}\det(a\mathbf{u} + b\mathbf{v}, c\mathbf{x} + d\mathbf{y}) &= a\det(\mathbf{u}, c\mathbf{x} + d\mathbf{y}) + b\det(\mathbf{v}, c\mathbf{x} + d\mathbf{y}) \\ &= ac\det(\mathbf{u}, \mathbf{x}) + bc\det(\mathbf{v}, \mathbf{x}) + ad\det(\mathbf{u}, \mathbf{y}) \\ &\quad + bd\det(\mathbf{v}, \mathbf{y})\end{aligned}$$

PROOF.– The line $\mathcal{D}(\mathbf{m}, \vec{u})$ passing through $\mathbf{m}$ and carried by $\vec{u}$ separates the plane into two half-planes: one that satisfies $\det(\mathbf{z} - \mathbf{m}, \vec{u}) \geq 0$ and another that verifies $\det(\mathbf{z} - \mathbf{m}, \vec{u}) \leq 0$. This line cuts the segment $[\mathbf{ab}]$ if $\mathbf{a}$ and $\mathbf{b}$ are not in the same half-plane (see Figure 6.6), in other words if $\det(\mathbf{a} - \mathbf{m}, \vec{u}) \cdot \det(\mathbf{b} - \mathbf{m}, \vec{u}) \leq 0$.

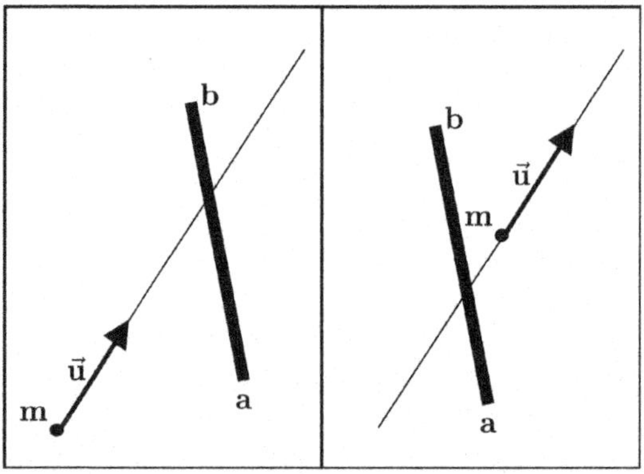

Figure 6.6. *The line* $\mathcal{D}(\mathbf{m}, \vec{u})$ *cuts the segment* $[\mathbf{ab}]$

In the left figure, the ray $\mathcal{E}(\mathbf{m}, \vec{u})$ cuts the segment $[\mathbf{ab}]$, which is not the case in the right figure. The inequality is, therefore, insufficient for stating that $\mathcal{E}(\mathbf{m}, \vec{u})$ cuts $[\mathbf{ab}]$. Let us assume that $\det(\mathbf{a} - \mathbf{m}, \vec{u}) . \det(\mathbf{b} - \mathbf{m}, \vec{u}) \leq$

0 (i.e. $\mathcal{D}(\mathbf{m},\vec{\mathbf{u}})$ cuts $[\mathbf{ab}]$). The points of $\mathcal{E}(\mathbf{m},\vec{\mathbf{u}})$ satisfy $\mathbf{z} = \mathbf{m} + \alpha\vec{\mathbf{u}}$, $\alpha \geq 0$. The point $\mathbf{z}$ belongs to the segment $[\mathbf{ab}]$ if $\mathbf{m} + \alpha\vec{\mathbf{u}} - \mathbf{a}$ and $\mathbf{b} - \mathbf{a}$ are collinear, in other words when α satisfies $\det(\mathbf{m} + \alpha\vec{\mathbf{u}} - \mathbf{a}, \mathbf{b} - \mathbf{a}) = 0$ (see Figure 6.7).

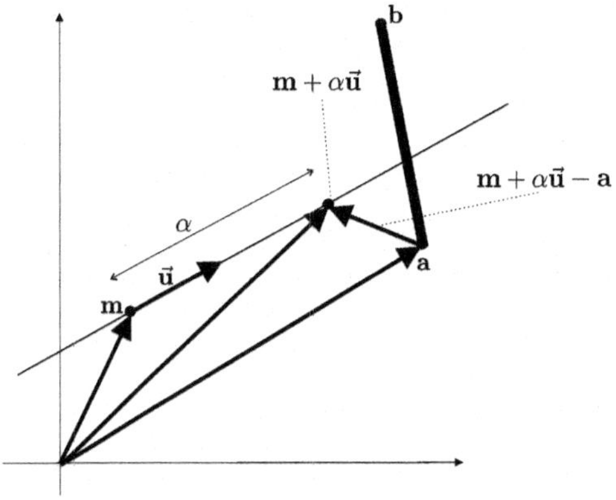

Figure 6.7. *If the point $\mathbf{m} + \alpha\vec{\mathbf{u}}$ is on the segment $[\mathbf{a},\mathbf{b}]$, then α corresponds to the directional distance*

Since the determinant is a multilinear form, we have:

$$\det(\mathbf{m} - \mathbf{a}, \mathbf{b} - \mathbf{a}) + \alpha \det(\vec{\mathbf{u}}, \mathbf{b} - \mathbf{a}) = 0$$

This yields:

$$\alpha = \frac{\det(\mathbf{a} - \mathbf{m}, \mathbf{b} - \mathbf{a})}{\det(\vec{\mathbf{u}}, \mathbf{b} - \mathbf{a})}$$

If $\alpha \geq 0$, then α represents the distance d from $\mathbf{m}$ to the next segment $\vec{\mathbf{u}}$. If $\alpha < 0$, then the radius of the telemeter will never reach the segment. The condition $\alpha \geq 0$ corresponds to the second inequality that we had to demonstrate.

2) In the following, the coordinates of the center $\mathbf{m}$ of the robot are denoted by (x, y) and $\vec{\mathbf{u}}$ represents a unit vector corresponding to the direction of the laser. For the kth sensor, we have:

$$\vec{\mathbf{u}} = \begin{pmatrix} \cos\left(\frac{k\pi}{4} + \theta\right) \\ \sin\left(\frac{k\pi}{4} + \theta\right) \end{pmatrix}, \quad k \in \{0, \ldots, 7\}$$

To have an expression of $\mathbf{f}(\mathbf{p})$, we need to calculate the distances returned by the telemeters. By using the theorem shown in question 1, we can deduce the following simulator $\mathbf{f}(\mathbf{p})$:

input: (x, y, θ)
for $i = 1$to 8
  $\overrightarrow{\mathbf{u}} := \left(\cos\left(\frac{(i-1)\pi}{4} + \theta\right); \sin\left(\frac{(i-1)\pi}{4} + \theta\right)\right);$
  $\mathbf{m} := (x\ y)^{\mathrm{T}}; \quad \ell_i := \infty;$
  for $j = 1$to n
    $\alpha := \frac{\det(\mathbf{a}_j - \mathbf{m}, \mathbf{b}_j - \mathbf{a}_j)}{\det(\overrightarrow{\mathbf{u}}, \mathbf{b}_j - \mathbf{a}_j)};$
    if $\left(\det\left(\mathbf{a}_j - \mathbf{m}, \overrightarrow{\mathbf{u}}\right) \cdot \det\left(\mathbf{b}_j - \mathbf{m}, \overrightarrow{\mathbf{u}}\right) \leq 0\right)$and $(\alpha \geq 0)$
      then $\ell_i := \min(\ell_i, \alpha);$
  next j
next i;
return $(\ell_1, \ldots, \ell_8)$

The MATLAB function below corresponds to this simulator:

```
function y=f(p)
y=inf(8,1);
for i=1:8,
u=[cos((i-1)*pi/4+p(3));sin((i-1)*pi/4+p(3))];
m=[p(1);p(2)];
for j=1:length(A),
a=A(:,j);b=B(:,j);
if det([a-m u])*det([b-m u]) <= 0
alpha=-det([b-a m-a])/det([b-a u]);
if alpha >= 0, y(i)=min(alpha,y(i)); end;
end; end; end; end
```

3) The following MATLAB program, which can also be found in `anneal.m`, proposes a minimization of the criterion $j(\mathbf{p}) = \|\mathbf{f}(\mathbf{p}) - \mathbf{y}\|$ using the simulated annealing method:

```
function j1=j(p), j1=norm(y-f(p)); end  % function to
minimize
A=[0 7 7 9 9 7 7 4 2 0 5 6 6 5; 0 0 2 2 4 4 7 7 5 5 2 2 3 3
];
B=[7 7 9 9 7 7 4 2 0 0 6 6 5 5; 0 2 2 4 4 7 7 5 5 0 2 3 3
2];
```

```
y=[6.4;3.6;2.3;2.1;1.7;1.6;3.0;3.1];
p0=[0;0;0];
T=10;
while (T>0.01)
p=p0+T*randn(3,1); draw(p,y);
if j(p)<j(p0), p0=p; end;
T=0.99*T;
end
```

This program performs a random search and keeps the parameter $\mathbf{p}$ which is the current best parameter. The variable T is the temperature that gives the search step. This temperature decreases exponentially with time. The algorithm generates the solution represented in Figure 6.8.

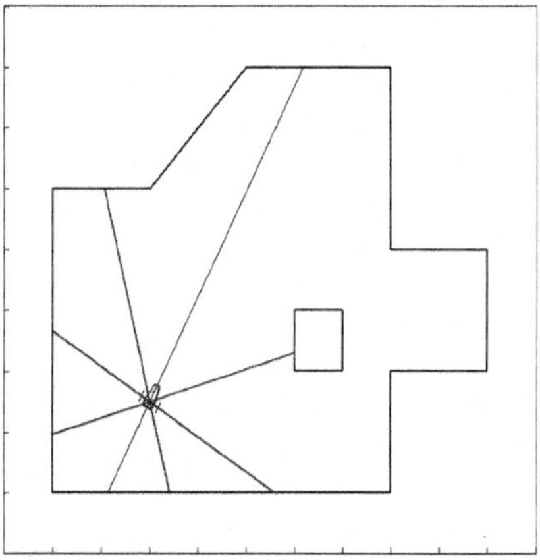

Figure 6.8. *The position of the robot as found from the eight distances given by the laser telemeters*

7

Kalman Filter

In Chapters 2 and 3, we have looked at tools for controlling robots in a nonlinear manner. For this purpose, we have assumed that the state vector was completely known. However, this is not the case in practice. This vector must be estimated from sensor measurements. In the case where the only unknown variables are associated with the position of the robot, Chapter 5 gives guidelines to find them. In the more general case, *filtering* or *state observation* seeks to reconstruct this state vector as well as possible from all the data measured on the robot throughout time by taking into account the state equations. The aim of this chapter is to show how such reconstruction is performed, within a stochastic context in which the system to observe is assumed to be linear. This is the purpose of the Kalman filter [KAL 60], which will be discussed in this chapter. The Kalman filter is used in numerous mobile robotics applications, even though the robots in question are strongly nonlinear. For such applications, the initial conditions are assumed to be relatively well known in order to allow a reliable linearization.

7.1. Covariance matrices

The Kalman filter is mainly based on the concept of covariance matrix which is important to grasp in order to understand the design and utilization of the observer. This section recalls the fundamental concepts surrounding covariance matrices.

7.1.1. *Definitions and interpretations*

Let us consider two random vectors $\mathbf{x} \in \mathbb{R}^n$ and $\mathbf{y} \in \mathbb{R}^m$. The mathematical expectations of $\mathbf{x}$ and $\mathbf{y}$ are denoted by $\bar{\mathbf{x}} = E(\mathbf{x})$, $\bar{\mathbf{y}} = E(\mathbf{y})$.

Let us define the *variations* of $\mathbf{x}$ and $\mathbf{y}$ by $\widetilde{\mathbf{x}} = \mathbf{x} - \bar{\mathbf{x}}$ and $\widetilde{\mathbf{y}} = \mathbf{y} - \bar{\mathbf{y}}$. The *covariance matrix* is given by:

$$\mathbf{\Gamma_{xy}} = E\left(\widetilde{\mathbf{x}} \cdot \widetilde{\mathbf{y}}^{\mathrm{T}}\right) = E\left(\left(\mathbf{x} - \bar{\mathbf{x}}\right)\left(\mathbf{y} - \bar{\mathbf{y}}\right)^{\mathrm{T}}\right)$$

The covariance matrix for $\mathbf{x}$ is defined by:

$$\mathbf{\Gamma_{x}} = \mathbf{\Gamma_{xx}} = E\left(\widetilde{\mathbf{x}} \cdot \widetilde{\mathbf{x}}^{\mathrm{T}}\right) = E\left(\left(\mathbf{x} - \bar{\mathbf{x}}\right)\left(\mathbf{x} - \bar{\mathbf{x}}\right)^{\mathrm{T}}\right)$$

The covariance matrix for $\mathbf{y}$ is:

$$\mathbf{\Gamma_{y}} = \mathbf{\Gamma_{yy}} = E\left(\widetilde{\mathbf{y}} \cdot \widetilde{\mathbf{y}}^{\mathrm{T}}\right) = E\left(\left(\mathbf{y} - \bar{\mathbf{y}}\right)\left(\mathbf{y} - \bar{\mathbf{y}}\right)^{\mathrm{T}}\right)$$

Let us note that $\mathbf{x}$, $\mathbf{y}$, $\widetilde{\mathbf{x}}$, $\widetilde{\mathbf{y}}$ are random vectors, whereas $\bar{\mathbf{x}}, \bar{\mathbf{y}}, \mathbf{\Gamma_x}, \mathbf{\Gamma_y}, \mathbf{\Gamma_{xy}}$ are deterministic. A covariance matrix $\mathbf{\Gamma_x}$ of a random vector $\mathbf{x}$ is always positive definite (we will write $\mathbf{\Gamma_x} \succ \mathbf{0}$), except in the degenerate case. In a computer, a random vector can be represented by a cloud of points associated with realizations. Let us consider the following MATLAB program:

```
x=2+randn(1000,1); e=randn(1000,1); y=2*x.^2+e; plot(x,y);
xbar=mean(x); ybar=mean(y);    xtilde=x-xbar; ytilde=y-ybar;
plot(xtilde,ytilde);
Gx=mean(xtilde.^2); Gy=mean(ytilde.^2);
Gxy=mean(xtilde.*ytilde);
```

This yields Figure 7.1, which gives us a representation of the random variables x, y (in (a)) and $\widetilde{x}, \widetilde{y}$ (in (b)). The program also gives us the estimations:

$$\bar{x} \simeq 1.99, \bar{y} \simeq 9.983, \Gamma_x \simeq 1.003, \Gamma_y \simeq 74.03, \Gamma_{xy} \simeq 8.082$$

where $\bar{x}, \bar{y}, \Gamma_x, \Gamma_y, \Gamma_{xy}$ correspond to `xbar`, `ybar`, `Gx`, `Gy`, `Gxy`.

Two random vectors $\mathbf{x}$ and $\mathbf{y}$ are linearly independent (or non-correlated or orthogonal) if $\mathbf{\Gamma_{xy}} = \mathbf{0}$. In Figure 7.2, the two point clouds correspond to non-correlated variables. Only Figure 7.2(b) corresponds to independent variables.

Figure 7.2(a) was generated by:

```
rho=10+randn(2000,1);                theta=2*pi*rand(2000,1);
x=rho.*sin(theta); y=rho.*cos(theta);
```

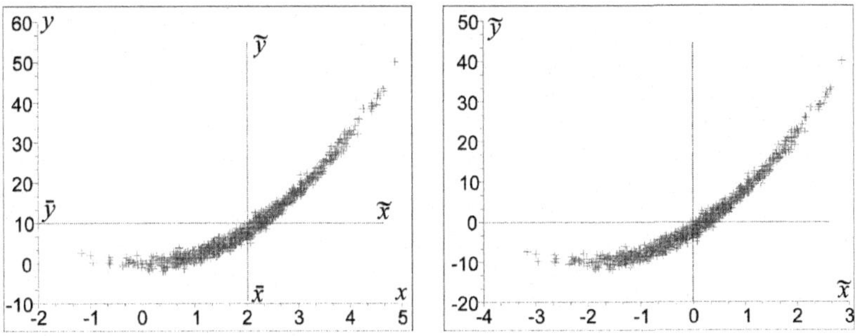

Figure 7.1. *Point cloud that represents a pair of two random variables*

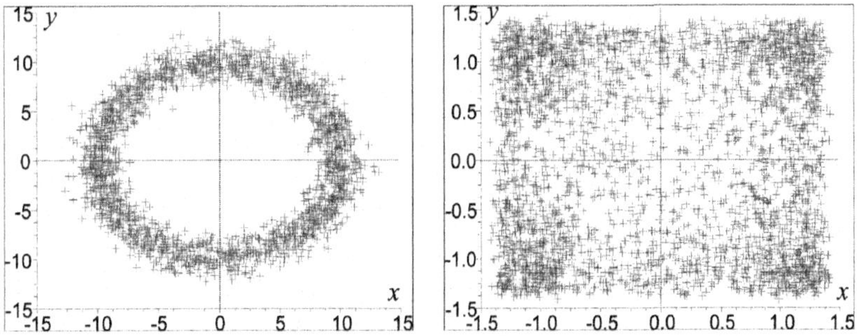

Figure 7.2. *a) Dependent but non-correlated variables* (x, y)*;*
b) independent variables

And Figure 7.2(b) was generated by:

```
x=atan(2*randn(3000,1)); y=atan(2*randn(3000,1));
```

Whiteness. A random vector $\mathbf{x}$ is called *white* if all of its components x_i are independent from one another. In such a case, the covariance vector $\Gamma_{\mathbf{x}}$ of $\mathbf{x}$ is diagonal.

7.1.2. *Properties*

Covariance matrices are symmetric and positive, in other words all of their eigenvalues are real and positive. The set of all covariance matrices of $\mathbb{R}^{n\times n}$ will be denoted by $\mathcal{S}^+(\mathbb{R}^n)$.

Decomposition. Every symmetric matrix $\mathbf{\Gamma}$ can be put into a diagonal form and has an orthonormal eigenvector basis. Therefore, we may write:

$$\mathbf{\Gamma} = \mathbf{R} \cdot \mathbf{D} \cdot \mathbf{R}^{-1}$$

where $\mathbf{R}$ is a rotation matrix (i.e. $\mathbf{R}^{\mathrm{T}}\mathbf{R} = \mathbf{I}$ and $\det \mathbf{R} = 1$). The matrix $\mathbf{R}$ corresponds to the eigenvectors and $\mathbf{D}$ is a diagonal matrix whose elements are the eigenvalues. For the matrices of $\mathcal{S}^+(\mathbb{R}^n)$, these eigenvalues are positive.

Square root. Every matrix $\mathbf{\Gamma}$ of $\mathcal{S}^+(\mathbb{R}^n)$ has a square root in $\mathcal{S}^+(\mathbb{R}^n)$. This square root will be denoted by $\mathbf{\Gamma}^{\frac{1}{2}}$. Following the eigenvalue correspondence theorem, the eigenvalues of $\mathbf{\Gamma}^{\frac{1}{2}}$ are the square roots of those of the eigenvalues of $\mathbf{\Gamma}$.

EXAMPLE 7.1.– Consider the following MATLAB script:

```
A=rand(3,3); S1=A*A'; [R,D]=eig(S1); S2=R*D*R';
A2=sqrtm(S2); S3=A2*A2'.
```

The matrix `D` is diagonal and the matrix `R` is a rotation matrix that contains the eigenvectors of `S1`. The three matrices `S1`, `S2` and `S3` are equal. This is not the case for matrices `A` and `A2` since only `A2` is symmetric. Here, `sqrt` returns the square root of `S2` and therefore `A2` is a covariance matrix.

Order. If $\mathbf{\Gamma}_1$ and $\mathbf{\Gamma}_2$ belong to $\mathcal{S}^+(\mathbb{R}^n)$, then $\mathbf{\Gamma} = \mathbf{\Gamma}_1 + \mathbf{\Gamma}_2$ also belongs to $\mathcal{S}^+(\mathbb{R}^n)$. This is equivalent to saying that $\mathcal{S}^+(\mathbb{R}^n)$ is a convex cone of $\mathbb{R}^{n \times n}$. Let us define the order relation:

$$\mathbf{\Gamma}_1 \leq \mathbf{\Gamma}_2 \Leftrightarrow \mathbf{\Gamma}_2 - \mathbf{\Gamma}_1 \in \mathcal{S}^+(\mathbb{R}^n)$$

It can be easily verified that it is reflexive, antisymmetric and transitive. If $\mathbf{\Gamma}_1 \leq \mathbf{\Gamma}_2$, then the a-level confidence ellipse (see the next section) of $\mathbf{\Gamma}_1$ is included in the one that corresponds to $\mathbf{\Gamma}_2$. The smaller the covariance matrix (in the sense of this order relation), the more precise it is. We will say that it is better or more precise.

7.1.3. *Confidence ellipse*

A random vector $\mathbf{x}$ of $\mathbb{R}^n$ can be characterized by the pair $(\bar{\mathbf{x}}, \mathbf{\Gamma}_{\mathbf{x}})$, to which we can associate an ellipse of $\mathbb{R}^n$ which encloses the consistent values for $\mathbf{x}$. In practice, for purely graphical reasons, we often only look at two components $\mathbf{w} = (x_i, x_j)$ of $\mathbf{x}$ (a computer screen is in fact two-dimensional). The average $\bar{\mathbf{w}}$ can be directly deduced from $\bar{\mathbf{x}}$ by extracting the ith and j^{th} components. The covariance matrix $\mathbf{\Gamma_w} \in \mathcal{S}^+ \left(\mathbb{R}^2\right)$ can also be obtained from $\mathbf{\Gamma_x} \in \mathcal{S}^+ \left(\mathbb{R}^n\right)$ by extracting the ith and j^{th} lines and columns. The *confidence ellipse* associated with $\mathbf{w}$ is described by the inequality:

$$\mathcal{E}_{\mathbf{w}} : (\mathbf{w} - \bar{\mathbf{w}})^{\mathrm{T}} \mathbf{\Gamma}_{\mathbf{w}}^{-1} (\mathbf{w} - \bar{\mathbf{w}}) \leq a^2$$

where a is an arbitrary positive real number. Therefore, if $\mathbf{w}$ is a Gaussian random vector, this ellipse corresponds to a contour line of the probability density for $\mathbf{w}$. Since $\mathbf{\Gamma}_{\mathbf{w}}^{-1} \succ \mathbf{0}$, it has a square root $\mathbf{\Gamma}_{\mathbf{w}}^{-\frac{1}{2}}$ which is also positive definite. Therefore, we may write:

$$\begin{aligned}
\mathcal{E}_{\mathbf{w}} &= \left\{ \mathbf{w} \mid (\mathbf{w} - \bar{\mathbf{w}})^{\mathrm{T}} \, \mathbf{\Gamma}_{\mathbf{w}}^{-\frac{1}{2}} \cdot \mathbf{\Gamma}_{\mathbf{w}}^{-\frac{1}{2}} (\mathbf{w} - \bar{\mathbf{w}}) \leq a^2 \right\} \\
&= \left\{ \mathbf{w} \mid \left\| \tfrac{1}{a} \cdot \mathbf{\Gamma}_{\mathbf{w}}^{-\frac{1}{2}} (\mathbf{w} - \bar{\mathbf{w}}) \right\| \leq 1 \right\} \\
&= \left\{ \mathbf{w} \mid \tfrac{1}{a} \cdot \mathbf{\Gamma}_{\mathbf{w}}^{-\frac{1}{2}} (\mathbf{w} - \bar{\mathbf{w}}) \in \mathcal{U} \right\}, \text{ where } \mathcal{U} \text{ is the unit disk} \\
&= \left\{ \mathbf{w} \mid \mathbf{w} \in \bar{\mathbf{w}} + a \mathbf{\Gamma}_{\mathbf{w}}^{\frac{1}{2}} \mathcal{U} \right\} \\
&= \bar{\mathbf{w}} + a \mathbf{\Gamma}_{\mathbf{w}}^{\frac{1}{2}} \mathcal{U}
\end{aligned}$$

The ellipse $\mathcal{E}_{\mathbf{w}}$ can, therefore, be defined as the image of the unit disk by the affine function $\mathbf{w}(\mathbf{s}) = \bar{\mathbf{w}} + a \cdot \mathbf{\Gamma}_{\mathbf{w}}^{\frac{1}{2}} \mathbf{s}$, as shown in Figure 7.3.

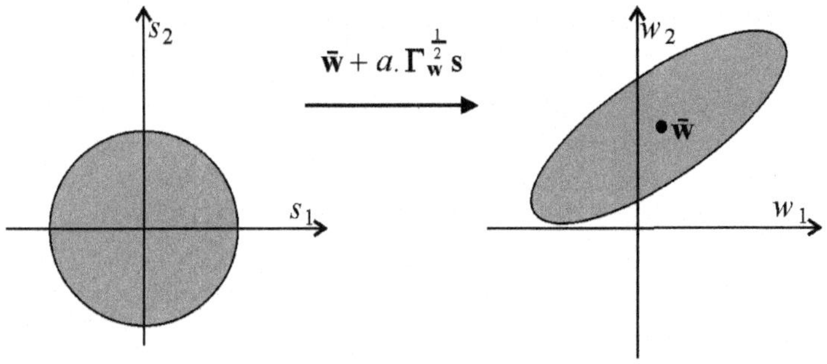

Figure 7.3. *A confidence ellipse is the image of the unit circle by an affine function*

Recall that for a centered, normed Gaussian random vector $\mathbf{s}$, the random variable $z = \mathbf{s}^{\mathrm{T}}\mathbf{s}$ follows a χ^2 law. In two dimensions, this probability density is given by:

$$\pi_z(z) = \begin{cases} \frac{1}{2}\exp\left(-\frac{z}{2}\right) & \text{if } z \geq 0 \\ 0 \text{ otherwise} \end{cases}$$

Thus, for a given $a > 0$, we have:

$$\begin{array}{rl} \eta \stackrel{\text{def}}{=} \text{prob}\left(\|\mathbf{s}\| \leq a\right) = \text{prob}\left(\mathbf{s}^{\mathrm{T}}\mathbf{s} \leq a^2\right) &= \text{prob}\left(z \leq a^2\right) \\ = \int_0^{a^2} \frac{1}{2}\exp\left(-\frac{z}{2}\right) dz &= 1 - e^{-\frac{1}{2}a^2} \end{array}$$

And therefore:

$$a = \sqrt{-2\ln\left(1-\eta\right)}$$

This relation allows us to calculate the threshold a that we need to choose in order to have a probability of being in the ellipse of η. We must, however, be careful as this probabilistic interpretation only makes sense in the Gaussian case. The following MATLAB function draws $\mathcal{E}_{\mathbf{w}}$ for a given probability η:

```
function draw_ellipse(wbar,Gw,eta);
s=0:0.01:2*pi;
w=wbar*ones(size(s))+sqrtm(-2*log(1-eta)*Gw)*[cos(s);
sin(s)];
plot(w(1,:),w(2,:));
```

7.1.4. *Generating Gaussian random vectors*

If we generate n centered Gaussian random numbers, we obtain the realization of a random vector whose center is $\bar{\mathbf{x}} = \mathbf{0}$ and whose covariance matrix $\mathbf{\Gamma}_{\mathbf{x}}$ is the identity matrix. In this section we will show, given a centered Gaussian random number generator allowing us to realize $\mathbf{x}$, how we can obtain a Gaussian random vector $\mathbf{y}$ of dimension n with an expectation and covariance matrix $\mathbf{\Gamma}_{\mathbf{y}}$. The main principle of this generation is based on the following theorem.

THEOREM 7.1.– If $\mathbf{x}, \alpha$ and $\mathbf{y}$ are three random vectors connected by the relation $\mathbf{y} = \mathbf{A}\mathbf{x} + \alpha + \mathbf{b}$ (where $\mathbf{A}$ and $\mathbf{b}$ are deterministic), and assuming that $\mathbf{x}, \alpha$ are independent and that α is centered, we have:

$$\begin{array}{ll} \bar{\mathbf{y}} &= \mathbf{A}\bar{\mathbf{x}} + \mathbf{b} \\ \mathbf{\Gamma}_{\mathbf{y}} &= \mathbf{A} \cdot \mathbf{\Gamma}_{\mathbf{x}} \cdot \mathbf{A}^{\mathrm{T}} + \mathbf{\Gamma}_{\alpha} \end{array} \qquad [7.1]$$

PROOF.– We have:

$$\bar{\mathbf{y}} = E(\mathbf{A}\mathbf{x} + \alpha + \mathbf{b}) = \mathbf{A}E(\mathbf{x}) + E(\alpha) + \mathbf{b} = \mathbf{A}\bar{\mathbf{x}} + \mathbf{b}$$

More over:

$$\begin{aligned}
\mathbf{\Gamma_y} &= E\left((\mathbf{y} - \bar{\mathbf{y}})(\mathbf{y} - \bar{\mathbf{y}})^{\mathrm{T}}\right) \\
&= E\left((\mathbf{A}\mathbf{x} + \alpha + \mathbf{b} - \mathbf{A}\bar{\mathbf{x}} - \mathbf{b})(\mathbf{A}\mathbf{x} + \alpha + \mathbf{b} - \mathbf{A}\bar{\mathbf{x}} - \mathbf{b})^{\mathrm{T}}\right) \\
&= E\left((\mathbf{A}\widetilde{\mathbf{x}} + \alpha)\cdot(\mathbf{A}\widetilde{\mathbf{x}} + \alpha)^{\mathrm{T}}\right) \\
&= \mathbf{A}\cdot\underbrace{E\left(\widetilde{\mathbf{x}}\cdot\widetilde{\mathbf{x}}^{\mathrm{T}}\right)}_{=\mathbf{\Gamma_x}}\cdot\mathbf{A}^{\mathrm{T}} + \mathbf{A}\cdot\underbrace{E\left(\widetilde{\mathbf{x}}\cdot\alpha^{\mathrm{T}}\right)}_{=0} + \underbrace{E\left(\alpha\cdot\widetilde{\mathbf{x}}^{\mathrm{T}}\right)}_{=0}\cdot\mathbf{A}^{\mathrm{T}} + \underbrace{E\left(\alpha\cdot\alpha^{\mathrm{T}}\right)}_{=\mathbf{\Gamma}_\alpha} \\
&= \mathbf{A}\cdot\mathbf{\Gamma_x}\cdot\mathbf{A}^{\mathrm{T}} + \mathbf{\Gamma}_\alpha
\end{aligned}$$

which concludes the proof. ■

Thus, if $\mathbf{x}$ is a centered, unit Gaussian white random noise (in other words, $\bar{\mathbf{x}} = \mathbf{0}$ and $\mathbf{\Gamma_x} = \mathbf{I}$), the random vector $\mathbf{y} = \mathbf{\Gamma_y}^{\frac{1}{2}}\mathbf{x} + \bar{\mathbf{y}}$ will have an expectation of $\bar{\mathbf{y}}$ and a covariance matrix equal to $\mathbf{\Gamma_y}$ (see Figure 7.4). To generate a Gaussian random vector with covariance matrix $\mathbf{\Gamma_y}$ and expectation $\bar{\mathbf{y}}$, we will use this property. Figure 7.4(b) was thus obtained by the script:

```
n=1000;Gy=[3,1;1,3]; ybar=[2;3]; x=randn(2,n);
y=ybar*ones(1,n)+sqrtm(Gy)*x; plot(y(1,:),y(2,:),'.');
```

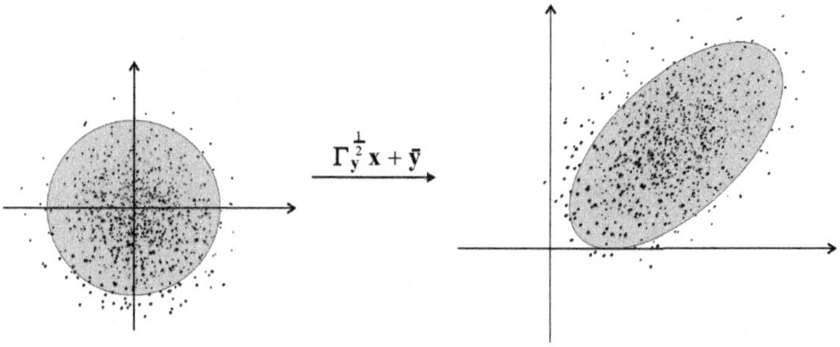

Figure 7.4. *The Gaussian random vector $\mathbf{y} : (\bar{\mathbf{y}}, \mathbf{\Gamma_y})$ is the image by an affine application of a unit Gaussian white random vector $\mathbf{x}$*

7.2. Unbiased orthogonal estimator

Let us consider two random vectors $\mathbf{x} \in \mathbb{R}^n$ and $\mathbf{y} \in \mathbb{R}^m$. Vector $\mathbf{y}$ corresponds to the measurement vector which is for the moment a random vector, and will only become available when the measurements have been made. The random vector $\mathbf{x}$ is the vector we need to estimate. An *estimator* is a function $\phi(\mathbf{y})$ that gives us an estimation of $\mathbf{x}$ given the knowledge of the measurement $\mathbf{y}$. Figure 7.5 shows a nonlinear estimator corresponding to $E(x|y)$.

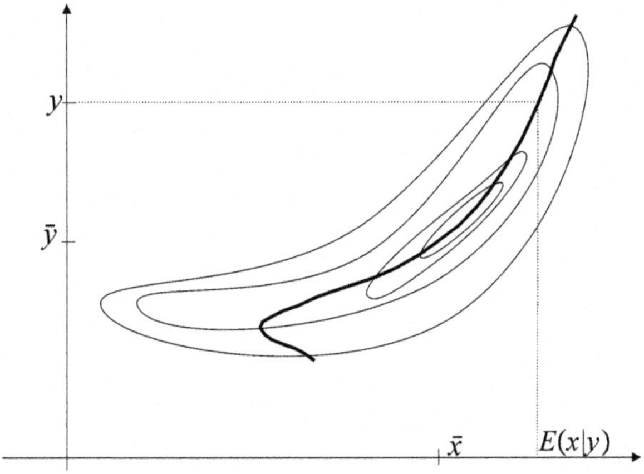

Figure 7.5. *Nonlinear estimator* $E(x|y)$

However, obtaining an analytic expression for such an estimator is generally not a simple task and it is preferable to limit ourselves to linear estimators. A *linear estimator* is a linear function of $\mathbb{R}^m \to \mathbb{R}^n$ of the form:

$$\hat{\mathbf{x}} = \mathbf{K}\mathbf{y} + \mathbf{b} \qquad [7.2]$$

where $\mathbf{K} \in \mathbb{R}^{n \times m}$ and $\mathbf{b} \in \mathbb{R}^n$. In this section, we will propose a method capable of finding a *good* $\mathbf{K}$ and a *good* $\mathbf{b}$ from the sole knowledge of the first-order moments $\bar{\mathbf{x}}, \bar{\mathbf{y}}$ and second-order moments $\mathbf{\Gamma_x}, \mathbf{\Gamma_x}, \mathbf{\Gamma_{xy}}$. The *estimation error* is:

$$\varepsilon = \hat{\mathbf{x}} - \mathbf{x}$$

The estimator is said to be *unbiased* if $E(\varepsilon) = \mathbf{0}$. It is *orthogonal* if $E\left(\varepsilon\tilde{\mathbf{y}}^{\mathrm{T}}\right) = \mathbf{0}$. This naming comes from the fact that the space of random variables of $\mathbb{R}$ can be equipped with a scalar product defined by $\langle a, b \rangle = E\left((a-\bar{a})\left(b-\bar{b}\right)\right)$ and that if this scalar product is zero, the two random variables a and b are called orthogonal. In the vectorial case (which is that of our section since ε and $\widetilde{\mathbf{y}}$ are vectors), we say that the two random vectors $\mathbf{a}$ and $\mathbf{b}$ are orthogonal if their components are, in other words $E\left((a_i-\bar{a}_i)\left(b_j-\bar{b}_j\right)\right) = 0$ for all (i,j), or equivalently $E\left((\mathbf{a}-\bar{\mathbf{a}})\left(\mathbf{b}-\bar{\mathbf{b}}\right)^{\mathrm{T}}\right) = \mathbf{0}$. Figure 7.6 represents the contour lines of a probability law for the pair (x, y). The line illustrates a linear estimator. Let us randomly pick a pair (x, y) while respecting its probability law. It is clear that the probability to be above the line is high, in other words the probability to have $\hat{x} < x$ is high, or even that $E(\varepsilon) < 0$. The estimator is thus biased. Figure 7.7 represents four different linear estimators. For estimator (a), $E(\varepsilon) < 0$ and for estimator (c), $E(\varepsilon) > 0$. For estimators (b) and (d), $E(\varepsilon) = 0$ and therefore the two estimators are unbiased. However, it is evident that estimator (b) is better. What differentiates these two is orthogonality. For (d), we have $E(\varepsilon\widetilde{y}) < 0$ (if $\widetilde{y} > 0$, ε tends to be negative, whereas if $\widetilde{y} < 0$, ε tends to be positive).

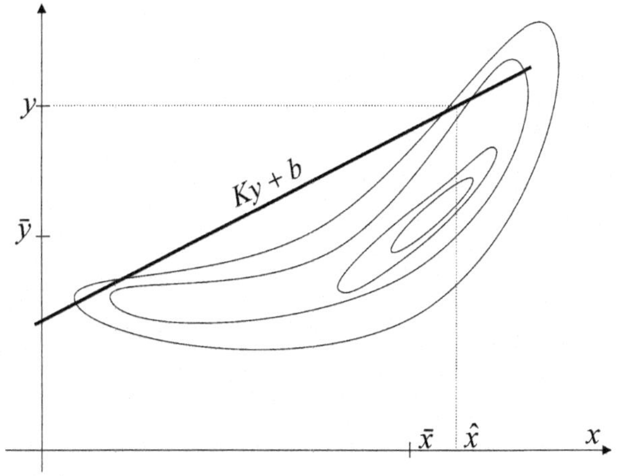

Figure 7.6. *Biased linear estimator*

THEOREM 7.2.– Consider two random vectors $\mathbf{x}$ and $\mathbf{y}$. A unique unbiased orthogonal estimator exists. It is given by:

$$\hat{\mathbf{x}} = \bar{\mathbf{x}} + \mathbf{K} \cdot (\mathbf{y} - \bar{\mathbf{y}}) \quad [7.3]$$

where:

$$\mathbf{K} = \mathbf{\Gamma_{xy}} \mathbf{\Gamma_y^{-1}} \qquad [7.4]$$

is referred to as the *Kalman gain*.

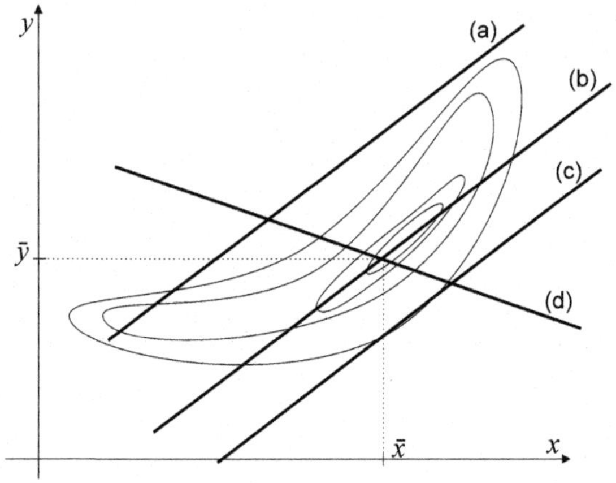

Figure 7.7. *Among these four linear estimators, estimator (b), which is unbiased and orthogonal, seems to be the best*

EXAMPLE 7.2.– Let us consider once again the example given in section 7.1.1. We obtain:

$$\hat{x} = \bar{x} + \Gamma_{xy}\Gamma_y^{-1} \cdot (y - \bar{y}) = 2 + 0.1 \cdot (y - 10)$$

The corresponding estimator is shown in Figure 7.8.

Proof of the theorem. We have:

$$\begin{aligned} E(\varepsilon) &= E(\hat{\mathbf{x}} - \mathbf{x}) \overset{[7.2]}{=} E(\mathbf{K}\mathbf{y} + \mathbf{b} - \mathbf{x}) \\ &= \mathbf{K}E(\mathbf{y}) + \mathbf{b} - E(\mathbf{x}) = \mathbf{K}\bar{\mathbf{y}} + \mathbf{b} - \bar{\mathbf{x}} \end{aligned}$$

The estimator is unbiased if $E(\varepsilon) = \mathbf{0}$, i.e.:

$$\mathbf{b} = \bar{\mathbf{x}} - \mathbf{K}\bar{\mathbf{y}} \qquad [7.5]$$

which gives us [7.3]. In this case:

$$\varepsilon = \hat{\mathbf{x}} - \mathbf{x} \overset{[7.3]}{=} \bar{\mathbf{x}} + \mathbf{K} \cdot (\mathbf{y} - \bar{\mathbf{y}}) - \mathbf{x} = \mathbf{K}\widetilde{\mathbf{y}} - \widetilde{\mathbf{x}} \quad [7.6]$$

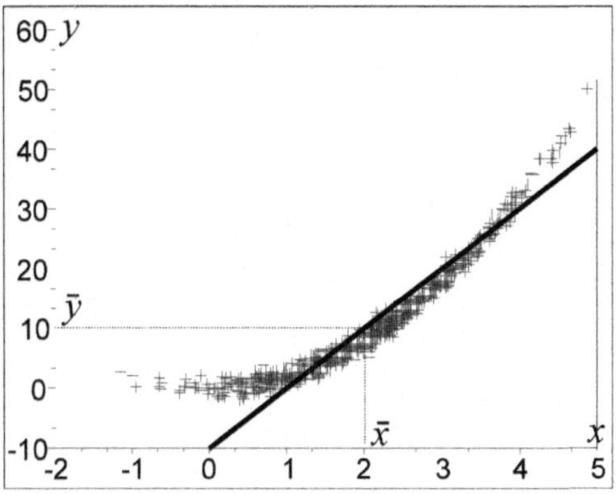

Figure 7.8. *Unbiased orthogonal linear estimator*

The estimator is orthogonal if:

$$\begin{aligned} E\left(\varepsilon \cdot \widetilde{\mathbf{y}}^{\mathrm{T}}\right) = \mathbf{0} &\overset{[7.6]}{\Leftrightarrow} E\left((\mathbf{K}\widetilde{\mathbf{y}} - \widetilde{\mathbf{x}}) \cdot \widetilde{\mathbf{y}}^{\mathrm{T}}\right) = \mathbf{0} \\ &\Leftrightarrow E\left(\mathbf{K}\widetilde{\mathbf{y}}\widetilde{\mathbf{y}}^{\mathrm{T}} - \widetilde{\mathbf{x}}\widetilde{\mathbf{y}}^{\mathrm{T}}\right) = \mathbf{0} \\ &\Leftrightarrow \mathbf{K}\mathbf{\Gamma}_{\mathbf{y}} - \mathbf{\Gamma}_{\mathbf{xy}} = \mathbf{0} \\ &\Leftrightarrow \mathbf{K} = \mathbf{\Gamma}_{\mathbf{xy}} \cdot \mathbf{\Gamma}_{\mathbf{y}}^{-1} \end{aligned}$$

which concludes the proof. ■

THEOREM 7.3.– The covariance matrix of the error associated with the unbiased orthogonal linear estimator is:

$$\mathbf{\Gamma}_{\varepsilon} = \mathbf{\Gamma}_{\mathbf{x}} - \mathbf{K} \cdot \mathbf{\Gamma}_{\mathbf{yx}} \quad [7.7]$$

PROOF.– The covariance matrix of ε in the unbiased case is written as:

$$\begin{aligned} \mathbf{\Gamma}_{\varepsilon} &= E\left(\varepsilon \cdot \varepsilon^{\mathrm{T}}\right) \overset{[7.6]}{=} E\left((\mathbf{K}\widetilde{\mathbf{y}} - \widetilde{\mathbf{x}}) \cdot (\mathbf{K}\widetilde{\mathbf{y}} - \widetilde{\mathbf{x}})^{\mathrm{T}}\right) \\ &= E\left((\mathbf{K}\widetilde{\mathbf{y}} - \widetilde{\mathbf{x}}) \cdot (\widetilde{\mathbf{y}}^{\mathrm{T}}\mathbf{K}^{\mathrm{T}} - \widetilde{\mathbf{x}}^{\mathrm{T}})\right) \\ &= E\left(\mathbf{K}\widetilde{\mathbf{y}}\widetilde{\mathbf{y}}^{\mathrm{T}}\mathbf{K}^{\mathrm{T}} - \widetilde{\mathbf{x}}\widetilde{\mathbf{y}}^{\mathrm{T}}\mathbf{K}^{\mathrm{T}} - \mathbf{K}\widetilde{\mathbf{y}}\widetilde{\mathbf{x}}^{\mathrm{T}} + \widetilde{\mathbf{x}}\widetilde{\mathbf{x}}^{\mathrm{T}}\right) \end{aligned}$$

Using the linearity of the expectation operator:

$$\mathbf{\Gamma}_\varepsilon = (\mathbf{K}\mathbf{\Gamma}_\mathbf{y} - \mathbf{\Gamma}_{\mathbf{xy}})\,\mathbf{K}^\mathrm{T} - \mathbf{K}\mathbf{\Gamma}_{\mathbf{yx}} + \mathbf{\Gamma}_\mathbf{x} \qquad [7.8]$$

However, following [7.4], in the orthogonal case $\mathbf{K}\mathbf{\Gamma}_\mathbf{y} - \mathbf{\Gamma}_{\mathbf{xy}} = 0$, which concludes the proof. ■

We will now present a theorem that shows that the unbiased orthogonal linear estimator is the best among all unbiased estimators. In order to understand this concept of *best*, we need to recall the inequalities on the covariance matrices (see section 7.1), which tells us that $\Gamma_1 \le \Gamma_2$ if and only if $\Delta = \Gamma_2 - \Gamma_1$ is a covariance matrix.

THEOREM 7.4.– No unbiased linear estimator exists allowing us to obtain a smaller covariance matrix on the error $\mathbf{\Gamma}_\varepsilon$ than the one given by the orthogonal estimator.

PROOF.– Every possible matrix $\mathbf{K}$ for our unbiased linear estimator is written in the form $\mathbf{K} = \mathbf{K}_0 + \mathbf{\Delta}$ with $\mathbf{K}_0 = \mathbf{\Gamma}_{\mathbf{xy}}\mathbf{\Gamma}_\mathbf{y}^{-1}$ and $\mathbf{\Delta}$ being an arbitrary matrix. Following [7.8], the covariance matrix for the error is:

$$\begin{aligned}
\mathbf{\Gamma}_\varepsilon &= ((\mathbf{K}_0 + \mathbf{\Delta})\,\mathbf{\Gamma}_\mathbf{y} - \mathbf{\Gamma}_{\mathbf{xy}})\,(\mathbf{K}_0 + \mathbf{\Delta})^\mathrm{T} - (\mathbf{K}_0 + \mathbf{\Delta})\,\mathbf{\Gamma}_{\mathbf{yx}} + \mathbf{\Gamma}_\mathbf{x} \\
&= (\mathbf{K}_0 + \mathbf{\Delta})\,(\underbrace{\mathbf{\Gamma}_\mathbf{y}\mathbf{K}_0^\mathrm{T}}_{=\mathbf{\Gamma}_{\mathbf{yx}}} + \mathbf{\Gamma}_\mathbf{y}\mathbf{\Delta}^\mathrm{T}) - (\underbrace{\mathbf{\Gamma}_{\mathbf{xy}}\mathbf{K}_0^\mathrm{T}}_{=\mathbf{K}_0\mathbf{\Gamma}_{\mathbf{yx}}} + \mathbf{\Gamma}_{\mathbf{xy}}\mathbf{\Delta}^\mathrm{T}) \\
&\quad - (\mathbf{K}_0\mathbf{\Gamma}_{\mathbf{yx}} + \mathbf{\Delta}\mathbf{\Gamma}_{\mathbf{yx}}) + \mathbf{\Gamma}_\mathbf{x} \\
&= \mathbf{K}_0\mathbf{\Gamma}_{\mathbf{yx}} + \mathbf{\Delta}\mathbf{\Gamma}_{\mathbf{yx}} + \underbrace{\mathbf{K}_0\mathbf{\Gamma}_\mathbf{y}}_{=\mathbf{\Gamma}_{\mathbf{xy}}}\mathbf{\Delta}^\mathrm{T} + \mathbf{\Delta}\mathbf{\Gamma}_\mathbf{y}\mathbf{\Delta}^\mathrm{T} - \mathbf{K}_0\mathbf{\Gamma}_{\mathbf{yx}} - \mathbf{\Gamma}_{\mathbf{xy}}\mathbf{\Delta}^\mathrm{T} \\
&\quad -\mathbf{K}_0\mathbf{\Gamma}_{\mathbf{yx}} - \mathbf{\Delta}\mathbf{\Gamma}_{\mathbf{yx}} + \mathbf{\Gamma}_\mathbf{x} \\
&= -\mathbf{K}_0\mathbf{\Gamma}_{\mathbf{yx}} + \mathbf{\Delta}\mathbf{\Gamma}_\mathbf{y}\mathbf{\Delta}^\mathrm{T} + \mathbf{\Gamma}_\mathbf{x}
\end{aligned}$$

Since $\mathbf{\Delta}\mathbf{\Gamma}_\mathbf{y}\mathbf{\Delta}^\mathrm{T}$ is always positive symmetric, the covariance matrix $\mathbf{\Gamma}_\varepsilon$ is minimal for $\mathbf{\Delta} = \mathbf{0}$, i.e. for $\mathbf{K} = \mathbf{\Gamma}_{\mathbf{xy}}\mathbf{\Gamma}_\mathbf{y}^{-1}$, which corresponds to the orthogonal unbiased estimator. ■

7.3. Application to linear estimation

Let us assume that $\mathbf{x}$ and $\mathbf{y}$ are connected by the relation:

$$\mathbf{y} = \mathbf{C}\mathbf{x} + \beta$$

where β is a centered random vector non-correlated with $\mathbf{x}$. The covariance matrices of $\mathbf{x}$ and β are denoted by $\mathbf{\Gamma_x}$ and $\mathbf{\Gamma}_\beta$. Let us utilize the results obtained in the previous section in order to find the best unbiased linear estimator for $\mathbf{x}$ (refer to [WAL 14] for more details on linear estimation). We have:

$$\begin{array}{lll}\bar{\mathbf{y}} & = & \mathbf{C}\bar{\mathbf{x}} + \bar{\beta} = \mathbf{C}\bar{\mathbf{x}} \\ \mathbf{\Gamma_y} & \overset{[7.1]}{=} & \mathbf{C}\mathbf{\Gamma_x}\mathbf{C}^\mathrm{T} + \mathbf{\Gamma}_\beta \\ \mathbf{\Gamma_{xy}} & = & E\left(\widetilde{\mathbf{x}} \cdot \widetilde{\mathbf{y}}^\mathrm{T}\right) = E\left(\widetilde{\mathbf{x}} \cdot \left(\mathbf{C}\widetilde{\mathbf{x}} + \widetilde{\beta}\right)^\mathrm{T}\right) \\ & = & E\left(\widetilde{\mathbf{x}} \cdot \widetilde{\mathbf{x}}^\mathrm{T}\mathbf{C}^\mathrm{T} + \widetilde{\mathbf{x}} \cdot \widetilde{\beta}^\mathrm{T}\right) \\ & = & E\left(\widetilde{\mathbf{x}} \cdot \widetilde{\mathbf{x}}^\mathrm{T}\right)\mathbf{C}^\mathrm{T} + \underbrace{E\left(\widetilde{\mathbf{x}} \cdot \widetilde{\beta}^\mathrm{T}\right)}_{=0} = \mathbf{\Gamma_x}\mathbf{C}^\mathrm{T}\end{array} \qquad [7.9]$$

Consequently, the best unbiased estimator for $\mathbf{x}$ and covariance matrix of the error can be obtained from $\mathbf{\Gamma_x}, \mathbf{\Gamma}_\beta, \mathbf{C}, \bar{\mathbf{x}}$ by using the following formulas:

$$\begin{array}{llll}\text{(i)} & \hat{\mathbf{x}} & \overset{[7.3]}{=} & \bar{\mathbf{x}} + \mathbf{K}\widetilde{\mathbf{y}} & \text{(estimation)} \\ \text{(ii)} & \mathbf{\Gamma}_\varepsilon & \overset{[7.7]}{=} & \mathbf{\Gamma_x} - \mathbf{K}\mathbf{C}\mathbf{\Gamma_x} & \text{(covariance of the error)} \\ \text{(iii)} & \widetilde{\mathbf{y}} & \overset{[7.9]}{=} & \mathbf{y} - \mathbf{C}\bar{\mathbf{x}} & \text{(innovation)} \\ \text{(iv)} & \mathbf{\Gamma_y} & \overset{[7.9]}{=} & \mathbf{C}\mathbf{\Gamma_x}\mathbf{C}^\mathrm{T} + \mathbf{\Gamma}_\beta & \text{(covariance of the innovation)} \\ \text{(v)} & \mathbf{K} & \overset{[7.4,7.9]}{=} & \mathbf{\Gamma_x}\mathbf{C}^\mathrm{T}\mathbf{\Gamma_y}^{-1} & \text{(Kalman gain)}\end{array} \qquad [7.10]$$

NOTE 7.1.– Figure 7.5 shows a situation in which it could be advantageous not to use a linear estimator. Here, the chosen estimator corresponds to $\hat{x} = E(x|y)$. In the particular case where the pair $(\mathbf{x}, \mathbf{y})$ is Gaussian, the estimator $\hat{\mathbf{x}} = E(\mathbf{x}|\mathbf{y})$ corresponds to the unbiased orthogonal estimator. In this case, we have, following [7.10]:

$$\begin{array}{ll}E(\mathbf{x}|\mathbf{y}) & = \bar{\mathbf{x}} + \mathbf{\Gamma_{xy}}\mathbf{\Gamma_y}^{-1}(\mathbf{y} - \bar{\mathbf{y}}) \\ E\left(\varepsilon \cdot \varepsilon^\mathrm{T}|\mathbf{y}\right) & = E\left((\hat{\mathbf{x}} - \mathbf{x})(\hat{\mathbf{x}} - \mathbf{x})^\mathrm{T}|\mathbf{y}\right) = \mathbf{\Gamma_x} - \mathbf{\Gamma_{xy}}\mathbf{\Gamma_y}^{-1}\mathbf{\Gamma_{yx}}\end{array}$$

7.4. Kalman filter

This section presents the Kalman filter (refer to [DEL 93] for more information). Let us consider the system described by the following state equations:

$$\left\{\begin{array}{ll}\mathbf{x}_{k+1} & = \mathbf{A}_k\mathbf{x}_k + \mathbf{u}_k + \alpha_k \\ \mathbf{y}_k & = \mathbf{C}_k\mathbf{x}_k + \beta_k\end{array}\right.$$

where α_k and β_k are the random, independent Gaussian noises white in time. By white in time, we mean that the vectors α_{k_1} and α_{k_2} (or β_{k_1} and β_{k_2}) are independent of each other if $k_1 \neq k_2$. The Kalman filter alternates between two phases: *correction* and *prediction*. To understand the mechanism of the filter, let us position ourselves at time k and assume that we have already processed the measurements $\mathbf{y}_0, \mathbf{y}_1, \ldots, \mathbf{y}_{k-1}$. At this stage, the state vector is a random vector that we will denote by $\mathbf{x}_{k|k-1}$ (since we are at time k and the measurements have been processed until $k-1$). This random vector is represented by an estimation denoted by $\hat{\mathbf{x}}_{k|k-1}$ and a covariance matrix $\mathbf{\Gamma}_{k|k-1}$.

Correction. Let us take the measurement $\mathbf{y}_k$. The random vector representing the state is now $\mathbf{x}_{k|k}$, which is different from $\mathbf{x}_{k|k-1}$ since $\mathbf{x}_{k|k}$ has knowledge of the measurement $\mathbf{y}$. The expectation $\hat{\mathbf{x}}_{k|k}$ and the covariance matrix $\mathbf{\Gamma}_{k|k}$ associated with $\mathbf{x}_{k|k}$ are given by equations [7.10]. Therefore, we have:

$$\begin{array}{llll}
\text{(i)} & \hat{\mathbf{x}}_{k|k} & = \hat{\mathbf{x}}_{k|k-1} + \mathbf{K}_k \cdot \widetilde{\mathbf{y}}_k & \text{(corrected estimation)} \\
\text{(ii)} & \mathbf{\Gamma}_{k|k} & = \mathbf{\Gamma}_{k|k-1} - \mathbf{K}_k \cdot \mathbf{C}_k \mathbf{\Gamma}_{k|k-1} & \text{(corrected covariance)} \\
\text{(iii)} & \widetilde{\mathbf{y}}_k & = \mathbf{y}_k - \mathbf{C}_k \hat{\mathbf{x}}_{k|k-1} & \text{(innovation)} \\
\text{(iv)} & \mathbf{S}_k & = \mathbf{C}_k \mathbf{\Gamma}_{k|k-1} \mathbf{C}_k^{\mathrm{T}} + \mathbf{\Gamma}_{\beta_k} & \text{(covariance of the innovation)} \\
\text{(v)} & \mathbf{K}_k & = \mathbf{\Gamma}_{k|k-1} \mathbf{C}_k^{\mathrm{T}} \mathbf{S}_k^{-1} & \text{(Kalman gain)}
\end{array} \quad [7.11]$$

Prediction. Given the measurements $\mathbf{y}_0, \mathbf{y}_1, \ldots, \mathbf{y}_k$, the random vector representing the state is now $\mathbf{x}_{k+1|k}$. Let us calculate its expectation $\hat{\mathbf{x}}_{k+1|k}$ and covariance matrix $\mathbf{\Gamma}_{k+1|k}$. Since:

$$\mathbf{x}_{k+1} = \mathbf{A}_k \mathbf{x}_k + \mathbf{u}_k + \alpha_k$$

we have, following [7.1]:

$$\hat{\mathbf{x}}_{k+1|k} = \mathbf{A}_k \hat{\mathbf{x}}_{k|k} + \mathbf{u}_k \quad [7.12]$$

and:

$$\mathbf{\Gamma}_{k+1|k} = \mathbf{A}_k \cdot \mathbf{\Gamma}_{k|k} \cdot \mathbf{A}_k^{\mathrm{T}} + \mathbf{\Gamma}_{\alpha_k} \quad [7.13]$$

Kalman filter. The complete Kalman filter is given by the following equations:

$$\begin{array}{lll}
\hat{\mathbf{x}}_{k+1|k} & \overset{[7.12]}{=} \mathbf{A}_k \hat{\mathbf{x}}_{k|k} + \mathbf{u}_k & \text{(predicted estimation)} \\
\mathbf{\Gamma}_{k+1|k} & \overset{[7.13]}{=} \mathbf{A}_k \cdot \mathbf{\Gamma}_{k|k} \cdot \mathbf{A}_k^{\mathrm{T}} + \mathbf{\Gamma}_{\alpha_k} & \text{(predicted covariance)}
\end{array}$$

$$\begin{array}{lcll}
\hat{\mathbf{x}}_{k|k} & \overset{[7.10,\text{i}]}{=} & \hat{\mathbf{x}}_{k|k-1}+\mathbf{K}_k \cdot \widetilde{\mathbf{y}}_k & \text{(corrected estimation)} \\
\mathbf{\Gamma}_{k|k} & \overset{[7.10,\text{ii}]}{=} & (\mathbf{I}-\mathbf{K}_k\mathbf{C}_k)\,\mathbf{\Gamma}_{k|k-1} & \text{(corrected covariance)} \\
\widetilde{\mathbf{y}}_k & \overset{[7.10,\text{iii}]}{=} & \mathbf{y}_k - \mathbf{C}_k\hat{\mathbf{x}}_{k|k-1} & \text{(innovation)} \\
\mathbf{S}_k & \overset{[7.10,\text{iv}]}{=} & \mathbf{C}_k\mathbf{\Gamma}_{k|k-1}\mathbf{C}_k^{\text{T}} + \mathbf{\Gamma}_{\beta_k} & \text{(covariance of the innovation)} \\
\mathbf{K}_k & \overset{[7.10,\text{v}]}{=} & \mathbf{\Gamma}_{k|k-1}\mathbf{C}_k^{\text{T}}\mathbf{S}_k^{-1} & \text{(Kalman gain)}
\end{array}$$

Figure 7.9 shows the fact that the Kalman filter stores the vector $\hat{\mathbf{x}}_{k+1|k}$ and the matrix $\mathbf{\Gamma}_{k+1|k}$. Its inputs are $\mathbf{y}_k$, $\mathbf{u}_k$, $\mathbf{A}_k$, $\mathbf{C}_k$, Γ_{α_k} and Γ_{β_k}. The quantities $\hat{\mathbf{x}}_{k|k}$, $\mathbf{\Gamma}_{k|k}$, $\widetilde{\mathbf{y}}_k$, $\mathbf{S}_k$, $\mathbf{K}_k$ are auxiliary variables.

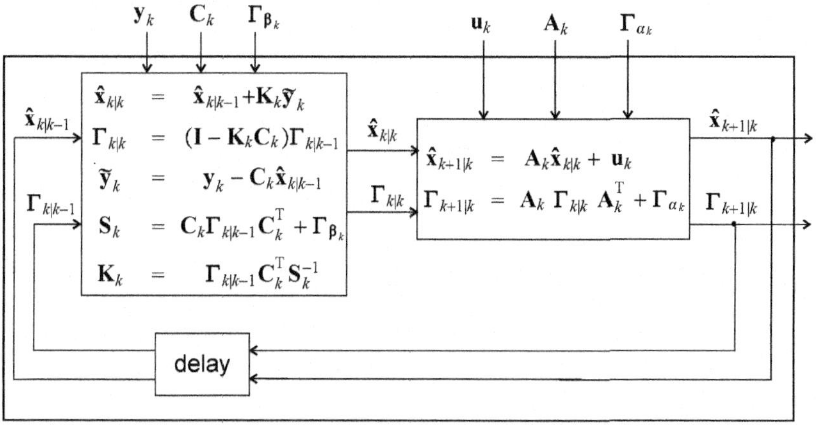

Figure 7.9. *The Kalman filter is composed of a corrector followed by a predictor*

The following MATLAB function implements the Kalman filter. In this program, we have the following correspondences: x0↔ $\hat{\mathbf{x}}_{k|k-1}$, G1↔ $\mathbf{\Gamma}_{k|k-1}$, x1↔ $\hat{\mathbf{x}}_{k+1|k}$, G1↔ $\mathbf{\Gamma}_{k+1|k}$, xup↔ $\hat{\mathbf{x}}_{k|k}$, Gup↔ $\mathbf{\Gamma}_{k|k}$ (the term up refers to *update*, in other words correction).

```
function [x1,G1]=kalman(x0,G0,u,y,Galpha,Gbeta,A,C);
S=C*G0*C'+Gbeta;
K=G0*C'*inv(S);
ytilde=y-C*x0;
xup=x0+K*ytilde;
Gup=G0-K*C*G0;
```

```
x1=A*xup + u;
G1=A*Gup*A'+Galpha;
end
```

NOTE 7.2.– Due to numerical problems, the covariance of the innovation $\mathbf{S}_k$ can sometimes loose its positivity. If such a problem arises, it is preferable to replace the corrected covariance equation with:

$$\mathbf{\Gamma}_{k|k} = \sqrt{(\mathbf{I} - \mathbf{K}_k\mathbf{C}_k)\,\mathbf{\Gamma}_{k|k-1}\mathbf{\Gamma}_{k|k-1}^{\mathrm{T}}\,(\mathbf{I} - \mathbf{K}_k\mathbf{C}_k)^{\mathrm{T}}}$$

which will always be positive definite, even when the matrix $\mathbf{\Gamma}_{k|k-1}$ is not. The Kalman filter equations will then be more stable in the sense that a slight error on the positive character of the covariance matrices is removed at the next iteration.

NOTE 7.3.– When no measurement is available, the Kalman filter operates in *predictor* mode. In order to be able to use the `kalman` function, $\mathbf{y}$, $\mathbf{\Gamma}_\beta$, $\mathbf{C}$ have to become empty quantities. However, they have to have correct dimensions in order to allow, in MATLAB, us to perform matrix operations. The function call will then be as follows:

```
[xhat,Gx]=kalman(xhat,Gx,u,eye(0,1),Galpha,eye(0,0),A,
eye(0,length(x)))
```

7.5. Kalman smoother

The Kalman filter is causal. This means that the estimation $\hat{\mathbf{x}}_{k|k-1}$ only takes into account the past. The *smoothing* process consists of a state estimation when all the measurements (future, present and past) are available. Let us denote by N the maximum time k. This time can correspond, for instance, to the end date of a mission performed by the robot and for which we are trying to estimate its path. In order to perform smoothing, we simply need to rerun a Kalman filter in the backward direction and merge, for each k, the information from the future with that of the past. An optimized version referred to as *Kalman smoother* can then be applied by adding the following equations to those of the Kalman filter:

$$\begin{aligned} \mathbf{J}_k &= \mathbf{\Gamma}_{k|k} \cdot \mathbf{A}_k^{\mathrm{T}} \cdot \mathbf{\Gamma}_{k+1|k}^{-1} \\ \hat{\mathbf{x}}_{k|N} &= \hat{\mathbf{x}}_{k|k} + \mathbf{J}_k \left(\hat{\mathbf{x}}_{k+1|N} - \hat{\mathbf{x}}_{k+1|k}\right) \\ \mathbf{\Gamma}_{k|N} &= \mathbf{\Gamma}_{k|k} + \mathbf{J}_k \left(\mathbf{\Gamma}_{k+1|N} - \mathbf{\Gamma}_{k+1|k}\right) \mathbf{J}_k^{\mathrm{T}} \end{aligned} \qquad [7.14]$$

In order to perform the smoothing process, we first need to run the Kalman filter for k ranging from 0 to N, then run equations [7.14] backward for k ranging form N to 0. Note that all the quantities $\hat{\mathbf{x}}_{k+1|k}$, $\hat{\mathbf{x}}_{k|k}$, $\mathbf{\Gamma}_{k+1|k}$, $\mathbf{\Gamma}_{k|k}$, $\hat{\mathbf{x}}_{k|N}$ are stored in the lists `x_forw{k}`, `x_up{k}`, `G_forw{k}`, `G_up{k}`, `x_back{k}`, `G_back{k}`, where `forw`, `up`, `back`, respectively, mean *forward*, *update* and *backward*. The quantities $\mathbf{u}_k, \mathbf{\Gamma}_\alpha(k), \mathbf{y}_k, \mathbf{\Gamma}_\beta(k), \mathbf{A}_k$ are also stored in lists. The direct, or *forward* part of the smoother corresponding to the Kalman filter, is given by the script:

```
x_forw{1}=...; G_forw{1}=...; % initialization
for k=1:kmax,
[x_forw{k+1},G_forw{k+1},x_up{k},G_up{k}]
=kalman(x_forw{k},G_forw{k},u{k},y{k},Galpha{k},Gbeta{k},
A{k},C{k});
end;
```

As for the *backward* part of the smoother, this is given by:

```
x_back{kmax}=x_up{kmax};
G_back{kmax}=G_up{kmax};
for k=kmax-1:-1:1,
J=G_up{k}*A'/G_forw{k+1};
x_back{k}=x_up{k}+J*(x_back{k+1}-x_forw{k+1});
G_back{k}=G_up{k}+J*(G_back{k+1}-G_forw{k+1})*J';
end;
```

Note that, given the specificity of MATLAB to index lists starting from 1, the initial condition corresponds to $k = 1$.

7.6. Exercises

EXERCISE 7.1.– Gaussian distribution

The probability distribution of a random Gaussian vector $\mathbf{x}$ is fully characterized by its expectation $\bar{\mathbf{x}}$ and covariance matrix $\mathbf{\Gamma}_\mathbf{x}$. More precisely, it is given by:

$$\pi_\mathbf{x}(\mathbf{x}) = \frac{1}{\sqrt{(2\pi)^n \det(\mathbf{\Gamma}_\mathbf{x})}} \cdot \exp\left(-\frac{1}{2}(\mathbf{x}-\bar{\mathbf{x}})^\mathrm{T} \mathbf{\Gamma}_\mathbf{x}^{-1} (\mathbf{x}-\bar{\mathbf{x}})\right)$$

1) Draw the graph and contour lines of $\pi_{\mathbf{x}}$ with:

$$\bar{\mathbf{x}} = \begin{pmatrix} 1 \\ 2 \end{pmatrix} \text{ and } \mathbf{\Gamma}_{\mathbf{x}} = \begin{pmatrix} 1 & 0 \\ 0 & 1 \end{pmatrix}$$

2) We define the random vector:

$$\mathbf{y} = \begin{pmatrix} \cos\frac{\pi}{6} & -\sin\frac{\pi}{6} \\ \sin\frac{\pi}{6} & \cos\frac{\pi}{6} \end{pmatrix} \begin{pmatrix} 1 & 0 \\ 0 & 3 \end{pmatrix} \mathbf{x} + \begin{pmatrix} 2 \\ -5 \end{pmatrix}$$

Draw the graph and contour lines of $\pi_{\mathbf{y}}$. Discuss.

EXERCISE 7.2.– Confidence ellipses

Let us generate six covariance matrices in MATLAB as follows:

```
A1=[1 0;0 3]; A2=[cos(pi/4) -sin(pi/4);sin(pi/4)
cos(pi/4)];
G1=eye(2,2); G2=3*eye(2,2); G3=A1*G2*A1'+G1; G4=A2*G3*A2';
G5=G4+G3; G6=A2*G5*A2';
```

Here, A2 corresponds to a rotation matrix of angle $\frac{\pi}{4}$. Then, we draw the six confidence ellipses at 90 % associated with these matrices by centering them around 0. Thus, we obtain Figure 7.10.

1) Associate each covariance matrix with its confidence ellipse in the figure.

2) Verify the result by regenerating these ellipses in MATLAB.

EXERCISE 7.3.– Confidence ellipse: prediction

1) Generate a cloud of $n = 1\,000$ points in MATLAB representing a random Gaussian vector centered at $\mathbb{R}^2$ whose covariance matrix is the identity matrix. Deduce from the latter a cloud of points for the random vector $\mathbf{x}$ such that:

$$\bar{\mathbf{x}} = \begin{pmatrix} 1 \\ 2 \end{pmatrix} \text{ and } \mathbf{\Gamma}_{\mathbf{x}} = \begin{pmatrix} 3 & 1 \\ 1 & 3 \end{pmatrix}$$

Use a two-line, n-column matrix to store the clouds.

2) Draw the confidence ellipses for the probabilities $\eta \in \{0.9, 0.99, 0.999\}$.

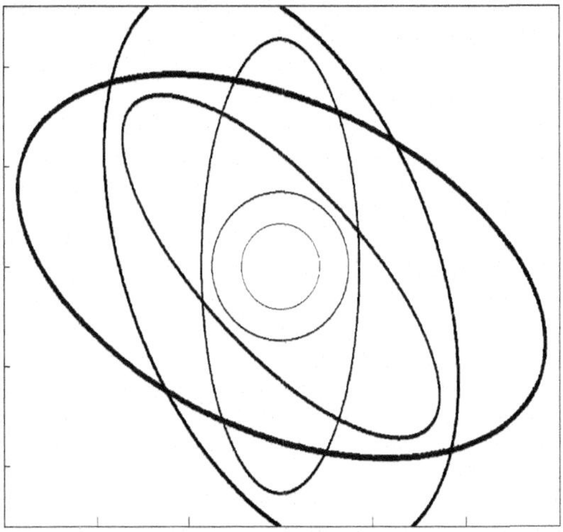

Figure 7.10. *Confidence ellipses associated with the six covariance matrices*

3) Find an estimation of $\bar{\mathbf{x}}$ and $\Gamma_{\mathbf{x}}$ from the cloud of $\mathbf{x}$.

4) This distribution represents the knowledge we have of the initial conditions of a system (for instance, a robot) described by state equations of the form:

$$\dot{\mathbf{x}} = \begin{pmatrix} 0 & 1 \\ -1 & 0 \end{pmatrix} \mathbf{x} + \begin{pmatrix} 2 \\ 3 \end{pmatrix} u$$

where the input $u(t) = \sin(t)$ is known. Write a program that illustrates the evolution of this particle cloud with time. Use a sampling period of $\delta = 0.01$ sec.

5) Represent this evolution using only the confidence ellipses.

EXERCISE 7.4.– Confidence ellipse: correction

1) As in the previous exercise, generate a Gaussian point cloud of $n = 1\,000$ points associated with the random vector $\mathbf{x}$ with:

$$\bar{\mathbf{x}} = \begin{pmatrix} 1 \\ 2 \end{pmatrix} \text{ and } \Gamma_{\mathbf{x}} = \begin{pmatrix} 3 & 1 \\ 1 & 3 \end{pmatrix}$$

Use a two-line, n-column matrix to store the clouds. Verify the coherence of the random generator by comparing it with MATLAB's `mvnrnd` generator (where *mvn* refers to *multivariate normal distribution*).

2) Find an unbiased and orthogonal linear estimator which allows us to find x_1 from x_2. Draw this estimator.

3) Same question as above, but one that allows us to find x_2 from x_1. Draw the estimator and discuss the difference with the previous question.

EXERCISE 7.5.– Covariance matrix propagation

Consider three centered random vectors $\mathbf{a}, \mathbf{b}, \mathbf{c}$ with covariance matrices equal to the identity matrix. These three vectors are independent of each other. Let $\mathbf{x}, \mathbf{y}$ be two random vectors defined as follows:

$$\begin{aligned} \mathbf{x} &= \mathbf{A}\,\mathbf{a} - \mathbf{b} \\ \mathbf{y} &= \mathbf{C}\,\mathbf{x} + \mathbf{c} \end{aligned}$$

where $\mathbf{A}, \mathbf{C}$ are matrices that are known.

1) Give the expression of the mathematical expectations $\bar{\mathbf{x}}, \bar{\mathbf{y}}$ and the covariance matrices $\mathbf{\Gamma_x}, \mathbf{\Gamma_y}$ of these two vectors, as a function of $\mathbf{A}$ and $\mathbf{C}$.

2) We form the vector $\mathbf{v} = (\mathbf{x}\ ,\ \mathbf{y})$. Calculate the mathematical expectation $\bar{\mathbf{v}}$ and covariance matrix $\Gamma_{\mathbf{v}}$ for $\mathbf{v}$.

3) Deduce from the previous question the covariance matrix of the random vector $\mathbf{z} = \mathbf{y} - \mathbf{x}$. We assume of course that $\mathbf{x}$ and $\mathbf{y}$ are of same dimension.

4) We measure $\mathbf{y}$, which means that now the random vector $\mathbf{y}$ becomes deterministic and is well known. Give an estimation $\hat{\mathbf{x}}$ for $\mathbf{x}$ using an unbiased and orthogonal linear estimator.

EXERCISE 7.6.– Brownian noise

We consider a random stationary, discretized, white and centered random signal. This signal is denoted by $x(t_k)$ with $k \in \mathbb{N}$. More precisely, for every $t_k = k\delta$, the random variables $x(t_k)$ with variance σ_x^2 are independent of each other. A Brownian noise is defined as the integral of a white noise. In our case, we form the Brownian noise as follows:

$$y(t_k) = \delta \cdot \sum_{i=0}^{k} x(t_k)$$

1) Calculate, as a function of time, the variance $\sigma_y^2(t_k)$ of the signal $y(t_k)$. How does the standard deviation $\sigma_y(t_k)$ evolve as a function of δ and as a function of t_k? Discuss. Validate the result with a MATLAB simulation.

2) We now tend δ toward 0. What standard deviation σ_x do we have to choose as a function of δ in order for the variances $\sigma_y^2(t_k)$ to remain unchanged? Illustrate this with a MATLAB program that generates Brownian noises $y(t)$ that are insensitive to sampling period changes.

EXERCISE 7.7.– Solving three equations using a linear estimator

The linear estimator can be used to solve problems that can be translated as linear equations. Let us consider as an illustration the system:

$$\begin{cases} 2x_1 + 3x_2 = 8 \\ 3x_1 + 2x_2 = 7 \\ x_1 - x_2 \quad = 0 \end{cases}$$

Since we have more equations than unknowns, the linear estimator must find some sort of compromise between all of these equations. Let us assume that the errors ε_i over the ith equation are centered and with variances: $\sigma_1^2 = 1$, $\sigma_2^2 = 4$ and $\sigma_3^2 = 4$. Solve the system by using a linear estimator and find the associated covariance matrix.

EXERCISE 7.8.– Estimating the parameters of an electric motor using a linear estimator

Let us consider a DC motor whose parameters have been estimated with a least squares method (see exercise 6.3). Recall that in that example, the angular speed Ω of a DC motor verifies the relation:

$$\Omega = x_1 U + x_2 T_r$$

where U is the input voltage, T_r is the resistive torque and $\mathbf{x} = (x_1, x_2)$ is the vector of parameters that we need to estimate. The following table recalls the measurements made on the motor for various experimental conditions:

U(V)	4	10	10	13	15
T_r(Nm)	0	1	5	5	3
Ω(rad/ sec)	5	10	8	14	17

We assume that the variance of the measurement error is equal to 9 and does not depend on the experimental conditions. Moreover, we assume that we know *a priori* that $x_1 \simeq 1$ and $x_2 \simeq -1$ with a variance of 4. Estimate the parameters of the motor and find the associated covariance matrix.

EXERCISE 7.9.– Trochoid

1) A point mass (placed on a wheel) is moving following a trochoid of the form:

$$\begin{cases} x(t) = p_1 t - p_2 \sin t \\ y(t) = p_1 - p_2 \cos t \end{cases}$$

where x corresponds to the abscissa and y corresponds to the altitude of the mass. We measure y for various instants t:

t(sec)	1	2	3	7
y(m)	0.38	3.25	4.97	−0.26

The measurement errors have a standard deviation of 10 cm. By using an unbiased orthogonal filter, calculate an estimation for p_1 and p_2.

2) Draw the estimated path of the mass in MATLAB.

EXERCISE 7.10.– Solving three equations using a Kalman filter

Let us consider once again the linear equations of exercise 7.7:

$$\begin{cases} 2x_1 + 3x_2 = 8 + \beta_1 \\ 3x_1 + 2x_2 = 7 + \beta_2 \\ x_1 - x_2 \quad = 0 + \beta_3 \end{cases}$$

where $\beta_1, \beta_2, \beta_3$ are the three independent, centered random variables with respective variances $1, 4, 4$.

1) Solve this system in MATLAB by calling the Kalman filter three times. Give an estimation of the solution and find the covariance matrix of the error.

2) Draw the confidence ellipses associated with each call.

3) Compare these with the results obtained for exercise 7.7.

EXERCISE 7.11.– Three-step Kalman filter

Let us consider the discrete-time system:

$$\begin{cases} \mathbf{x}_{k+1} &= \mathbf{A}_k \mathbf{x}_k + \mathbf{u}_k + \alpha_k \\ y_k &= \mathbf{C}_k \mathbf{x}_k + \beta_k \end{cases}$$

with $k \in \{0, 1, 2\}$. The values for the quantities $\mathbf{A}_k, \mathbf{C}_k, \mathbf{u}_k, y_k$ are given by:

k	$\mathbf{A}_k$	$\mathbf{u}_k$	$\mathbf{C}_k$	y_k
0	$\begin{pmatrix} 0.5 & 0 \\ 0 & 1 \end{pmatrix}$	$\begin{pmatrix} 8 \\ 16 \end{pmatrix}$	$(1\ 1)$	7
1	$\begin{pmatrix} 1 & -1 \\ 1 & 1 \end{pmatrix}$	$\begin{pmatrix} -6 \\ -18 \end{pmatrix}$	$(1\ 1)$	30
2	$\begin{pmatrix} 1 & -1 \\ 1 & 1 \end{pmatrix}$	$\begin{pmatrix} 32 \\ -8 \end{pmatrix}$	$(1\ 1)$	-6

Let us assume that the signals α_k and β_k are white Gaussian signals with a unitary covariance matrix, in other words:

$$\Gamma_\alpha = \begin{pmatrix} 1 & 0 \\ 0 & 1 \end{pmatrix} \text{ and } \mathbf{\Gamma}_\beta = 1$$

The initial state vector is unknown and is represented by an estimation $\hat{\mathbf{x}}_{0|-1}$ and a covariance matrix $\mathbf{\Gamma}_{0|-1}$. We will take:

$$\hat{\mathbf{x}}_{0|-1} = \begin{pmatrix} 0 \\ 0 \end{pmatrix}, \ \mathbf{\Gamma}_{0|-1} = \begin{pmatrix} 100 & 0 \\ 0 & 100 \end{pmatrix}$$

Draw the confidence ellipses with center $\hat{\mathbf{x}}_{k|k}$ and covariance matrix $\Gamma_{k|k}$ obtained by the Kalman filter, in MATLAB.

EXERCISE 7.12.– Estimating the parameters of an electric motor

Let us consider once more the DC motor with angular speed Ω (see exercises 6.3–7.8). We have:

$$\Omega = x_1 U + x_2 T_r$$

where U is the input voltage, T_r is the resistive torque and $\mathbf{x} = (x_1, x_2)$ is the vector of parameters to estimate. The following table presents the measurements obtained for various experimental conditions:

k	0	1	2	3	4
U(V)	4	10	10	13	15
T_r(Nm)	0	1	5	5	3
Ω(rad/ sec)	5	10	11	14	17

We still assume that the variance of the measurement error is equal to 9 and that $x_1 \simeq 1$ and $x_2 \simeq -1$ with a variance of 4. Using the Kalman filter, calculate an estimation of the parameters x_1, x_2 and give the associated covariance matrix.

EXERCISE 7.13.– Localization from wall distance measurements

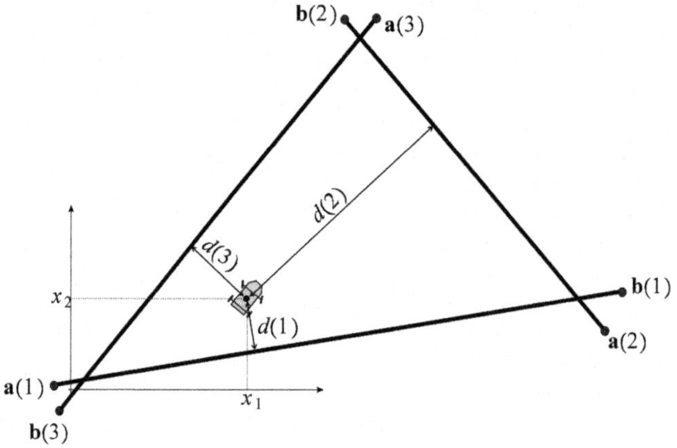

Figure 7.11. *The robot must locate itself by measuring its distance to the three walls*

Consider a punctual robot positioned at $\mathbf{x} = (x_1, x_2)$. This robot measures its distance to the three walls, as shown in Figure 7.11. The ith wall corresponds to a line defined by two points $\mathbf{a}(i)$ and $\mathbf{b}(i)$. The distance to the ith wall is:

$$d(i) = \det(\mathbf{u}(i), \mathbf{x} - \mathbf{a}(i)) + \beta_i$$

with $\mathbf{u}(i) = \frac{\mathbf{b}(i)-\mathbf{a}(i)}{\|\mathbf{b}(i)-\mathbf{a}(i)\|}$. Each distance is measured with a centered error β_i with variance 1 and all the errors are independent of each other. Before taking any measurements, the robot assumes that it is in position $\bar{\mathbf{x}} = (1, 2)$ with the associated covariance matrix given by $100 \cdot \mathbf{I}$ where $\mathbf{I}$ is the identity matrix.

1) Give, as a function of the $\mathbf{a}(i), \mathbf{b}(i), d(i)$, an estimation of the robot's position as well as the covariance matrix for the error. For this, you can use the expression of the unbiased orthogonal linear estimator or equivalently the expression of the Kalman filter in correction mode.

2) The coordinates of the points as well as the distances are given by:

i	1	2	3
$\mathbf{a}(i)$	$\begin{pmatrix} 2 \\ 1 \end{pmatrix}$	$\begin{pmatrix} 15 \\ 5 \end{pmatrix}$	$\begin{pmatrix} 3 \\ 12 \end{pmatrix}$
$\mathbf{b}(i)$	$\begin{pmatrix} 15 \\ 5 \end{pmatrix}$	$\begin{pmatrix} 3 \\ 12 \end{pmatrix}$	$\begin{pmatrix} 2 \\ 1 \end{pmatrix}$
$d(i)$	2	5	4

Write a MATLAB program that gives us the required estimation.

EXERCISE 7.14.– Temperature estimation

The temperature in a room has to verify (after temporal discretization) the state equation:

$$\begin{cases} x_{k+1} = x_k + \alpha_k \\ y_k \quad\;\, = x_k + \beta_k \end{cases}$$

We assume that the state noise α_k and the measurement noise β_k are independent and Gaussian with covariance $\Gamma_\alpha = 4$ and $\Gamma_\beta = 3$.

1) Give the expression of the Kalman filter that allows us to estimate the temperature x_k from the measurement y_k. From this, deduce an expression of $\hat{x}_{k+1|k}$ and $\Gamma_{k+1|k}$ as a function of $\hat{x}_{k|k-1}, \Gamma_{k|k-1}, y_k$.

2) For large enough k, we may assume that $\Gamma_{k+1|k} = \Gamma_{k|k-1} = \Gamma_\infty$. We then obtain the so-called *asymptotic* Kalman filter. Give the expression of the asymptotic Kalman filter. How would you characterize the precision of this filter?

3) Going back to the non-asymptotic case, but now assuming that $\Gamma_{\alpha_k} = 0$, what is the value of Γ_∞? Discuss.

EXERCISE 7.15.– Blind walker

We consider a blind walker moving on a horizontal line. Its movement is described by the discretized state equation:

$$\begin{cases} x_1(k+1) = x_1(k) + x_2(k) \cdot u(k) \\ x_2(k+1) = x_2(k) + \alpha_2(k) \end{cases}$$

where $x_1(k)$ is the position of the walker, $x_2(k)$ is the length of a step (referred to as *scale factor*) and $u(k)$ is the number of steps per time unit. We measure the quantity $u(k)$. Thus, at each unit of time, the walker moves a distance of $x_2(k)\,u(k)$. At the initial moment, we know that x_1 is zero and that x_2 is close to 1. $x_2(0)$ will be represented by a Gaussian distribution whose mean is equal to 1 and whose standard deviation is 0.02. The scale factor x_2 evolves slowly by means of $\alpha_2(k)$ that we will assume to be centered, white and of standard deviation 0.01.

1) We apply an input $u(k) = 1$ for $k = 0, \ldots, 9$ and $u(k) = -1$ for $k = 10, \ldots, 19$. Write a MATLAB program that implements a predictive Kalman filter capable of estimating the position $x_1(k)$.

2) Draw the confidence ellipses associated with the probability $\eta = 0.99$. How does the uncertainty evolve for x_1 as a function of k ?

3) As a function of k, draw the determinant of the covariance matrix $\mathbf{\Gamma}_\mathbf{x}$. Discuss.

EXERCISE 7.16.– Simple pendulum

Let us consider the pendulum in Figure 7.12. The input of this system is the torque u exerted on the pendulum.

Its state representation is assumed to be:

$$\begin{pmatrix} \dot{x}_1 \\ \dot{x}_2 \end{pmatrix} = \begin{pmatrix} x_2 \\ -\sin x_1 + u \end{pmatrix}$$

This is of course a normalized model in which the coefficients (mass, gravity and length) have all been set to 1.

1) We would like the position of the pendulum $x_1(t)$ to be equal to a setpoint $w(t)$ that varies with time. Using a feedback linearization method,

suggest a state feedback controller such that the error $e = w - x_1$ converges toward 0 at $\exp(-t)$ (which means that we place the poles at -1). For this, the expression of u as a function of $\mathbf{x}, w, \dot{w}, \ddot{w}$ must be written.

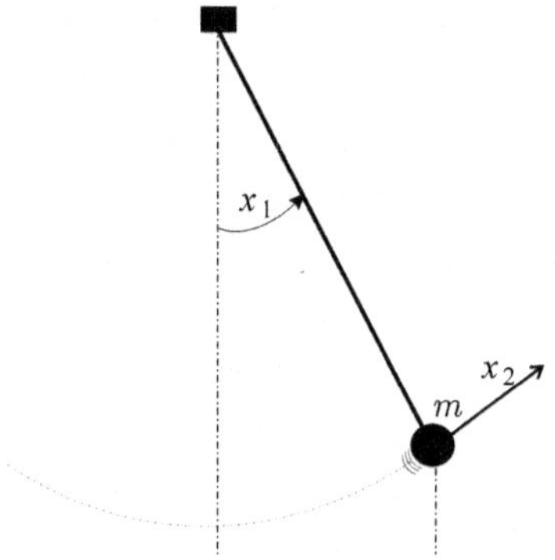

Figure 7.12. *Simple pendulum with state vector* $\mathbf{x} = (x_1, x_2)$

2) We would like the angle of the pendulum x_1 to be equal to $\sin t$ once the transient regime has passed. What expression do we need to choose for the control $u(t)$? Give the expression of u as a function of $\mathbf{x}$ and t.

3) In order to implement the control proposed in the previous question, it is necessary for the complete state to be available. However, we only have a gyro placed on the axis of the pendulum that gives us a measurement of x_2 every $\delta = 0.01$ s, with a white Gaussian error of standard deviation 0.1rad/ sec. Therefore, we must reconstruct the state $\mathbf{x}$ of the pendulum to implement our control. This can be done using a state observer. Here, we also propose using a Kalman filter to realize this observer. Before doing so, the system must be linearized around the point $\mathbf{x} = (0, 0)$ and discretized with a sampling period of δ. Find the arguments $(y_k, u_k, \mathbf{\Gamma}_{\alpha_k}, \Gamma_{\beta_k}, \mathbf{A}_k, \mathbf{C}_k)$ to give the Kalman at each iteration k so that the latter generates an estimation $\hat{\mathbf{x}}$ of our state.

4) Give the entire system in the form of a block-diagram in which the inputs and outputs of each subsystem appear (pendulum, observer and controller).

EXERCISE 7.17.– State estimation of the inverted rod pendulum

We consider an inverted rod pendulum whose state equations are given by:

$$\begin{pmatrix} \dot{x}_1 \\ \dot{x}_2 \\ \dot{x}_3 \\ \dot{x}_4 \end{pmatrix} = \begin{pmatrix} x_3 \\ x_4 \\ \dfrac{m \sin x_2 (g \cos x_2 - \ell x_4^2) + u}{M + m \sin^2 x_2} \\ \dfrac{\sin x_2 ((M+m) g - m \ell x_4^2 \cos x_2) + \cos x_2 u}{\ell (M + m \sin^2 x_2)} \end{pmatrix}$$

$$\text{and} \quad y = \begin{pmatrix} x_1 \\ x_2 \end{pmatrix}$$

Here, we have taken as state vector $\mathbf{x} = \left(x, \theta, \dot{x}, \dot{\theta}\right)$, where the input u is the force exerted on the cart of mass M, x is the position of the cart and θ is the angle between the pendulum and the vertical direction. We will assume here that only the position of the cart x and the angle θ of the pendulum are measured.

1) Linearize this system around the state $\mathbf{x} = \mathbf{0}$.

2) Suggest a state feedback controller of the form $u = -\mathbf{K} \cdot \mathbf{x} + h\, w$ that stabilizes the system. Use a pole placement method to achieve this (the `place` instruction in MATLAB). All the poles will be equal to -2. For the precompensator h, take a setpoint w that corresponds to the desired position for the cart. Following [JAU 15], we must take:

$$h = -\left(\mathbf{E} \cdot (\mathbf{A} - \mathbf{B} \cdot \mathbf{K})^{-1} \cdot \mathbf{B}\right)^{-1}$$

where $\mathbf{E}$ is the setpoint matrix given by:

$$\mathbf{E} = (1\,0\,0\,0)$$

Simulate the system controlled by this state feedback.

3) In order to preform output feedback, we need an estimation $\hat{\mathbf{x}}$ of the state vector $\mathbf{x}$. For this, we will use a Kalman filter (see Figure 7.13):

Discretize the system using steps of $dt = 0.01$ sec, then propose a Kalman filter for observing the state.

4) Implement this filter in MATLAB. We will take a centered Gaussian white noise for the state with variance $\sigma_x^2 = dt \cdot 0.01$ for the variables x, θ.

For the measurement noise, we will take a centered Gaussian white noise with variance $\sigma_y^2 = 0.01^2$. Study the robustness of the observer when the measurement noise is increased. On a separate figure, draw the confidence ellipses in the space (x, θ) and verify whether their true value (x^*, θ^*) is within the ellipse.

5) An extended Kalman filter can be obtained by replacing, in the prediction step of the Kalman filter, the instruction:

$$\hat{\mathbf{x}}_{k+1|k} = \mathbf{A}_k \hat{\mathbf{x}}_{k|k} + \mathbf{B}_k \mathbf{u}_k$$

by:

$$\hat{\mathbf{x}}_{k+1|k} = \hat{\mathbf{x}}_{k|k} + \mathbf{f}\left(\hat{\mathbf{x}}_{k|k}, \mathbf{u}_k\right) \cdot dt$$

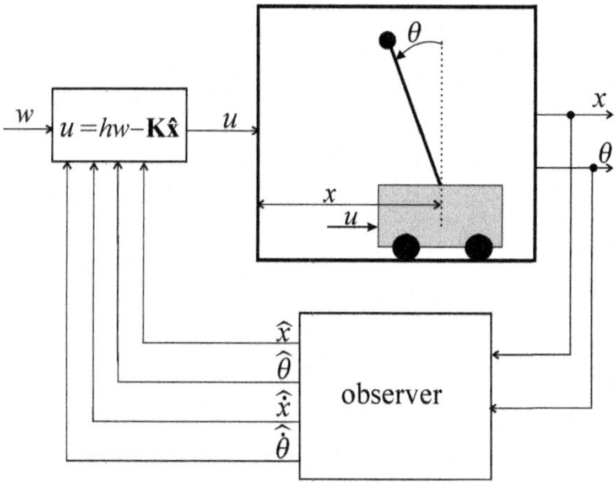

Figure 7.13. *Kalman filter used to estimate the state of the inverted rod pendulum*

Here, we have replaced the prediction performed on the linearized model by a prediction performed on the initial nonlinear model that is closer to reality. Propose an implementation of this extended Kalman filter.

EXERCISE 7.18.– Dead reckoning

Dead reckoning corresponds to the problem of localization in which only proprioceptive sensors are available. This type of navigation was used by early navigators who were trying to locate themselves during long journeys. They were able to do this in a very approximative way by measuring the heading of the boat, the speed at various instants and integrating all the corresponding variations in position over the entire journey. In a more general context, we may consider that using a state observer in prediction mode and without correction (in the particular case in which the state is the position of the robot) corresponds to dead reckoning. Let us consider the robot represented in Figure 7.14 and whose state equations are:

$$\begin{pmatrix} \dot{x} \\ \dot{y} \\ \dot{\theta} \\ \dot{v} \\ \dot{\delta} \end{pmatrix} = \begin{pmatrix} v \cos \delta \cos \theta \\ v \cos \delta \sin \theta \\ \frac{v \sin \delta}{3} + \alpha_\theta \\ u_1 + \alpha_v \\ u_2 + \alpha_\delta \end{pmatrix}$$

where $\alpha_\theta, \alpha_v, \alpha_\delta$ are the independent continuous-time Gaussian white noises. In a more rigorous way, these are random distributions with infinite power, but once they are discretized, the mathematical difficulties disappear.

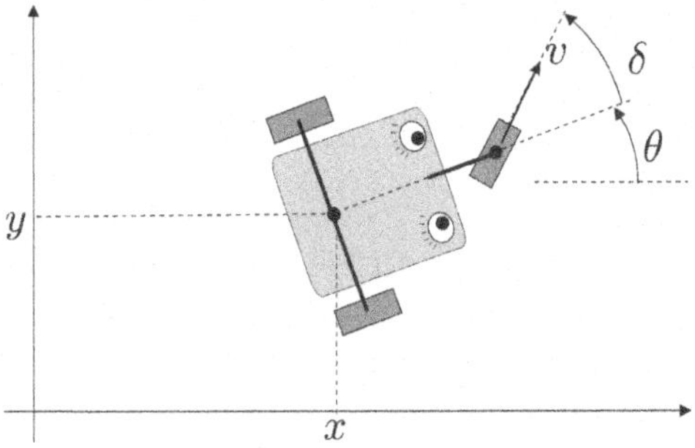

Figure 7.14. *Dead reckoning for a tricycle robot*

The robot is equipped with a compass that returns θ with high precision and an angle sensor that returns the angle δ of the front wheel.

1) Discretize this system with an Euler method. Simulate this system in MATLAB for an arbitrary input $\mathbf{u}(t)$ and initial vector. For the variance of the discretized noises $\alpha_\theta, \alpha_v, \alpha_\delta$ we will take $0.01 \cdot dt$, where dt is the discretization step.

2) Express this localization problem in a linear and discretized form.

3) Using a Kalman filter, predict the position of the robot as well as the associated covariance matrix.

4) How does the localization program change if we assume that, using odometers, the robot is capable of measuring its speed v with a variance of 0.01?

EXERCISE 7.19.– Goniometric localization

Let us consider once again a robot vehicle described by the state equations:

$$\begin{pmatrix} \dot{x}_1 \\ \dot{x}_2 \\ \dot{x}_3 \\ \dot{x}_4 \\ \dot{x}_5 \end{pmatrix} = \begin{pmatrix} x_4 \cos x_5 \cos x_3 \\ x_4 \cos x_5 \sin x_3 \\ \frac{x_4 \sin x_5}{3} \\ u_1 \\ u_2 \end{pmatrix}$$

The vector (x_1, x_2) represents the coordinates of the center of the robot, x_3 is the heading of the robot, x_4 is its speed and x_5 is the angle of its front wheels. The robot is surrounded by point landmarks $\mathbf{m}(1), \mathbf{m}(2), \ldots$ whose positions are known. The robot can only detect these landmarks $\mathbf{m}(i)$ if the distance to them is sufficiently small (smaller than 15 m). In such a case, the robot measures the angle δ_i with high precision. We will also assume that the robot knows the angles x_3 and x_5 at all times, without any error. Finally, it measures its speed x_4 with an error of variance 1. Figure 7.15 shows a situation in which two landmarks $\mathbf{m}(1)$ and $\mathbf{m}(2)$ are detected by the robot.

In order for the robot to locate itself, we would like to use a Kalman filter. For this, we need linear equations, which we do not have here. Since x_3 and x_5 are known, the nonlinearity can be based on a temporal dependency. Let us take for this $\mathbf{z} = (x_1, x_2, x_4)$.

1) Show that $\mathbf{z}$ satisfies a linear state evolution equation. Find the associated observation equation.

2) Find a discretization for the evolution of $\mathbf{z}$ in order to feed a Kalman filter.

3) Implement a simulator in MATLAB with the robot surrounded by the following four landmarks:

$$\mathbf{a}(1) = \begin{pmatrix} 0 \\ 25 \end{pmatrix}, \ \mathbf{a}(2) = \begin{pmatrix} 15 \\ 30 \end{pmatrix}, \ \mathbf{a}(3) = \begin{pmatrix} 30 \\ 15 \end{pmatrix}, \ \mathbf{a}(4) = \begin{pmatrix} 15 \\ 20 \end{pmatrix}$$

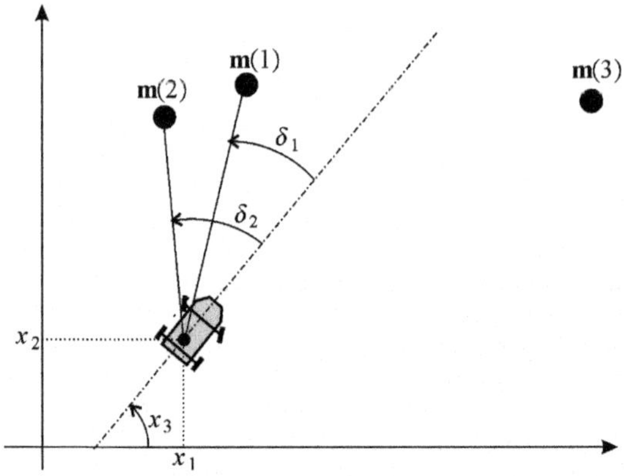

Figure 7.15. *Goniometric localization*

As stated above, the robot can only goniometrically detect landmarks once they are close.

4) Implement a Kalman filter for the localization. The initial state will be assumed unknown.

5) We now have two robots $\mathcal{R}_a$ and $\mathcal{R}_b$ capable of communicating wirelessly while measuring the landmark angles (see Figure 7.16). When the distances are small (i.e. smaller than 20 m), the robots can measure the angles φ_a and φ_b with high precision using cameras (see Figure 7.16). Suggest a centralized Kalman filter for the localization of the two robots.

EXERCISE 7.20.– Following a boat with two radars

The movement of a boat that we are seeking to follow is described by the state equations:

$$\begin{cases} p_x(k+1) = p_x(k) + dt \cdot v_x(k) \\ v_x(k+1) = v_x(k) - dt \cdot v_x(k) + \alpha_x(k) \\ p_y(k+1) = p_y(k) + dt \cdot v_y(k) \\ v_y(k+1) = v_y(k) - dt \cdot v_y(k) + \alpha_y(k) \end{cases}$$

where $dt = 0.01$ and α_x and α_y are the Gaussian white noises with variance matrix dt. The state vector is, therefore, $\mathbf{x} = (p_x, v_x, p_y, v_y)$.

1) Write a MATLAB program that simulates the boat.

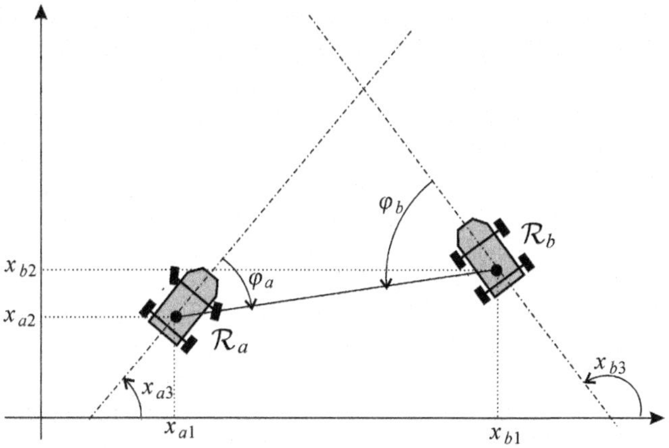

Figure 7.16. *Goniometric localization for two communicating robots*

2) Two radars placed at $\mathbf{a} : (a_x, a_y) = (0, 0)$ and $\mathbf{b} : (b_x, b_y) = (1, 0)$ measure the square of the distance to the boat. The observation equation is:

$$\mathbf{y}_k = \underbrace{\begin{pmatrix} (p_x(k) - a_x)^2 + (p_y(k) - a_y)^2 \\ (p_x(k) - b_x)^2 + (p_y(k) - b_y)^2 \end{pmatrix}}_{\mathbf{g}(\mathbf{x}_k)} + \beta_k$$

where $\beta_1(k)$ and $\beta_2(k)$ are the independent unit Gaussian white noises. Adjust the simulation in order to visualize the radars and generate the measurement vector $\mathbf{y}(k)$.

3) Linearize this observation equation around the current estimation $\hat{\mathbf{x}}_k$ of the state vector $\mathbf{x}_k$. Deduce from this an equation of the form $\mathbf{z}_k = \mathbf{C}_k \mathbf{x}_k$ where $\mathbf{z}_k = \mathbf{h}(\mathbf{y}_k, \hat{\mathbf{x}}_k)$ takes the role of the measurement taken at time k.

4) Implement a Kalman filter that allows the localization of the boat.

EXERCISE 7.21.– Robot localization in a pool

Consider an underwater robot moving within a rectangular pool of length $2R_y$ and width $2R_y$. A sonar placed just above the robot rotates with a constant angular speed. The depth z is easily obtained using a pressure sensor and therefore we will assume this quantity to be known. The robot is weighted in such a way that the bank and elevation angles may be assumed zero. In our context, localizing the robot means estimating the coordinates (x, y) of the robot. The origin of the coordinate system will be middle of the pool. For this localization, we will assume that the angle α of the sonar is measured relative to the body of the robot, the heading angle θ is measured with a compass and the tangential a_T and normal a_N accelerations with accelerometers. Every 0.1 s, the sonar returns the length ℓ of the sonar beam. Figure 7.17 represents the length $\ell(t)$ of the sonar beam, obtained by simulation when the sonar performs seven rotations around itself while the robot is moving.

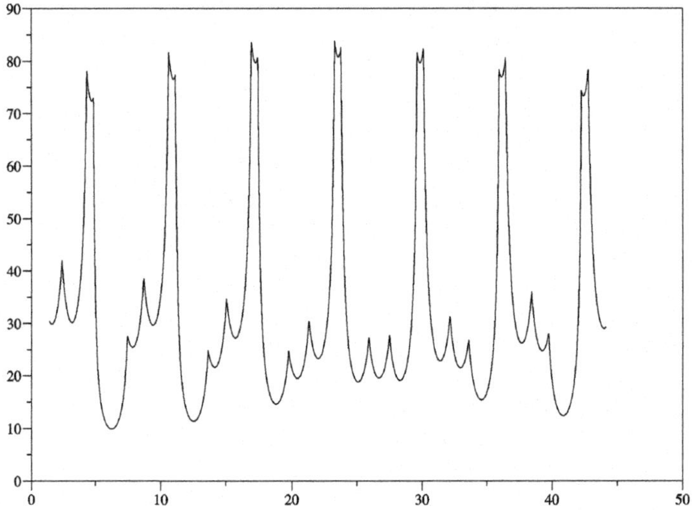

Figure 7.17. *Telemetric measurements collected by the robot*

1) Given the signal collected by the sonar, suggest a robust method for detecting local minima that correspond to the situation where the sonar is perpendicularly pointing toward one of the four walls of the pool.

2) Let $v_x = \dot{x}$, $v_y = \dot{y}$. Show that we can write the following state equations to connect the measurements:

$$\begin{cases} \dot{x} &= v_x \\ \dot{y} &= v_y \\ \dot{v}_x &= a_T \cos\theta - a_N \sin\theta \\ \dot{v}_y &= a_T \sin\theta + a_N \cos\theta \end{cases}$$

3) Let us now assume that we have collected (either by simulation or real experience) for each $t \in [0, t_{\max}]$, data relative to the accelerations (a_T, a_N),

to the heading angle θ and to the angle of the sonar α. Propose a recursive method based on the Kalman filter in order to localize the robot.

EXERCISE 7.22.– Simultaneous localization and mapping (SLAM)

The *Redermor* (underwater robot built by Groupe d'Etude Sous-Marine de l'Atlantique (GESMA), Brest) performed a 2-h mission in the Douarnenez bay (see Figure 7.18). During its mission, it collected data from its inertial unit (which gives us the Euler angles ϕ, θ, ψ), its Doppler log (which gives us the robot's speed $\mathbf{v}_r$ in the robot's coordinate system), its pressure sensor (which gives us the robot's depth p_z) and its altitude sensor (sonar that gives us the altitude a), with a sampling period of $dt = 0.1$ sec. These data can be found in the file `slam_data.txt`. The file is composed of 59,996 lines (one line per sampling period) and nine columns which are, respectively:

$$(t, \varphi, \theta, \psi, v_x, v_y, v_z, p_z, a)$$

where p_z is the depth of the robot and a is its altitude (in other words, its distance to the seabed).

Figure 7.18. *The* Redermor, *built by Groupe d'Etude Sous-Marine de l'Atlantique (GESMA), right before diving into the water*

1) Given that the robot started from a position $\mathbf{p} = (0, 0, 0)$, at time $t = 0$, and using an Euler method, deduce an estimation for the path. Use for this, the state equation:

$$\dot{\mathbf{p}}(t) = \mathbf{R}(\varphi(t), \theta(t), \psi(t)) \cdot \mathbf{v}_r(t)$$

with $\mathbf{R}(\varphi, \theta, \psi)$ the Euler matrix (see [1.7] in section 1.2.1) whose expression we recall below:

$$\mathbf{R}(\varphi, \theta, \psi) = \begin{pmatrix} \cos\psi & -\sin\psi & 0 \\ \sin\psi & \cos\psi & 0 \\ 0 & 0 & 1 \end{pmatrix} \begin{pmatrix} \cos\theta & 0 & \sin\theta \\ 0 & 1 & 0 \\ -\sin\theta & 0 & \cos\theta \end{pmatrix} \begin{pmatrix} 1 & 0 & 0 \\ 0 & \cos\varphi & -\sin\varphi \\ 0 & \sin\varphi & \cos\varphi \end{pmatrix}$$

2) The angles ψ, θ, φ are measured with a standard deviation of $(2 \times 10^{-4}, 2 \times 10^{-4}, 5 \times 10^{-3})$. The components of $\mathbf{v}_r$ are measured with a standard deviation of $\sigma_v = 1\ \text{ms}^{-1}$. We may assume that the robot satisfies the equation:

$$\mathbf{p}_{k+1} = \mathbf{p}_k + (dt \cdot \mathbf{R}(k)) \cdot \bar{\mathbf{v}}_r(k) + \alpha_k$$

where α_k is a white noise and $\bar{\mathbf{v}}_r(k)$ is a measurement of the average speed over the corresponding sampling period. Show that a realistic covariance matrix for α_k is:

$$\mathbf{\Gamma}_\alpha = dt^2 \sigma_v^2 \cdot \begin{pmatrix} 1 & 0 & 0 \\ 0 & 1 & 0 \\ 0 & 0 & 1 \end{pmatrix}$$

3) Using the Kalman filter as predictor, calculate the precision with which the robot knows its position at each moment $t = k \cdot dt$. Give, as a function of t, the standard deviation of the error over the position. What will this become after 1 h? After 2 h? Verify your calculations experimentally in MATLAB by implementing a Kalman predictor.

4) During its mission, the robot may detect several landmarks with its lateral sonar (here, these will be mines). When the robot detects a landmark, it will be on its right side and in a plane perpendicular to the robot. The seabed is assumed to be flat and horizontal. The following table shows the detection times, the numbers i of the landmarks and the distance r_i between the robot and landmark:

t	1 054	1 092	1 374	1 748	3 038	3 688	4 024	4 817	5 172	5 232	5 279	5 688
i	1	2	1	0	1	5	4	3	3	4	5	1
$r_i(t)$	52.42	12.47	54.40	52.68	27.73	26.98	37.90	36.71	37.37	31.03	33.51	15.05

SLAM seeks to use these repeated detections to improve the precision of the estimation of its path. For this, we form a large state vector $\mathbf{x}$, of dimension $3 + 2 \cdot 6 = 15$ that contains the position of the robot $\mathbf{p}$ as well as the vector $\mathbf{q}$ of dimension 12 containing the coordinates (as x and y) of the six landmarks. Let us note that since the landmarks are immobile, we have $\dot{\mathbf{q}} = \mathbf{0}$. Give the MATLAB function `[y,C,Gbeta]=g(k)` that corresponds to the observation. This function returns the measurement vector $\mathbf{y}$, matrix $\mathbf{C}(k)$ and covariance matrix of the measurement noise. As for the standard deviation of the measurement noise β_k, we will take 0.1 for that of the depth and 1 for that of the robot-landmark distance.

5) Using a Kalman filter, find the position of the landmarks together with the associated uncertainty. Show how the robot was able to readjust its position.

6) Use the Kalman smoother to improve the precision over the landmark positions by taking into account the past as well as the future.

EXERCISE 7.23.– *A priori* SLAM

An underwater robot carries a navigation system (inertial unit and Doppler log) that gives its position as x, y with a drift of 100 m per hour. This means that if the robot knows its position with a precision of r meters at time t, then an hour before and an hour later it knows its position with an error smaller than $r + 100$ m.

In the beginning, our robot locates itself by GPS with a precision of 10 m and proceeds to dive in a zone of flat seabed. It swims around for 8 h at constant depth (that it can measure with a pressure sensor). When it resurfaces, it locates itself again using the GPS, again with a precision of 10 m. Each hour, the robot passes above a remarkable landmark (for instance, small rock with a particular form) that can be detected by a camera placed under the robot. The following table indicates the number of the detected landmarks as a function of time, expressed in hours:

t(hour)	1	2	3	4	5	6	7
landmark	1	2	1	3	2	1	4

We may deduce from this table that the robot encounters four remarkable landmarks in total and that it encounters landmark 1 three times, at times $t = 1H, t = 3H$ and $t = 6H$. With what precision is it able to localize landmarks $1, 2, 3$ and 4?

7.7. Corrections

CORRECTION FOR EXERCISE 7.1.– (Gaussian distribution)

1) In order to draw the graph of $\pi_{\mathbf{x}}$, we write the following lines of code:

```
Mx1 = -5:0.1:5; Mx2 = -5:0.1:5; [X1,X2] =
meshgrid(Mx1,Mx2);
xbar=[1;2]; Gx=[1 0;0 1]; invGx=inv(Gx)
dX1=X1-xbar(1); dX2=X2-xbar(2);
Q=invGx(1,1)*(dX1.^2)+2*(invGx(1,2)*dX1.*dX2)
+(invGx(2,2)*dX2.^2)
Z=(1/(2*pi*sqrt(det(Gx))*exp(-(1/2)*Q);
contour3(X1,X2,Z,20,'black');
surface(X1,X2,Z);
```

We then obtain the results shown in Figure 7.19 (the script can be found in gauss.m).

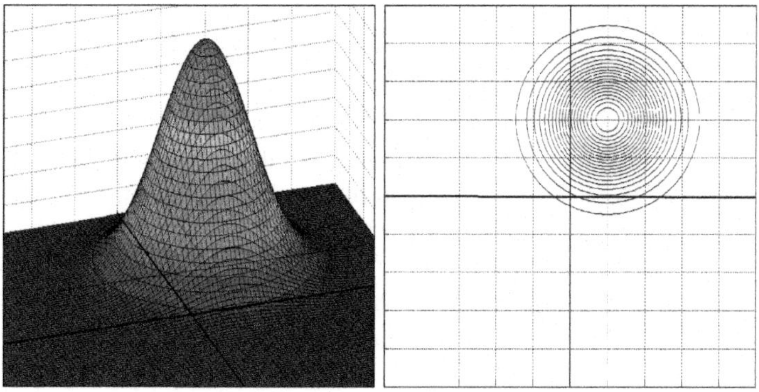

Figure 7.19. *Probability density of a non-centered normed Gaussian random vector*

2) We obtain the characteristics of the law $\pi_{\mathbf{y}}$ using the following relations:

$$\begin{aligned} \bar{\mathbf{y}} &= \mathbf{A}\bar{\mathbf{x}} + \mathbf{b} \\ \mathbf{\Gamma}_{\mathbf{y}} &= \mathbf{A} \cdot \mathbf{\Gamma}_{\mathbf{x}} \cdot \mathbf{A}^{\mathrm{T}} \end{aligned}$$

This gives in MATLAB:

```
A=[cos(pi/6) -sin(pi/6);sin(pi/6) cos(pi/6)]*[1 0;0,3];
Gy=A*Gx*A'; ybar=A*xbar+[2;-5];
```

Figure 7.20 shows the obtained result.

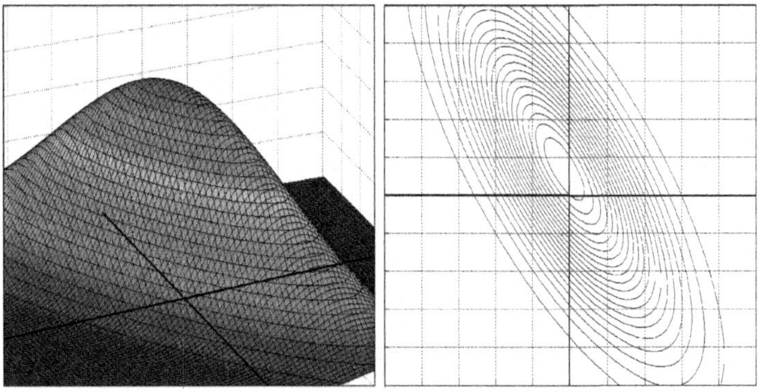

Figure 7.20. *Density of the random vector* $\mathbf{y}$ *after transformation of* $\mathbf{x}$ *by a linear application*

Note that, in accordance with the linear relation that connects $\mathbf{x}$ and $\mathbf{y}$, $\pi_{\mathbf{y}}$ can be obtained from $\pi_{\mathbf{x}}$ by an expansion of 2 with respect to x_1 followed by a rotation of $\frac{\pi}{6}$ and finally translation. The Gaussian character is preserved.

CORRECTION FOR EXERCISE 7.2.– (Confidence ellipses)

The matrices `G1`,. . . ,`G6`, are drawn in Figure 7.21. Given the instructions `G1=eye(2,2)` and `G2=3*eye(2,2)`, the ellipses associated with `G1` and `G2` are circles and `G2` is obtained from `G1` by a homothety with ratio $\sqrt{3}$. Since `G3=A1*G2*A1'+G1` with `A1=[1 0;0 3]`, we can obtain `G3` by an expansion with respect to x_2 with ratio 3 followed by an inflation of the obtained ellipse, given the addition of `G1`. To obtain `G4`, we subject `G3` to a rotation of angle $\frac{\pi}{4}$ (since `G4=A2*G3*A2'`). As `G5=G4+G3`, `G5` encloses ellipses `G4` and `G3`. Once again, `G6` is obtained by a rotation of angle $\frac{\pi}{4}$ of `G5`. The associated program can be found in `sixcov.m`.

CORRECTION FOR EXERCISE 7.3.– (Confidence ellipse: prediction)

1)
```
n=1000; Gx=[3,1;1,3]; xbar=[2;3]; b=randn(2,n);
x=xbar*ones(1,n)+sqrtm(Gx)*b; plot(x(1,:),x(2,:),'.');
```

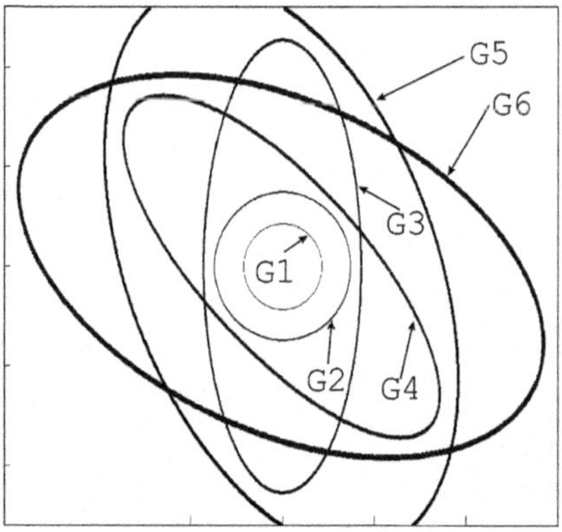

Figure 7.21. *Confidence ellipses, from the thinnest to the thickest*

2) The following functions have to be called:

```
draw_ellipse(xbar,Gx,0.9); draw_ellipse(xbar,Gx,0.99);
draw_ellipse(xbar,Gx,0.999);
```

3)

```
xhat=[mean(x(1,:));mean(x(2,:))];
xtilde=x-xbar*ones(1,n);
g11=mean(xtilde(1,:).*xtilde(1,:));
g12=mean(xtilde(1,:).*xtilde(2,:));
g22=mean(xtilde(2,:).*xtilde(2,:));
Ghat=[g11,g12;g12,g22];
```

4) The simulation program in question is:

```
dt=0.01; A=[0 1;-1 0]; B=[2;3];
for t=0:dt:5
Ad=(eye(2,2)+dt*A); ud=dt*sin(t)*B
x=Ad*x+ud*ones(1,n);
end
```

5) Given Euler's formula around $t = k\delta$, we have the approximation:

$$\begin{aligned}\mathbf{x}(k+1) &= \mathbf{x}(k) + \delta \cdot \left(\begin{pmatrix} 0 & 1 \\ -1 & 0 \end{pmatrix} \mathbf{x}(k) + \begin{pmatrix} 2 \\ 3 \end{pmatrix} u(t) \right) \\ &= \mathbf{A}\mathbf{x}(k) + \mathbf{B}u(k)\end{aligned}$$

with:

$$\mathbf{A} = \begin{pmatrix} 1 & \delta \\ -\delta & 1 \end{pmatrix} \text{ and } \mathbf{B} = \begin{pmatrix} 2\delta \\ 3\delta \end{pmatrix}$$

Therefore, the matrix of the ellipse is calculated as follows:

$$\mathbf{\Gamma}_{k+1} = \mathbf{A}\mathbf{\Gamma}_k \mathbf{A}^{\mathrm{T}}$$

and its center $\hat{\mathbf{x}}$ by:

$$\hat{\mathbf{x}}(k+1) = \mathbf{A}\hat{\mathbf{x}}(k) + \mathbf{B}u(k)$$

Therefore, we need to add the following instructions to the program loop: `xbar=Ad*xbar+ud` and `Gx=Ad*Gx*Ad'`. Of course, these instructions require a negligible amount of time compared to what we would need to do if were manipulating the particle cloud (see question 4). In a linear Gaussian context, therefore, we will directly manipulate the covariance matrices and the averages, which will lead us to the Kalman filter. In a nonlinear context, manipulating the particle cloud will often be preferred. The entire program associated with this exercise can be found in `predicov.m`.

CORRECTION FOR EXERCISE 7.4.– (Confidence ellipse: correction)

1) `n=1000; Gx=[3,1;1,3]; xbar=[1;2]; b=randn(2,n); x=xbar*ones(1,n)+sqrtm(Gx)*b; plot(x(1,:),x(2,:),'.');`

To generate this same cloud more quickly, we could have also written `x=mvnrnd(xbar,Gx,n)'`.

2) We have $\hat{x}_1 = \bar{x}_1 + K \cdot (x_2 - \bar{x}_2)$ with $K = \Gamma_{x_1 x_2}\Gamma_{x_2}^{-1} = \frac{1}{3}$. In other words:

$$\hat{x}_1 = 1 + \frac{1}{3} \cdot (x_2 - 2)$$

3) We have $\hat{x}_2 = \bar{x}_2 + K \cdot (x_1 - \bar{x}_1)$ with $K = \Gamma_{x_1 x_2}\Gamma_{x_1}^{-1} = \frac{1}{3}$. In other words:

$$\hat{x}_2 = 2 + \frac{1}{3} \cdot (x_1 - 1)$$

To correctly understand the non-invertibility of estimators, we represent them graphically as shown in Figure 7.22. We then invoke the estimator in one direction and then in the other direction. We do not return to the same point. The program associated with this exercise is given in `corrcov.m`.

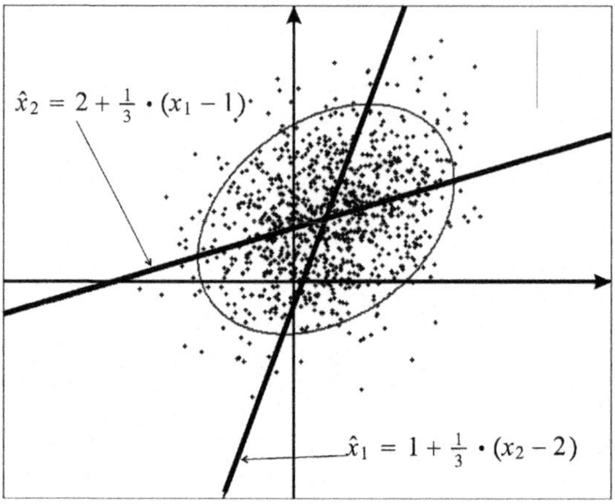

Figure 7.22. *The two lines in bold represent the two estimators $\hat{x}_1(x_2)$ and $\hat{x}_2(x_1)$*

CORRECTION FOR EXERCISE 7.5.– (Covariance matrix propagation)

1) We have:

$$\bar{\mathbf{x}} = E(\mathbf{x}) = E(\mathbf{A}\,\mathbf{a} - \mathbf{b}) = \mathbf{A}E(\mathbf{a}) - E(\mathbf{b}) = \mathbf{0}$$

since the random variables $\mathbf{a}$ and $\mathbf{b}$ are centered. Similarly:

$$\bar{y} = E(\mathbf{y}) = E(\mathbf{C}\,\mathbf{x} + \mathbf{c}) = \mathbf{C}E(\mathbf{x}) + E(\mathbf{c}) = \mathbf{0}$$

since $\mathbf{c}$ is centered. For the covariance matrices, we have:

$$\begin{aligned}
\mathbf{\Gamma}_{\mathbf{x}} &= E\left((\mathbf{x} - \bar{\mathbf{x}})\,(\mathbf{x} - \bar{\mathbf{x}})^{\mathrm{T}}\right) = E\left(\mathbf{x}\,\mathbf{x}^{\mathrm{T}}\right) \\
&= E\left((\mathbf{A}\,\mathbf{a} - \mathbf{b})\,(\mathbf{A}\,\mathbf{a} - \mathbf{b})^{\mathrm{T}}\right) \\
&= E\left(\mathbf{A}\,\mathbf{a}\,\mathbf{a}^{\mathrm{T}}\mathbf{A}^{\mathrm{T}} - \mathbf{A}\,\mathbf{a}\,\mathbf{b}^{\mathrm{T}} - \mathbf{b}\,\mathbf{a}^{\mathrm{T}}\mathbf{A}^{\mathrm{T}} + \mathbf{b}\,\mathbf{b}^{\mathrm{T}}\right) \\
&= \mathbf{A}\underbrace{E\left(\mathbf{a}\,\mathbf{a}^{\mathrm{T}}\right)}_{=\mathbf{I}}\mathbf{A}^{\mathrm{T}} - \mathbf{A}\underbrace{E\left(\mathbf{a}\,\mathbf{b}^{\mathrm{T}}\right)}_{=\mathbf{0}} - \underbrace{E\left(\mathbf{b}\,\mathbf{a}^{\mathrm{T}}\right)}_{=\mathbf{0}}\mathbf{A}^{\mathrm{T}} + \underbrace{E\left(\mathbf{b}\,\mathbf{b}^{\mathrm{T}}\right)}_{=\mathbf{I}} \\
&= \mathbf{A}\cdot\mathbf{A}^{\mathrm{T}} + \mathbf{I}
\end{aligned}$$

Similarly:

$$\mathbf{\Gamma_y} = \mathbf{C}\,\mathbf{\Gamma_x}\,\mathbf{C}^{\mathrm{T}} + \mathbf{\Gamma_c} = \mathbf{C}\,\left(\mathbf{A}\,\mathbf{A}^{\mathrm{T}} + \mathbf{I}\right)\,\mathbf{C}^{\mathrm{T}} + \mathbf{I}$$

2) We have of course $\bar{\mathbf{v}} = (\bar{\mathbf{x}}, \bar{\mathbf{y}}) = \mathbf{0}$. This gives:

$$\mathbf{\Gamma_v} = E\left(\mathbf{v}\,\mathbf{v}^{\mathrm{T}}\right) = \begin{pmatrix} E\left(\mathbf{x}\,\mathbf{x}^{\mathrm{T}}\right) & E\left(\mathbf{x}\,\mathbf{y}^{\mathrm{T}}\right) \\ E\left(\mathbf{y}\,\mathbf{x}^{\mathrm{T}}\right) & E\left(\mathbf{y}\,\mathbf{y}^{\mathrm{T}}\right) \end{pmatrix}$$

However:

$$\begin{aligned} \mathbf{\Gamma_{xy}} &= E\left(\mathbf{x}\,\mathbf{y}^{\mathrm{T}}\right) = E\left(\mathbf{x}\,\left(\mathbf{C}\,\mathbf{x} + \mathbf{c}\right)^{\mathrm{T}}\right) = E\left(\mathbf{x}\mathbf{x}^{\mathrm{T}}\mathbf{C}^{\mathrm{T}} + \mathbf{x}\mathbf{c}^{\mathrm{T}}\right) \\ &= \mathbf{\Gamma_x}\mathbf{C}^{\mathrm{T}} + \mathbf{\Gamma_{xc}} = \mathbf{\Gamma_x}\mathbf{C}^{\mathrm{T}} \end{aligned}$$

since $\mathbf{x}$ and $\mathbf{c}$ are independent (and thus $\mathbf{\Gamma_{xc}} = \mathbf{0}$). It follows that:

$$\mathbf{\Gamma_v} = \begin{pmatrix} \mathbf{\Gamma_x} & \mathbf{\Gamma_x}\mathbf{C}^{\mathrm{T}} \\ \mathbf{C}\,\mathbf{\Gamma_x} & \mathbf{\Gamma_y} \end{pmatrix}$$

3) We have:

$$\mathbf{z} = \left(-\mathbf{I}\ \ \mathbf{I}\right) \cdot \mathbf{v}$$

Therefore:

$$\begin{aligned} \mathbf{\Gamma_z} &= \left(-\mathbf{I}\ \ \mathbf{I}\right)\,\mathbf{\Gamma_v}\,\begin{pmatrix} -\mathbf{I} \\ \mathbf{I} \end{pmatrix} = \left(-\mathbf{I}\ \ \mathbf{I}\right)\begin{pmatrix} \mathbf{\Gamma_x} & \mathbf{\Gamma_x}\mathbf{C}^{\mathrm{T}} \\ \mathbf{C}\,\mathbf{\Gamma_x} & \mathbf{\Gamma_y} \end{pmatrix}\begin{pmatrix} -\mathbf{I} \\ \mathbf{I} \end{pmatrix} \\ &= \left(-\mathbf{I}\ \ \mathbf{I}\right)\begin{pmatrix} -\mathbf{\Gamma_x} + \mathbf{\Gamma_x}\mathbf{C}^{\mathrm{T}} \\ -\mathbf{C}\,\mathbf{\Gamma_x} + \mathbf{\Gamma_y} \end{pmatrix} = \mathbf{\Gamma_x} - \mathbf{\Gamma_x}\mathbf{C}^{\mathrm{T}} - \mathbf{C}\,\mathbf{\Gamma_x} + \mathbf{\Gamma_y} \end{aligned}$$

4) Following the expression of the linear estimator, we have $\hat{\mathbf{x}} = \bar{\mathbf{x}} + \mathbf{K}\widetilde{\mathbf{y}}$ with $\mathbf{K} = \mathbf{\Gamma_{xy}}\mathbf{\Gamma_y}^{-1}.= \mathbf{\Gamma_x}\mathbf{C}^{\mathrm{T}}\mathbf{\Gamma_y}^{-1}$ and $\widetilde{\mathbf{y}} = \mathbf{y} - \mathbf{C}\bar{\mathbf{x}}$. Therefore:

$$\hat{\mathbf{x}} = \bar{\mathbf{x}} + \left(\mathbf{\Gamma_x}\mathbf{C}^{\mathrm{T}}\mathbf{\Gamma_y}^{-1}\right)\left(\mathbf{y} - \mathbf{C}\bar{\mathbf{x}}\right) = \mathbf{\Gamma_x}\mathbf{C}^{\mathrm{T}}\mathbf{\Gamma_y}^{-1} \cdot \mathbf{y}$$

since $\bar{\mathbf{x}} = \mathbf{0}$.

CORRECTION FOR EXERCISE 7.6.– (Brownian noise)

1) Let us first recall that if a and b are two independent centered random variables with variances σ_a^2 and σ_b^2, and if α is a deterministic real number then:

$$\begin{aligned}\sigma_{a+b}^2 &= E\left((a+b)^2\right) = E\left(a^2+b^2+2ab\right) = \sigma_a^2+\sigma_b^2+2E\left(ab\right)\\ &= \sigma_a^2+\sigma_b^2\\ \sigma_{\alpha a}^2 &= E\left((\alpha a)^2\right) = \alpha^2\sigma_a^2\end{aligned}$$

It follows that:

$$\sigma_y^2\left(t_k\right) = \delta^2 \cdot \sum_{j=0}^{k} \sigma_x^2\left(t_j\right) = \delta^2 \cdot k\sigma_x^2$$

However, $t_k = k\delta$ and therefore $\sigma_y^2\left(t_k\right) = \delta t_k \sigma_x^2$. Thus, we can see that the standard deviation $\sigma_y\left(t_k\right) = \sqrt{\delta}\sqrt{t_k}\sigma_x$ tends toward zero when $\delta \to 0$ and that this error evolves with $\sqrt{t}$ (due to the random walk). This phenomenon comes from the fact that the errors compensate for each other, even more so as δ is small. In order to illustrate this phenomenon in MATLAB, we write a simulation function as follows:

```
function [T,X,Y]=Simu(delta,sig_x,tmax)
T=0:delta:tmax;
kmax=length(T);
X=sig_x*randn(1,kmax); Y=0*X;
for k=1:kmax-1, Y(k+1)=Y(k)+delta*X(k); end
end
```

We then apply this function for $\delta = 0.1$, $\delta = 0.01$ and $\delta = 0.001$. The obtained result is represented in Figure 7.23. Each figure corresponds to $t \in [0, 100]$ and $y \in [-7, 7]$. Note that when the sampling period δ decreases, the Brownian noise becomes smaller due to the compensation effect. The figures were generated using the program below:

```
for i=1:100,
delta=0.1; % 0.01, 0.001
[T,X,Y]=Simu(delta,1,100);
plot(T,Y);
end
```

2) We have:

$$\sigma_x(\delta) = \frac{1}{\sqrt{\delta}} \cdot \frac{\sigma_y(t_k)}{\sqrt{t_k}}$$

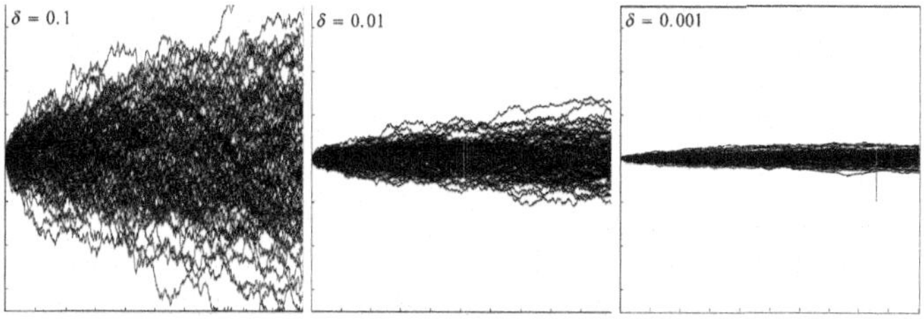

Figure 7.23. *When δ approaches 0, the Brownian noise (integral of the white noise) becomes smaller*

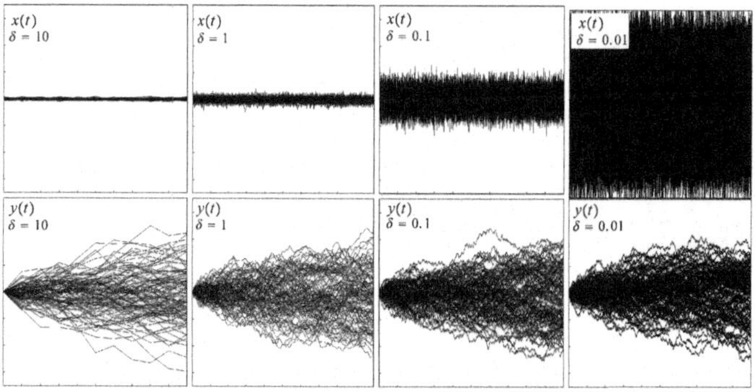

Figure 7.24. *White noise with its corresponding Brownian noise for different values of δ*

In order to maintain $\sigma_y(t_k)$ independent of δ, the standard deviation σ_x for $x(t_k)$ must, therefore, increase to infinity in order to decrease the compensation effect. When we reach the limit, in other words $\delta = 0$, the signal $x(t_k)$ has an infinite standard deviation. It is for this reason that a continuous-time centered white random signal (in other words, $\delta = 0$) has to have an infinite power in order to have influence over an integrator or, more generally, over any physical system. We can observe this phenomenon in physics when we study the Brownian motion of particles moving within a limited space with an infinite speed. In such a case, $y(t)$ corresponds to the movement of the particle and $x(t)$ corresponds to its speed. Figure 7.24 gives the white noise for different values of δ, together with their integrals. The scales are identical to those chosen for Figure 7.23. We can see that when δ tends toward 0, the power of the white noise $x(t)$ must be increased in order to obtain a similar Brownian noise. The figures were generated with the following program, which can also be found in `brownien.m`:

```
for i=1:100,
delta=10; % or 1, 0.1, 0.01
[T,X,Y]=Simu(delta,0.2/sqrt(delta),100);
plot(T,X); plot(T,Y);
end
```

CORRECTION FOR EXERCISE 7.7.– (Solving three equations using a linear estimator)

We translate the problem as follows:

$$\underbrace{\begin{pmatrix} 8 \\ 7 \\ 0 \end{pmatrix}}_{\mathbf{y}} = \underbrace{\begin{pmatrix} 2 & 3 \\ 3 & 2 \\ 1 & -1 \end{pmatrix}}_{\mathbf{C}} \underbrace{\begin{pmatrix} x_1 \\ x_2 \end{pmatrix}}_{\mathbf{x}} + \underbrace{\begin{pmatrix} \beta_1 \\ \beta_2 \\ \beta_3 \end{pmatrix}}_{\beta}$$

We take:

$$\bar{\mathbf{x}} = \begin{pmatrix} 0 \\ 0 \end{pmatrix}, \mathbf{\Gamma_x} = \begin{pmatrix} 1000 & 0 \\ 0 & 1000 \end{pmatrix}, \ \mathbf{\Gamma}_\beta = \begin{pmatrix} 1\,0\,0 \\ 0\,4\,0 \\ 0\,0\,4 \end{pmatrix}$$

which is equivalent to saying that *a priori* the vector $\mathbf{x}$ is more or less within the interval $[-33, 33]$, that the β_i are independent of each other and that the first equation is twice as accurate. We obtain:

$$\begin{aligned}
\tilde{\mathbf{y}} &= \mathbf{y} - \mathbf{C}\bar{\mathbf{x}} = \begin{pmatrix} 8 \\ 7 \\ 0 \end{pmatrix} \\
\mathbf{\Gamma_y} = \mathbf{C}\mathbf{\Gamma_x}\mathbf{C}^\mathrm{T} + \mathbf{\Gamma}_\beta &= \begin{pmatrix} 13001 & 12000 & -1000 \\ 12\,000 & 13\,004 & 1000 \\ -1000 & 1000 & 2004 \end{pmatrix} \\
\mathbf{K} &= \mathbf{\Gamma_x}\mathbf{C}^\mathrm{T}\mathbf{\Gamma_y}^{-1} = \begin{pmatrix} -0.09 & 0.288 & 0.311 \\ 0.355 & -0.155 & -0.24 \end{pmatrix}
\end{aligned}$$

$$\hat{\mathbf{x}} \ = \bar{\mathbf{x}} + \mathbf{K}\widetilde{\mathbf{y}} = \begin{pmatrix} 1.311 \\ 1.756 \end{pmatrix}$$

$$\mathbf{\Gamma}_\varepsilon = \mathbf{\Gamma}_\mathbf{x} - \mathbf{K}\mathbf{C}\mathbf{\Gamma}_\mathbf{x} = \begin{pmatrix} 0.722 & -0.517 \\ -0.54 & 0.44 \end{pmatrix}$$

Thus, we can represent the solution of our linear system by $\hat{\mathbf{x}}$ and the matrix $\mathbf{\Gamma}_\varepsilon$.

CORRECTION FOR EXERCISE 7.8.– (Estimating the parameters of an electric motor using a linear estimator)

Let us apply the formulas in [7.10] with:

$$\bar{\mathbf{x}} = \begin{pmatrix} 1 \\ -1 \end{pmatrix}, \mathbf{\Gamma}_\mathbf{x} = \begin{pmatrix} 4\,0 \\ 0\,4 \end{pmatrix}; \mathbf{C} = \begin{pmatrix} 4\ 0 \\ 10\ 1 \\ 10\ 5 \\ 13\ 5 \\ 15\ 3 \end{pmatrix}, \mathbf{\Gamma}_\beta = \begin{pmatrix} 9\,0\,0\,0\,0 \\ 0\,9\,0\,0\,0 \\ 0\,0\,9\,0\,0 \\ 0\,0\,0\,9\,0 \\ 0\,0\,0\,0\,9 \end{pmatrix}$$

$$\text{and } \mathbf{y} = \begin{pmatrix} 5 \\ 10 \\ 8 \\ 14 \\ 17 \end{pmatrix}$$

The corresponding MATLAB program is the following:

```
y=[5;10;8;14;17];
C=[4 0;10 1;10 5; 13 5;15 3];
xbar=[1;-1];
Gx=4*eye(2,2);
Gbeta=9*eye(5,5);
ytilde=y-C*xbar
Gy=C*Gx*C'+Gbeta;
K=Gx*C'*inv(Gy);
xhat=xbar+K*ytilde
Ge=Gx-K*C*Gx
```

We obtain:

$$\widetilde{\mathbf{y}} = \mathbf{y} - \mathbf{C}\bar{\mathbf{x}} = \begin{pmatrix} 1 \\ 1 \\ 3 \\ 6 \\ 5 \end{pmatrix}$$

$$\mathbf{\Gamma_y} = \mathbf{C}\mathbf{\Gamma_x}\mathbf{C}^{\mathrm{T}} + \mathbf{\Gamma}_\beta = \begin{pmatrix} 73 & 160 & 160 & 208 & 240 \\ 160 & 413 & 420 & 540 & 612 \\ 160 & 420 & 509 & 620 & 660 \\ 208 & 540 & 620 & 785 & 840 \\ 240 & 612 & 660 & 840 & 945 \end{pmatrix}$$

$$\mathbf{K} = \mathbf{\Gamma_x}\mathbf{C}^{\mathrm{T}}\mathbf{\Gamma_y}^{-1} = \begin{pmatrix} 0.027 & 0.0491 & -0.0247 & -0.0044 & 0.046 \\ -0.0739 & -0.118 & 0.148 & 0.092 & -0.077 \end{pmatrix}$$

$$\hat{\mathbf{x}} = \bar{\mathbf{x}} + \mathbf{K}\widetilde{\mathbf{y}} = \begin{pmatrix} 1.2 \\ -0.58 \end{pmatrix}$$

$$\mathbf{\Gamma}_\varepsilon = \mathbf{\Gamma_x} - \mathbf{K}\mathbf{C}\mathbf{\Gamma_x} = \begin{pmatrix} 0.062 & -0.166 \\ -0.179 & 0.593 \end{pmatrix}$$

CORRECTION FOR EXERCISE 7.9.– (Trochoid)

1) We have:

$$\bar{\mathbf{p}} = \begin{pmatrix} 0 \\ 0 \end{pmatrix}, \Gamma_{\mathbf{p}} = \begin{pmatrix} 10^2 & 0 \\ 0 & 10^2 \end{pmatrix}$$

$$\mathbf{y} = \begin{pmatrix} 0.38 \\ 3.25 \\ 4.97 \\ -0.26 \end{pmatrix}, \mathbf{C} = \begin{pmatrix} 1 - \cos(1) \\ 1 - \cos(2) \\ 1 - \cos(3) \\ 1 - \cos(7) \end{pmatrix}, \Gamma_\beta = \begin{pmatrix} 0.01 & 0 & 0 & 0 \\ 0 & 0.01 & 0 & 0 \\ 0 & 0 & 0.01 & 0 \\ 0 & 0 & 0 & 0.01 \end{pmatrix}$$

and we apply the linear estimator. The corresponding MATLAB script is the following:

```
y=[0.38;3.25;4.97;-0.26];t=[1;2;3;7];
C=[ones(size(t)),-cos(t)];
pbar=[0;0];Gp=1000*eye(2,2);Gbeta=0.01*eye(4,4);
ytilde=y-C*pbar; Gy=C*Gp*C'+Gbeta;
```

```
K=Gp*C'*inv(Gy);
phat=pbar+K*ytilde; Ge=Gp-K*C*Gp;
```

We obtain:

$$\hat{\mathrm{p}} = \begin{pmatrix} 2.001 \\ 2.999 \end{pmatrix} \text{ and } \mathbf{\Gamma}_\varepsilon = \begin{pmatrix} 0.0025 & -0.0001 \\ -0.0001 & 0.0050 \end{pmatrix}$$

2) To draw the estimated path of the mass, we write:

```
t1=0:0.01:20; x1=phat(1)*t1-phat(2)*sin(t1);
y1=phat(1)-phat(2)*cos(t1); plot(x1,y1);
```

Figure 7.25 is then obtained.

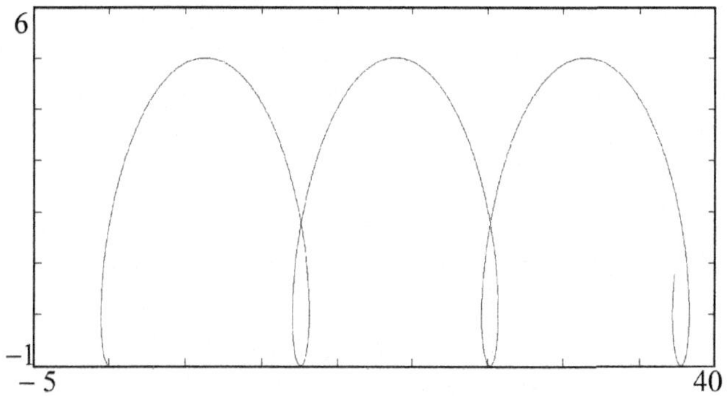

Figure 7.25. *Estimated path for the mass*

CORRECTION FOR EXERCISE 7.10.– (Solving three equations using a Kalman filter)

1) The required MATLAB script is the following:

```
Galpha=zeros(2,2); A=eye(2,2);
C0=[2 3]; C1=[3 2]; C2=[1 -1]; u=0;
xhat0=[0;0]; Gx0=1000*eye(2,2);
[xhat1,Gx1]=kalman(xhat0,Gx0,u,8,Galpha,1,A,C0)
[xhat2,Gx2]=kalman(xhat1,Gx1,u,7,Galpha,4,A,C1)
[xhat3,Gx3]=kalman(xhat2,Gx2,u,0,Galpha,4,A,C2)
```

2) If we draw the three associated covariance matrices, we realize that at each new iteration, the covariance matrix contracts until it is concentrated around a single point, which corresponds to the solution.

3) As expected, the obtained results are identical to those obtained in exercise 7.7.

CORRECTION FOR EXERCISE 7.11.– (Three-step Kalman filter)

Figure 7.26 gives the confidence ellipses obtained by the Kalman filter.

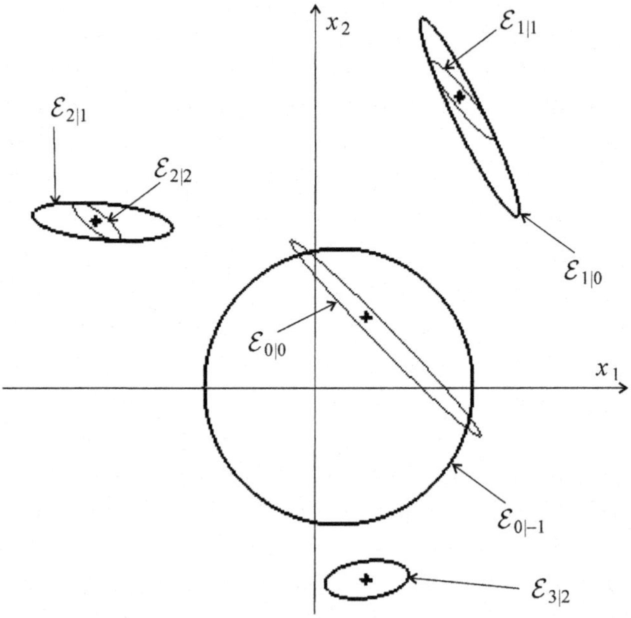

Figure 7.26. *Graphical illustration of the Kalman filter on a simple example*

The corresponding MATLAB program, also available in the file `kalm3steps.m`, is the following:

```
A0=[0.5 0;0 1]; A1=[1 -1;1 1]; A2=[1 -1;1 1];
u0=[8;16]; u1=[-6;-18]; u2=[32;-8];
C0=[1 1];C1=[1 1];C2=[1 1];
```

```
y0=7; y1=30; y2=-6;
Galpha=1*eye(2,2);
Gbeta=1*eye(1,1);
xhat0=[0;0]; Gx0=100*eye(2,2);
[xhat1,Gx1]=kalman(xhat0,Gx0,u0,y0,Galpha,Gbeta,A0,C0);
[xhat2,Gx2]=kalman(xhat1,Gx1,u1,y1,Galpha,Gbeta,A1,C1);
[xhat3,Gx3]=kalman(xhat2,Gx2,u2,y2,Galpha,Gbeta,A2,C2);
```

CORRECTION FOR EXERCISE 7.12.– (Estimating the parameters of an electric motor)

In order to use the Kalman filter, we will take the state equations:

$$\begin{cases} \mathbf{x}_{k+1} = \underbrace{\begin{pmatrix} 1 & 0 \\ 0 & 1 \end{pmatrix}}_{\mathbf{A}_k} \mathbf{x}_k + \mathbf{u}_k + \alpha_k \\ y_k \quad = \underbrace{\begin{pmatrix} U(k) & T_r(k) \end{pmatrix}}_{\mathbf{C}_k} \mathbf{x}_k + \beta_k \end{cases}$$

The associated MATLAB program is:

```
y=[5;10;11;14;17];
C=[4 0;10 1;10 5; 13 5;15 3]
xhat=[1;-1];Gx=4*eye(2,2);
Galpha=zeros(2,2); Gbeta=9;
A=eye(2,2); u=zeros(2,1);
for k=1:5,
[xhat,Gx]=kalman(xhat,Gx,u,y(k),Galpha,Gbeta,A,C(k,:));
end;
```

This program can be found in the file `kalmotor.m`. The obtained estimation is then:

$$\bar{\mathbf{x}} = \begin{pmatrix} 1.13 \\ -0.14 \end{pmatrix} \text{ and } \mathbf{\Gamma}_{\mathbf{x}} = \begin{pmatrix} 0.06 & -0.17 \\ -0.17 & 0.6 \end{pmatrix}$$

CORRECTION FOR EXERCISE 7.13.– (Localization from wall distance measurements)

1) We have:

$$d\left(i\right)=-u_{2}\left(i\right)\cdot x_{1}+u_{1}\left(i\right)\cdot x_{2}+\ u_{2}\left(i\right)\cdot a_{1}\left(i\right)-u_{1}\left(i\right)\cdot a_{2}\left(i\right)+\beta_{i}$$

By taking $y_i = d(i) - \bar{d}(i)$ and $\bar{d}(i) = u_2(i) \cdot a_1(i) - u_1(i) \cdot a_2(i)$, we obtain

$$y_i = \begin{pmatrix} -u_2(i) & u_1(i) \end{pmatrix} \mathbf{x} + \beta_i$$

Thus, we form the quantities:

$$\mathbf{y}=\begin{pmatrix} d(1)-\bar{d}(i) \\ d(2)-\bar{d}(i) \\ d(3)-\bar{d}(i) \end{pmatrix}, \mathbf{C}=\begin{pmatrix} -u_2(1) & u_1(1) \\ -u_2(2) & u_1(2) \\ -u_2(3) & u_1(3) \end{pmatrix},$$

$$\mathbf{\Gamma}_{\beta}=\mathbf{I}_3,\ \mathbf{\Gamma}_{\mathbf{x}}=100\cdot\mathbf{I}_2,\ \bar{\mathbf{x}}=\begin{pmatrix} 1 \\ 2 \end{pmatrix}$$

We then apply the equations of the orthogonal linear estimator:

$$\begin{array}{lll} \hat{\mathbf{x}}=\bar{\mathbf{x}}+\mathbf{K}\cdot\widetilde{\mathbf{y}} & \widetilde{\mathbf{y}}=\mathbf{y}-\mathbf{C}\bar{\mathbf{x}} & \mathbf{\Gamma}_{\mathbf{y}}=\mathbf{C}\mathbf{\Gamma}_{\mathbf{x}}\mathbf{C}^{\mathrm{T}}+\mathbf{\Gamma}_{\beta} \\ \mathbf{\Gamma}_{\varepsilon}=(\mathbf{I}-\mathbf{K}\mathbf{C})\,\mathbf{\Gamma}_{\mathbf{x}} & \mathbf{K}=\mathbf{\Gamma}_{\mathbf{x}}\mathbf{C}^{\mathrm{T}}\mathbf{\Gamma}_{\mathbf{y}}^{-1} & \end{array}$$

2) The requested program is:

```
A=[2 15 3; 1 5 12]; B=[15 3 2; 5 12 1];C=[];dbar=[];
for i=1:3,
u=(B(:,i)-A(:,i))/norm(B(:,i)-A(:,i));
C=[C;[-u(2),u(1)]]; dbar=[dbar;det([u,-A(:,i)])];
end
d=[2;5;4]; y=d-dbar;
x0=[1;2];G0=100*eye(2,2);u=0;Galpha=0*G0;Gbeta=eye(3,3);
[x1,G1]=kalman(x0,G0,u,y,Galpha,Gbeta,eye(2,2),C);
```

The estimation thus calculated is represented in Figure 7.27 together with its confidence ellipse.

CORRECTION FOR EXERCISE 7.14.– (Temperature estimation)

1) The Kalman filter is given by:

$$\begin{cases} \hat{x}_{k+1|k} = \hat{x}_{k|k} \\ \Gamma_{k+1|k} = \Gamma_{k|k} + \Gamma_\alpha \\ \hat{x}_{k|k} = \hat{x}_{k|k-1} + K_k \widetilde{y}_k \\ \Gamma_{k|k} = (1 - K_k)\,\Gamma_{k|k-1} \\ \widetilde{y}_k = y_k - \hat{x}_{k|k-1} \\ S_k = \Gamma_{k|k-1} + \Gamma_\beta \\ K_k = \Gamma_{k|k-1} S_k^{-1} \end{cases}$$

in other words:

$$\begin{cases} \hat{x}_{k+1|k} = \hat{x}_{k|k-1} + \frac{\Gamma_{k|k-1}}{\Gamma_{k|k-1}+\Gamma_\beta}\left(y_k - \hat{x}_{k|k-1}\right) \\ \quad = \hat{x}_{k|k-1} + \frac{\Gamma_{k|k-1}}{\Gamma_{k|k-1}+3}\left(y_k - \hat{x}_{k|k-1}\right) \\ \Gamma_{k+1|k} = \left(1 - \frac{\Gamma_{k|k-1}}{\Gamma_{k|k-1}+\Gamma_\beta}\right)\Gamma_{k|k-1} + \Gamma_\alpha = \left(1 - \frac{\Gamma_{k|k-1}}{\Gamma_{k|k-1}+3}\right)\Gamma_{k|k-1} + 4 \end{cases}$$

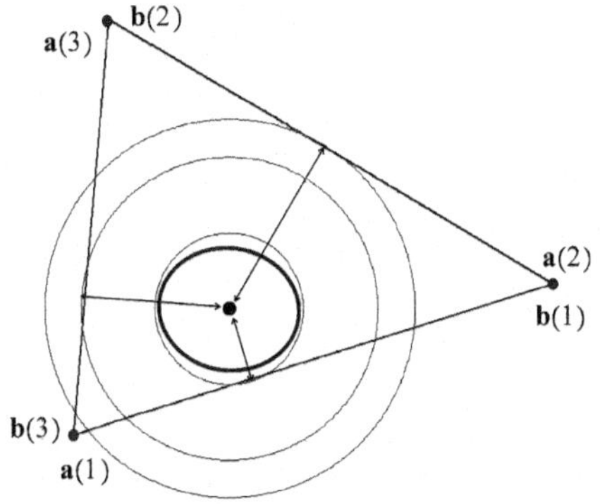

Figure 7.27. *Confidence ellipse (in bold) associated with the localization of the robot*

2) For $k \to \infty$, we have $\Gamma_{k+1|k} - \Gamma_{k|k-1} \to 0$, i.e. $\Gamma_{k+1|k} \to \Gamma_\infty$. Therefore:

$$\Gamma_\infty = \left(1 - \frac{\Gamma_\infty}{\Gamma_\infty + \Gamma_\beta}\right)\Gamma_\infty + \Gamma_\alpha$$

i.e.:

$$\Gamma_\infty^2 - \Gamma_\alpha\Gamma_\infty - \Gamma_\alpha\Gamma_\beta = 0$$

A single positive solution exists. It is given by:

$$\Gamma_\infty = \frac{\Gamma_\alpha + \sqrt{\Gamma_\alpha^2 + 4\Gamma_\alpha\Gamma_\beta}}{2} = \frac{4 + \sqrt{16 + 4 \cdot 4 \cdot 3}}{2} = 6$$

and therefore the asymptotic filter is expressed as:

$$\hat{x}_{k+1} = \hat{x}_k + \frac{2}{3}\left(y_k - \hat{x}_k\right)$$

The precision of the estimation is given by the variance $\Gamma_\infty = 6$. We will have the temperature with a precision of $\pm\sqrt{6}$ deg.

3) In the situation where $\Gamma_{\alpha_k} = 0$, we obtain $\Gamma_\infty = 0$. This means that after a sufficiently long amount of time, the non-asymptotic Kalman filter returns the correct temperature, without any uncertainties.

CORRECTION FOR EXERCISE 7.15.– (Blind walker)

1) We have a linear evolution of the state since:

$$\mathbf{x}(k+1) = \begin{pmatrix} 1 & u(k) \\ 0 & 1 \end{pmatrix} \mathbf{x}(k)$$

We initialize the filter by:

$$\hat{\mathbf{x}} = \begin{pmatrix} 0 \\ 1 \end{pmatrix} \text{ and } \Gamma_x = \begin{pmatrix} 0 & 0 \\ 0 & 0.02^2 \end{pmatrix}$$

The program that simulates our system and estimates the state using a Kalman filter is the following (see also in `blindwalk.m`):

```
xhat=[0;1]; Gx=diag([0,0.02^2]);
Galpha=diag([0;0.01^2]);
x=[0;1+0.02*randn(1)];
for k=0:19,
if (k<10), u=1; else u=-1; end
Ak=[1 u;0 1];
[xhat,Gx]=kalman(xhat,Gx,0,eye(0,1),Galpha,eye(0,0),
Ak,eye(0,2));
alpha=mvnrnd([0;0],Galpha)';
x=Ak*x+alpha;
end;
```

2) The ellipses are given on top of Figure 7.28. Since we have no exteroceptive measurements, we notice that these ellipses are growing. However, as the standard deviation shows as a function of time, the projection of these ellipses according to x_1 forms intervals whose size may decrease. Indeed, the accumulated uncertainties of the scale factor on the forward journey are partly recovered on the way back. If we take 10 steps forward and 10 steps back with equal length, we return to the initial position, regardless of the length of the individual steps.

NOTE 7.4.– In a commercial underwater robot using dead reckoning for navigation (localization using a Kalman predictor), if the scale factors are not well known a decreasing uncertainty can often be observed on the return journey. This is the same phenomenon as the one described in this exercise with the walker.

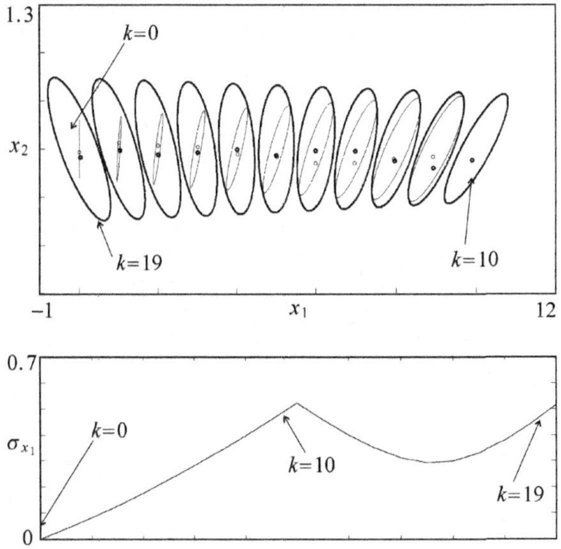

Figure 7.28. *Top: confidence ellipses for k from 0 to 19; bottom: standard deviation for x_1*

CORRECTION FOR EXERCISE 7.16.– (Simple pendulum)

1) We have:

$$\ddot{x}_1 = -\sin x_1 + u$$

therefore, by taking $u = \sin x_1 + v$, we obtain:

$$\ddot{x}_1 = v$$

We then take a proportional-derivative controller:

$$v = (w - x_1) + 2(\dot{w} - \dot{x}) + \ddot{w} = (w - x_1) + 2(\dot{w} - x_2) + \ddot{w}$$

The controller is, therefore:

$$u = \sin x_1 + (w - x_1) + 2(\dot{w} - x_2) + \ddot{w}$$

2) We take $w(t) = \sin t$. Thus, $\dot{w}(t) = \cos t$ and $\ddot{w} = -\sin t$. Consequently:

$$u = \sin x_1 + (\sin t - x_1) + 2(\cos t - x_2) - \sin t$$

3) Linearizing the pendulum yields:

$$\begin{pmatrix} \dot{x}_1 \\ \dot{x}_2 \end{pmatrix} = \begin{pmatrix} x_2 \\ -\sin x_1 + u \end{pmatrix} \simeq \begin{pmatrix} x_2 \\ -x_1 + u \end{pmatrix}$$
$$y = x_2$$

And we can, therefore, tell the Kalman filter that the system it is observing is described by the equations:

$$\mathbf{x}_{k+1} = \underbrace{\begin{pmatrix} 1 & \delta \\ -\delta & 1 \end{pmatrix}}_{\mathbf{A}_k} \mathbf{x}_k + \underbrace{\begin{pmatrix} 0 \\ \delta \end{pmatrix}}_{\mathbf{B}_k} u_k + \alpha_k$$
$$y_k = \underbrace{\begin{pmatrix} 0 & 1 \end{pmatrix}}_{\mathbf{C}_k} \mathbf{x}_k + \beta_k$$

For the covariance matrices of the state, we need to take something in the order of δ^2 that depends on the prediction precision (external perturbation such as the wind, approximation due to linearization, negligible friction, etc.). We may take, for example:

$$\Gamma_{\alpha_k} = \begin{pmatrix} \delta^2 & 0 \\ 0 & \delta^2 \end{pmatrix} \text{ and } \Gamma_{\beta_k} = 0.1^2$$

but other values are also possible. The adjustment of the covariance matrices of a Kalman filter is a delicate problem that is often performed after several experiments.

CORRECTION FOR EXERCISE 7.17.– (State estimation of the inverted rod pendulum)

1) Let us linearize this system around $\mathbf{x} = \mathbf{0}$ using the Taylor–Young method. We have:

$$\begin{cases} \frac{m \sin x_2 (g \cos x_2 - \ell x_4^2) + u}{M + m \sin^2 x_2} = \frac{m\,(x_2+\varepsilon)\,(g\,(1+\varepsilon) - \ell\varepsilon) + u}{M + m\,(x_2+\varepsilon)^2} \\ = \frac{m\,(x_2+\varepsilon)\,(g+\varepsilon) + u}{M+\varepsilon} = \frac{m g x_2 + u}{M} + \varepsilon \\ \frac{\sin x_2((M+m)g - m\ell x_4^2 \cos x_2) + \cos x_2 u}{\ell(M + m \sin^2 x_2)} \\ = \frac{(x_2+\varepsilon)\,((M+m)g - m\ell\varepsilon\,(1+\varepsilon)) + (1+\varepsilon)\,u}{\ell(M + m\,(x_2+\varepsilon)^2)} = \frac{x_2(M+m)g + u}{\ell M} + \varepsilon \end{cases}$$

Thus, we obtain the linearized system:

$$\begin{cases} \dot{\mathbf{x}} = \underbrace{\begin{pmatrix} 0 & 0 & 1 & 0 \\ 0 & 0 & 0 & 1 \\ 0 & \frac{mg}{M} & 0 & 0 \\ 0 & \frac{(M+m)g}{M\ell} & 0 & 0 \end{pmatrix}}_{\mathbf{A}} \mathbf{x} + \underbrace{\begin{pmatrix} 0 \\ 0 \\ \frac{1}{M} \\ \frac{1}{M\ell} \end{pmatrix}}_{\mathbf{B}} u \\ y = \underbrace{\begin{pmatrix} 1 & 0 & 0 & 0 \\ 0 & 1 & 0 & 0 \end{pmatrix}}_{\mathbf{C}} \mathbf{x} \end{cases}$$

2) The gain $\mathbf{K}$ is obtained by solving the system:

$$\det\left(\mathbf{A} - \mathbf{B}\mathbf{K}\right) = (s+2)^4$$

The result can be calculated by the MATLAB instruction:

```
K = place(A,B,[-2 -2.01 -2.02 -2.03]);
```

where we have taken care to avoid multiple poles in order for the `place` function to work. The precompensator is obtained by the relation:

$$h = -\left(\mathbf{E} \cdot (\mathbf{A} - \mathbf{B} \cdot \mathbf{K})^{-1} \cdot \mathbf{B}\right)^{-1}$$

where:

$$\mathbf{E} = \begin{pmatrix} 1 \, 0 \, 0 \, 0 \end{pmatrix}$$

The corresponding MATLAB code is, therefore, the following:

```
m=1;M=5;l=1;g=9.81;
A=[0 0 1 0;0 0 0 1;0 m*g/M 0 0;0 (M+m)*g/(l*M) 0 0];
B=[0;0;1/M;1/(l*M)]; C=[1 0 0 0;0 1 0 0];
E=[1 0 0 0];
K=place(A,B,[-2 -2.01 -2.02 -2.03]);
H=-inv(E*inv(A-B*K)*B);
```

The MATLAB code for the simulation is of the form:

```
x=[0;0.02;0;0];
for t=0:dt:30,
w=1; u=-K*x+H*w;
x=x+f(x,u)*dt;
end
```

3) Euler discretization yields:

$$\mathbf{x}(k+1) = (\mathbf{I}+dt\ \mathbf{A}) \cdot \mathbf{x}(k) + dt\ \mathbf{B}\ u(k) + \alpha(k)$$

where the vector $\alpha(k) \in \mathbb{R}^4$ is the state noise that takes into account the errors due to modeling and discretization. The observation equation is:

$$\mathbf{y}(k) = \underbrace{\begin{pmatrix} 1\,0\,0\,0 \\ 0\,1\,0\,0 \end{pmatrix}}_{\mathbf{C}} \mathbf{x}(k) + \begin{pmatrix} \beta_1(k) \\ \beta_2(k) \end{pmatrix}$$

The Kalman filter is expressed in the form:

```
[xr,P]=kalman(xr,P,dt*B*u,y,Q_alpha,Q_beta,eye(4,4)+A*dt,C)
```

4) The program, which can also be found in the file `invpend.m`, is the following:

```
sigm2_y=dt*0.01;
Q_alpha=dt*0.0001*eye(4,4); % state noise
Q_beta=(sigm2_y)*eye(2,2); % measurement noise
```

```
x=[0;0.02;0;0]; % initial state of the system
xr=[0;0;0;0]; P=eye(4,4);% initial state of the Kalman
filter
for t=0:dt:10,
w=1; % setpoint
y=C*x+sqrt(sigm2_y)*randn(2,1);
u=-K*xr+H*w;
[xr,P]=kalman(xr,P,dt*B*u,y,Q_alpha,Q_beta,eye(4,4)+A*dt,C);
x=x+f(x,u)*dt;
end
```

CORRECTION FOR EXERCISE 7.18.– (Dead reckoning)

1) The difficulty lies in choosing the covariance matrix and drawing the random noise. The simulation is performed using a simple Euler method:

```
x=[0;0;pi/3;4;0.3];
for t=0:dt:1,
Galpha=dt*diag([0 0 0.01 0.01 0.01]);
alpha=mvnrnd([0 0 0 0 0],Galpha)';
x=x+f(x,[0;0])*dt+alpha;
end
```

The evolution function $\mathbf{f}(\mathbf{x}, \mathbf{u})$ is the following:

```
function xdot=f(x,u)
xdot=[x(4)*cos(x(5))*cos(x(3));
x(4)*cos(x(5))*sin(x(3));
x(4)*sin(x(5))/3; u(1);u(2)];
end;
```

2) In order to use the Kalman filter to predict the state of the system, we need to describe the dependencies between the state variables using a linear state equation. If we let $z = (x, y, v)$, we obtain:

$$\dot{\mathbf{z}} = \begin{pmatrix} 0 & 0 & \cos\delta\cos\theta \\ 0 & 0 & \cos\delta\sin\theta \\ 0 & 0 & 0 \end{pmatrix} \mathbf{z} + \begin{pmatrix} 0 \\ 0 \\ u_1 \end{pmatrix} + \begin{pmatrix} 0 \\ 0 \\ \alpha_2 \end{pmatrix}$$

And, after discretization by Euler's method:

$$\mathbf{z}_{k+1} = \underbrace{\begin{pmatrix} 1 & 0 & dt\cos\delta\cos\theta \\ 0 & 1 & dt\cos\delta\sin\theta \\ 0 & 0 & 1 \end{pmatrix}}_{=\mathbf{A}_k} \mathbf{z}_k + \underbrace{\begin{pmatrix} 0 \\ 0 \\ dt \cdot u_1(k) \end{pmatrix}}_{=\mathbf{u}_k} + \underbrace{\begin{pmatrix} 0 \\ 0 \\ dt \cdot \alpha_2 \end{pmatrix}}_{=\alpha_k}$$

3) The program, which can be found in the file `deadreckoning.m`, is:

```
x=[0;0;pi/3;4;0.3]; Galphax=dt*diag([0 0 0.01 0.01 0.01]);
zhat=[x(1);x(2);x(4)]; Gz=zeros(3,3); Galphaz=dt*diag([0.01
0.01 0.01]);
for t=0:dt:10,
alphax=mvnrnd([0 0 0 0 0],Galphax)';
ux=[0;0]; x=x+f(x,ux)*dt+alphax;
uz=[0;0;dt*ux(1)];
y=[]; C=[]; Gbeta=[]; % without odometer
Ak=[1 0 dt*cos(x(5))*cos(x(3)); 0 1 dt*cos(x(5))*sin(x(3));
0 0 1];
[zhat,Gz]=kalman(zhat,Gz,uz,y,Galphaz,Gbeta,Ak,C);
end.
```

4) If we would have had odometers capable of giving an approximation of the speed with a variance of 0.01, we would have had to take following values as observation:

```
y=x(4)+mvnrnd(0,0.1); C=[0 0 1]; Gbeta=0.1;
```

CORRECTION FOR EXERCISE 7.19.– (Goniometric localization)

1) We have:

$$\begin{pmatrix} \dot{z}_1 \\ \dot{z}_2 \\ \dot{z}_3 \end{pmatrix} = \begin{pmatrix} \dot{x}_1 \\ \dot{x}_2 \\ \dot{x}_4 \end{pmatrix} = \begin{pmatrix} x_4\cos x_5\cos x_3 \\ x_4\cos x_5\sin x_3 \\ u_1 \end{pmatrix}$$

$$= \begin{pmatrix} 0 & 0 & \cos x_5\cos x_3 \\ 0 & 0 & \cos x_5\sin x_3 \\ 0 & 0 & 0 \end{pmatrix} \begin{pmatrix} z_1 \\ z_2 \\ z_3 \end{pmatrix} + \begin{pmatrix} 0 \\ 0 \\ u_1 \end{pmatrix}$$

When the robot detects the landmark $\mathbf{m}(i) = (x_m(i), y_m(i))$ with an angle δ_i, we have:

$$(x_m(i) - x_1)\sin(x_3 + \delta_i) - (y_m(i) - x_2)\cos(x_3 + \delta_i) = 0$$

i.e.:

$$\underbrace{-x_m(i)\sin(x_3 + \delta_i) + y_m(i)\cos(x_3 + \delta_i)}_{\text{known}} = \underbrace{\begin{pmatrix} -\sin(x_3 + \delta_i) & \cos(x_3 + \delta_i) \end{pmatrix}}_{\text{known}} \begin{pmatrix} x_1 \\ x_2 \end{pmatrix} + \beta_i$$

where β_i is a noise that we can assume to be white Gaussian with variance 1. This noise allows us to take into account the uncertainties on the measured angles (mainly the δ_i). If $\{i_1, i_2, \ldots\}$ are the numbers of the landmarks detected by the robot, we have the observation equation:

$$\mathbf{y}(k) = \underbrace{\begin{pmatrix} 0 & 0 & 1 \\ -\sin(x_3 + \delta_{i_1}) & \cos(x_3 + \delta_{i_1}) & 0 \\ -\sin(x_3 + \delta_{i_2}) & \cos(x_3 + \delta_{i_2}) & 0 \\ \vdots & \vdots & \vdots \end{pmatrix}}_{\mathbf{C}(k)} \cdot \mathbf{z}(k)$$

Note that the dimension of $\mathbf{y}$ depends on k. The first equation is given by the odometers that give us the speed. The other equations correspond to the goniometric measurement of the landmarks.

2) An Euler discretization yields:

$$\mathbf{z}(k+1) = \underbrace{\begin{pmatrix} 1 & 0 & dt \cdot \cos x_5 \cdot \cos x_3 \\ 0 & 1 & dt \cdot \cos x_5 \cdot \sin x_3 \\ 0 & 0 & 1 \end{pmatrix}}_{\mathbf{A}(k)} \cdot \mathbf{z}(k) + \begin{pmatrix} 0 \\ 0 \\ dt \cdot u_1 \end{pmatrix} + \alpha(k)$$

which is linear.

3) To design this simulator, we need to recall the simulator in exercise 7.19, to which we need to add an observation function. In order to feed the

Kalman filter, this function must return a vector of measurements $\mathbf{y}$, the associated covariance matrix and also the observation matrix $\mathbf{C}(k)$. This observation function is:

```
function [y,Gbeta,Ck]=g(x)
Ck=[0,0,1]; y=x(4); beta=1; %odometers
for i=1:length(landmarks),
a=landmarks(:,i);
da=a-x(1:2);
delta=angle(da)-x(3);
if (norm(da)<15),
yi=-a(1)*sin(x(3)+delta)+a(2)*cos(x(3)+delta);
Cki=[-sin(x(3)+delta),cos(x(3)+delta),0];
y=[y;yi]; Ck=[Ck;Cki]; beta=[beta;1];
end;
end;
Gbeta=diag(beta);
y=y+mvnrnd(zeros(size(y)),Gbeta)';
end
```

The landmark matrix was initialized in the main program as follows:

```
landmarks=[0 15 30 15; 25 30 15 20];
```

4) The main program including the simulator and the Kalman filter is the following:

```
dt=0.05; u=[0;0];
x=[0;-20;pi/3;20;0.1]; % initial state
zhat=[0;0;0]; Gz=10^3*eye(3,3); % the initial state is
unknown
Galpha=dt*0.001*eye(3,3); % state noise
for t=0:dt:10,
[y,Gbeta,Ck]=g(x);
Ak=eye(3,3)+dt*cos(x(5))*[0 0 cos(x(3)); 0 0 sin(x(3)); 0 0
0 ];
uk=dt*[0;0;u(1)];
[zhat,Gz]=kalman(zhat,Gz,uk,y,Galpha,Gbeta,Ak,Ck);
alphax=0*x; alphax([1;2;4])=mvnrnd(zeros(3,1),Galpha)';
x=x+f(x,u)*dt+alphax;
end;
```

5) We may consider the two robots as a single system whose state vector is:

$$\mathbf{x} = (x_{a1}, x_{a2}, x_{a3}, x_{a4}, x_{a5}, x_{b1}, x_{b2}, x_{b3}, x_{b4}, x_{b5})$$

The vector:

$$\mathbf{z} = (x_{a1}, x_{a2}, x_{a4}, x_{b1}, x_{b2}, x_{b4})$$

satisfies a linear evolution equation given by:

$$\underbrace{\begin{pmatrix} \dot{x}_{a1} \\ \dot{x}_{a2} \\ \dot{x}_{a4} \\ \dot{x}_{b1} \\ \dot{x}_{b1} \\ \dot{x}_{b1} \end{pmatrix}}_{=\dot{\mathbf{z}}} = \begin{pmatrix} 0\;0 & \cos x_{a5} \cos x_{a3} & 0\;0 & 0 \\ 0\;0 & \cos x_{a5} \sin x_{a3} & 0\;0 & 0 \\ 0\;0 & 0 & 0\;0 & 0 \\ 0\;0 & 0 & 0\;0 & \cos x_{b5} \cos x_{b3} \\ 0\;0 & 0 & 0\;0 & \cos x_{b5} \sin x_{b3} \\ 0\;0 & 0 & 0\;0 & 0 \end{pmatrix} \underbrace{\begin{pmatrix} x_{a1} \\ x_{a2} \\ x_{a4} \\ x_{b1} \\ x_{b1} \\ x_{b1} \end{pmatrix}}_{=\mathbf{z}} + \begin{pmatrix} 0 \\ 0 \\ u_{a1} \\ 0 \\ 0 \\ u_{b1} \end{pmatrix}$$

When the two robots can see each other, we have the relation:

$$(x_{b1} - x_{a1}) \sin (x_{a3} + \varphi_a) - (x_{b2} - x_{a2}) \cos (x_{a3} + \varphi_a) = 0$$

i.e.:

$$0 = \begin{pmatrix} -\sin (x_{a3} + \varphi_a) & \cos (x_{a3} + \varphi_a) \; 0 \; \sin (x_{a3} + \varphi_a) \\ -\cos (x_{a3} + \varphi_a) & 0 \end{pmatrix} \cdot \mathbf{z} + \beta (k)$$

where $\beta (k)$ corresponds to a measurement noise that we can assume to be white Gaussian. The 0 on the left of the equality corresponds to the measurement. The associated observation function can be coded in the following manner:

```
function [yab,Gab,Cab]=gab(xa,xb)
da=xb(1:2)-xa(1:2); phi=atan(da)-xa(3);
yab=[];Gab=[];Cab=[];
if (norm(da)<20),
Cab=[-sin(xa(3)+phi),cos(xa(3)+phi),0,sin(xa(3)+phi),
-cos(xa(3)+phi),0];
Gab=1; yab=mvnrnd(0,Gab);
end; end
```

To this interrobot observation function, we need to add the detection of landmarks as already seen in the previous question. We may then merge the interrobot observation function and that of the landmarks into a single function as follows:

```
function [y,Gbeta,Ck]=gall(xa,xb)
[ya,Ga,Cak]=g(xa); [yb,Gb,Cbk]=g(xb);
[yab,Gab,Cabk]=gab(xa,xb);
y=[ya;yb;yab];
Gbeta=blkdiag(Ga,Gb,Gab);
Ck=[blkdiag(Cak,Cbk);Cabk];
end
```

We can then use a Kalman filter to perform the localization. The script, which can also be found in `gonio.m`, is given by:

```
ua=[0;0]; ub=[0;0]; % input for robots A and B
xa=[-13;-22;pi/3;15;0.1]; % initial state of robot A
xb=[20;-10;pi/3;18;0.2]; % initial state of robot B
zhat=zeros(6,1); Gz=10^3*eye(6,6); % initialization of the
filter
Galphaa=dt*diag([0.1,0.1,0.5]); Galphab=Galphaa;
Galpha=blkdiag(Galphaa,Galphab); % covariance for the state
noise
for t=0:dt:10,
[y,Gbeta,Ck]=gall(xa,xb); % observation
Aak=[1 0 dt*cos(xa(5))*cos(xa(3)); 0 1
dt*cos(xa(5))*sin(xa(3)); 0 0 1 ];
Abk=[1 0 dt*cos(xb(5))*cos(xb(3)); 0 1
dt*cos(xb(5))*sin(xb(3)); 0 0 1 ];
Ak=blkdiag(Aak,Abk);
uk=dt*[0;0;ua(1);0;0;ub(1)];
[zhat,Gz]=kalman(zhat,Gz,uk,y,Galpha,Gbeta,Ak,Ck);
alphaa=0*xa; alphaa([1;2;4])=mvnrnd(zeros(3,1),Galphaa)';
alphab=0*xb; alphab([1;2;4])=mvnrnd(zeros(3,1),Galphab)';
xa=xa+f(xa,ua)*dt+alphaa; xb=xb+f(xb,ub)*dt+alphab;
end;
```

CORRECTION FOR EXERCISE 7.20.– (Following a boat with two radars)

1) The simulator can be found in the program given in the correction of question 4.

2) Given that:

$$\mathbf{g}(\mathbf{x}) = \begin{pmatrix} (p_x - a_x)^2 + (p_y - a_y)^2 \\ (p_x - b_x)^2 + (p_y - b_y)^2 \end{pmatrix}$$

we have:

$$\frac{d\mathbf{g}}{d\mathbf{x}}(\hat{\mathbf{x}}) = \begin{pmatrix} 2(\hat{p}_x - a_x)\; 0\; 2(\hat{p}_y - a_y)\; 0 \\ 2(\hat{p}_x - b_x)\; 0\; 2(\hat{p}_y - b_y)\; 0 \end{pmatrix}$$

And therefore, the observation equation can be approximated by its tangent equation:

$$\mathbf{y} \simeq \mathbf{g}(\hat{\mathbf{x}}) + \frac{d\mathbf{g}}{d\mathbf{x}}(\hat{\mathbf{x}}) \cdot (\mathbf{x} - \hat{\mathbf{x}})$$

in other words:

$$\underbrace{\mathbf{y} - \mathbf{g}(\hat{\mathbf{x}}) + \frac{d\mathbf{g}}{d\mathbf{x}}(\hat{\mathbf{x}}) \cdot \hat{\mathbf{x}}}_{\mathbf{z}} \simeq \underbrace{\frac{d\mathbf{g}}{d\mathbf{x}}(\hat{\mathbf{x}})}_{\mathbf{C}} \cdot \mathbf{x}$$

3) To implement the Kalman filter, we take:

$$\mathbf{A}_k = \begin{pmatrix} 1\; dt & 0\; 0 \\ 0\; 1 - dt & 0\; 0 \\ 0\; 0 & 1\; dt \\ 0\; 0 & 0\; 1 - dt \end{pmatrix} \text{ and } \mathbf{C}_k = \begin{pmatrix} 2(\hat{p}_x - a_x)\; 0\; 2(\hat{p}_y - a_y)\; 0 \\ 2(\hat{p}_x - b_x)\; 0\; 2(\hat{p}_y - b_y)\; 0 \end{pmatrix}$$

4) The corresponding main MATLAB program, which can also be found in the file `radar.m`, is the following:

```
dt=0.01;a=[0;0]; b=[1;0]; x=[0;0;2;0];
Ak=[1 dt 0 0; 0 (1-dt) 0 0; 0 0 1 dt; 0 0 0 (1-dt)];
Galpha=dt*diag([0;1;0;1]); Gbeta=eye(2,2);
xhat=[1;0;3;0]; Gx=10000*eye(4,4);
for t=0:dt:10,
beta=mvnrnd([0;0],Gbeta)'; y=g(x)+beta;
Ck=2*[xhat(1)-a(1),0,xhat(3)-a(2),0;xhat(1)-b(1),0,xhat(3)
-b(2),0];
zk=y-g(xhat)+Ck*xhat;
[xhat,Gx]=kalman(xhat,Gx,0,zk,Galpha,Gbeta,Ak,Ck);
alpha=mvnrnd([0;0;0;0],Galpha)';
x=Ak*x+alpha;
end
```

The function g (x) is:

```
function y=g(x)
y=[norm(x([1,3])-a)^2;norm(x([1,3])-b)^2];
end
```

CORRECTION FOR EXERCISE 7.21.– (Robot localization in a pool)

1) When the sonar is facing one of the walls, the returned distance ℓ satisfies:

$$a = \ell \cdot \cos \beta \qquad [7.15]$$

where β is the angle between the wall normal and sonar beam and a is the distance between the sonar and wall. We will assume that the robot is immobile and that only the sonar rotates (this is the same as assuming that the tangential v and angular $\dot{\theta}$ speeds of the robot are negligible compared to the rotation speed of the sonar $\dot{\alpha}$). First, let us consider the situation in which, at time t, the sonar is in the normal axis of the wall. If τ is a sufficiently small positive real number, in other words such that at time $t - \tau$ the sonar points toward the same wall, then from relation [7.15] we would have:

$$a = \ell(t - \tau) \cdot \cos(-\dot{\alpha}\tau)$$

Note that the rotation speed of the sonar $\dot{\alpha}$ is assumed to be known and constant. Let us take $\tau = k\delta, k \in \{0, 1, 2, \ldots, N-1\}$, where δ is the amount of time between two pings of the sonar and N is an integer such that at time $t - N\delta$, the sonar necessarily points toward the wall that is orthogonal to the sonar beam at time t. Let us take:

$$\widetilde{a}_k = \ell(t - k\delta) \cdot \cos(-k\delta\dot{\alpha})$$

The quantity $\widetilde{a}_k$ should correspond to the distance a between the robot and wall it points toward. However, given the presence of measurement noise, it is preferable to obtain an estimation $\hat{a}$ of the distance a using an average:

$$\hat{a} = \frac{1}{N} \sum_{k=0}^{N-1} \widetilde{a}_k$$

We can verify the fact that the sonar is perpendicularly pointing toward a wall by verifying that the variance of the $\widetilde{a}_k$ is small, in other words:

$$\frac{1}{N}\sum_{k=1}^{N-1}\left(\widetilde{a}_k - \hat{a}\right)^2 < \varepsilon$$

where ε is a fixed threshold close to 0. This is the *variance test*. However, in practice, there are a lot of incorrect data. Therefore, we need to change our method. A more robust method than the one presented earlier consists of calculating the median instead of the mean. For this, the $\widetilde{a}_k$ have to be sorted in an increasing order. Next, the middle $\bar{a}$ of the list is taken and the elements of this list that are the furthest from $\bar{a}$ are removed. Let us say half of them are removed. These elements are easy to find since they are either in the beginning or at the end of the list. The average of the remaining elements is then calculated to obtain $\hat{a}$. The variance test is performed on the remaining elements in order to verify that the sonar is in normal direction to a wall. Note that if the robot has a reliable compass, the variance test becomes useless. Indeed, the compass gives us θ and the angle α is known, which allows us to know whether the sonar points or not toward the normal axis of a wall and also which wall it is.

2) For the last two equations, it is sufficient to note that the absolute acceleration is obtained from the measured acceleration $(a_{\mathrm{T}}, a_{\mathrm{N}})$ by the accelerometers from a simple rotation of angle θ:

$$\begin{pmatrix} \ddot{x} \\ \ddot{y} \end{pmatrix} = \begin{pmatrix} \cos\theta & -\sin\theta \\ \sin\theta & \cos\theta \end{pmatrix} \begin{pmatrix} a_{\mathrm{T}} \\ a_{\mathrm{N}} \end{pmatrix}$$

3) In order to use a Kalman filter, we first need to discretize time. Note that when $\alpha + \theta$ is a multiple of $\frac{\pi}{2}$, the sonar beam is facing one of the four corners of the pool (given that the latter is assumed to be rectangular). In such a case, the length measured might allow us to calculate either x or y. In our problem, the discrete time k is increased whenever the sonar beam is facing one of the pool walls, in other words that $k = E(\frac{\alpha+\theta}{\pi/2})$, where E denotes the integer part of a real number. An Euler discretization of our state equations is the following:

$$\begin{cases} x(k+1) &= x(k) + v_x(k) \cdot T(k) \\ y(k+1) &= y(k) + v_y(k) \cdot T(k) \\ v_x(k+1) &= v_x(k) + (a_{\mathrm{T}}(k)\cos\theta(k) - a_{\mathrm{N}}(k)\sin\theta(k)) \cdot T(k) \\ v_y(k+1) &= v_y(k) + (a_{\mathrm{T}}(k)\sin\theta(k) + a_{\mathrm{N}}(k)\cos\theta(k)) \cdot T(k) \end{cases}$$

where $T(k)$ is the time elapsed between two consecutive increments of k. In matrix form, these state equations become:

$$\begin{aligned}
\mathbf{x}(k+1) &= \begin{pmatrix} 1 & 0 & T(k) & 0 \\ 0 & 1 & 0 & T(k) \\ 0 & 0 & 1 & 0 \\ 0 & 0 & 0 & 1 \end{pmatrix} \mathbf{x}(k) \\
&+ \begin{pmatrix} 0 & 0 \\ 0 & 0 \\ -T(k)\sin\theta(k) & T(k)\cos\theta(k) \\ T(k)\cos\theta(k) & T(k)\sin\theta(k) \end{pmatrix} \mathbf{u}(k) \\
r(k) &= C(k) \cdot \mathbf{x}(k)
\end{aligned} \qquad [7.16]$$

with:

$$\begin{aligned}
\mathbf{x}(k) &= (x(k), y(k), v_x(k), v_y(k)) \\
\mathbf{u}(k) &= (a_{\mathrm{N}}(k), a_{\mathrm{T}}(k))
\end{aligned}$$

Concerning the measurement equation, we need to distinguish between four cases:

– *case 0*, wall on the right ($\theta(k) + \alpha(k) = 2k\pi$). In this case, we have a measurement of x: $r(k) = x(k) \simeq R_x - d(k)$, where $d(k)$ is the distance returned by the sonar. The observation matrix will, therefore, be $C(k) = \begin{pmatrix} 1 & 0 & 0 & 0 \end{pmatrix}$;

– *case 1*, wall on the bottom ($\theta(k) + \alpha(k) = 2k\pi + \frac{\pi}{2}$). In this case, we have a measurement of y: $r(k) = y(k) \simeq R_y - d(k)$ and therefore $C(k) = \begin{pmatrix} 0 & 1 & 0 & 0 \end{pmatrix}$;

– *case 2*, wall on the left ($\theta(k) + \alpha(k) = 2k\pi + \pi$). Once more, we have a measurement of x: $r(k) = x(k) \simeq -R_x + d(k)$. The observation matrix will be $C(k) = \begin{pmatrix} 1 & 0 & 0 & 0 \end{pmatrix}$;

– *case 3*, wall in front ($\theta(k) + \alpha(k) = 2k\pi + \frac{3\pi}{2}$). We have a measurement of y: $r(k) = y(k) \simeq -R_y + d(k)$. Thus, $C(k) = \begin{pmatrix} 0 & 1 & 0 & 0 \end{pmatrix}$.

If we know the value of $\theta + \alpha$, in order to find out which case i we are in, we need to solve:

$$\exists k \in \mathbb{N}, \theta(k) + \alpha(k) = 2k\pi + \frac{i\pi}{2}$$

in other words:

$$\exists k \in \mathbb{N}, \frac{2}{\pi}\left(\theta(k) + \alpha(k)\right) = i + 4k$$

Therefore, in order to have the most favorable case i, we calculate the integer closest to the quantity $\frac{2}{\pi}\left(\theta(k) + \alpha(k)\right)$, and we look at the remainder of the Euclidean division of this integer by 4. In MATLAB, this is done using the following calculation:

```
i=mod(round((thetak+alphak)*2/pi),4)
```

Let us note that the goal of the system in [7.16] is not to reproduce the dynamic behavior of the system on the control level, but to allow the utilization of a Kalman filter with the aim of estimating the position and speed of the robot. We have made sure to have a discrete system described by linear state equations of the form:

$$\begin{cases} \mathbf{x}(k+1) = \mathbf{A}(k)\mathbf{x}(k) + \mathbf{B}(k)\mathbf{u}(k) + \alpha(k) \\ \quad r(k) \quad = C(k)\mathbf{x}(k) + \beta(k) \end{cases}$$

We have just added two signals with noises α and β which are assumed to be white and with covariance matrices $\mathbf{\Gamma}_\alpha$ and $\mathbf{\Gamma}_\beta$ (note that here, $\mathbf{\Gamma}_\beta$ is a scalar). The goal of having these two matrices is to model the uncertainties of the model and the measurement noises. A Kalman filter will then be able to help with the localization. Figure 7.29 represents the robot at time t, the associated sonar beam and a series of confidence ellipses created by the Kalman filter. The big circle represents the initial confidence ellipse.

CORRECTION FOR EXERCISE 7.22.– (SLAM)

1) The simulation program using Euler's method is the following:

```
M=load('slam_data.txt'); t=M(:,1); phi=M(:,2);
theta=M(:,3); psi=M(:,4);
vr=M(:,5:7); depth=M(:,8); alt=M(:,9);
dt=0.1; kmax=length(M);
xhat=[0;0;0];
for k=1:kmax,
    xhat=xhat+dt*eulermat(phi(k),theta(k),psi(k))*vr(k,:)';
end
```

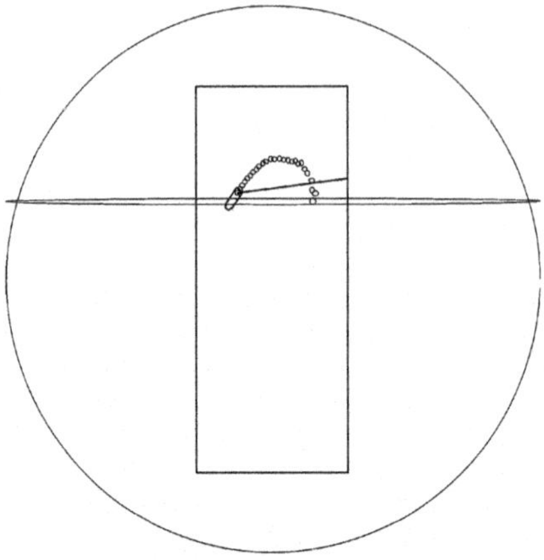

Figure 7.29. *Localization of the underwater robot using the Kalman filter*

2) The Euler angles are well known, given the inertial unit. Therefore, we have:

$$\begin{aligned}\mathbf{p}_{k+1} &= \mathbf{p}_k + dt \cdot \mathbf{R}(\varphi_k, \theta_k, \psi_k) \cdot (\bar{\mathbf{v}}_r(k) + \alpha_v(k)) \\ &= \mathbf{p}_k + \underbrace{dt \cdot \mathbf{R}(\varphi_k, \theta_k, \psi_k) \cdot \bar{\mathbf{v}}_r(k)}_{\mathbf{u}_k} + \underbrace{dt \cdot \mathbf{R}(\varphi_k, \theta_k, \psi_k) \cdot \alpha_v(k)}_{=\alpha_k}\end{aligned}$$

We can approximate α_k with a white noise of covariance matrix:

$$\begin{aligned}\mathbf{\Gamma}_\alpha &= dt^2 \cdot \mathbf{R}(\varphi_k, \theta_k, \psi_k)\mathbf{\Gamma}_{\alpha_v}\mathbf{R}^{\mathrm{T}}(\varphi_k, \theta_k, \psi_k) && \text{car } \alpha_k = dt \cdot \mathbf{R}(\varphi_k, \theta_k, \psi_k) \cdot \alpha_v(k) \\ &= dt^2\sigma_v^2\,\mathbf{I} && \text{car } \mathbf{R}\mathbf{R}^{\mathrm{T}} = \mathbf{I} \\ &= 10^{-2} \cdot \mathbf{I}\end{aligned}$$

Note that we have not taken into account the errors due to discretization nor the errors on the angles of the unit.

3) When the Kalman filter is used in predictor mode, we have:

$$\mathbf{\Gamma}_{k+1|k} = \mathbf{A}_k \cdot \mathbf{\Gamma}_{k|k} \cdot \mathbf{A}_k^{\mathrm{T}} + \mathbf{\Gamma}_{\alpha_k} = \mathbf{A}_k \cdot \mathbf{\Gamma}_{k|k-1} \cdot \mathbf{A}_k^{\mathrm{T}} + \mathbf{\Gamma}_\alpha$$

As no measurements are available in predictor mode, we have:

$$\mathbf{\Gamma}_{k|k} = \mathbf{\Gamma}_{k|k-1} = \mathbf{\Gamma}_k$$

and thus the equation becomes:

$$\mathbf{\Gamma}_{k+1} = \mathbf{A}_k \cdot \mathbf{\Gamma}_k \cdot \mathbf{A}_k^{\mathrm{T}} + \mathbf{\Gamma}_\alpha$$

Therefore:

$$\begin{aligned}
\mathbf{\Gamma}_1 &= \mathbf{A}_0 \cdot \mathbf{\Gamma}_0 \cdot \mathbf{A}_0^{\mathrm{T}} + \mathbf{\Gamma}_\alpha \\
\mathbf{\Gamma}_2 &= \mathbf{A}_1 \cdot \mathbf{\Gamma}_1 \cdot \mathbf{A}_1^{\mathrm{T}} + \mathbf{\Gamma}_\alpha = \mathbf{A}_1\mathbf{A}_0 \cdot \mathbf{\Gamma}_0 \cdot \mathbf{A}_0^{\mathrm{T}}\mathbf{A}_1^{\mathrm{T}} + \mathbf{A}_1\mathbf{\Gamma}_\alpha\mathbf{A}_1^{\mathrm{T}} + \mathbf{\Gamma}_\alpha \\
\mathbf{\Gamma}_3 &= \mathbf{A}_2 \cdot \mathbf{\Gamma}_2 \cdot \mathbf{A}_2^{\mathrm{T}} + \mathbf{\Gamma}_\alpha = \mathbf{A}_2\mathbf{A}_1\mathbf{A}_0 \cdot \mathbf{\Gamma}_0 \cdot \mathbf{A}_0^{\mathrm{T}}\mathbf{A}_1^{\mathrm{T}}\mathbf{A}_2^{\mathrm{T}} \\
&\quad + \mathbf{A}_2\mathbf{A}_1\mathbf{\Gamma}_\alpha\mathbf{A}_1^{\mathrm{T}}\mathbf{A}_2^{\mathrm{T}} + \mathbf{A}_2\mathbf{\Gamma}_\alpha\mathbf{A}_2^{\mathrm{T}} + \mathbf{\Gamma}_\alpha
\end{aligned}$$

However, in this context, the matrices $\mathbf{A}_i$ are equal to the identity matrix. Therefore, we have:

$$\mathbf{\Gamma}_k = k \cdot \mathbf{\Gamma}_\alpha = k dt^2 \sigma_v^2 \mathbf{I}$$

Note that this means that the covariance matrix increases linearly with time k. The standard deviation then increases by $\sqrt{k}$, which is a known phenomenon in random walk theory. Since $t = k \cdot dt$, the covariance of the predicted state is:

$$\mathbf{\Gamma}_{\mathbf{x}}(t) = \frac{t}{dt} \cdot \mathbf{\Gamma}_\alpha = \frac{t}{dt} dt^2 \sigma_v^2 \mathbf{I} = t \cdot dt \cdot \sigma_v^2 \cdot \mathbf{I}$$

which corresponds to a standard deviation (or drift) of:

$$\sigma_x(t) = \sigma_v \sqrt{t \cdot dt} = 0.3\sqrt{t}$$

After 1 h, the error is equal to $\sigma_x(3600) = 0.3 \cdot \sqrt{3600} = 18\ m$ and after 2 h, $\sigma_x(2 \cdot 3600) = 0.3 \cdot \sqrt{2 \cdot 3600} = 25\ m$.

The code corresponding to the predictor is given below:

```
M=dlmread('slam_data.txt');
t=M(:,1); phi=M(:,2); theta=M(:,3); psi=M(:,4);
vr=M(:,5:7);
depth=M(:,8); alt=M(:,9);
dt=0.1; kmax=size(M,1);
xhat=zeros(3,1);
Gx=diag([0,0,0]);
```

```
Galpha=[0.01*eye(3,3)];
A=eye(3,3);
for k=1:kmax,
u=[dt*eulermat(phi(k),theta(k),psi(k))*vr(k,:)'];
[xhat,Gx]=kalman(xhat,Gx,u,eye(0,1),Galpha,eye(0,0),
A,eye(0,3));
end;
```

Let us note that since we have no measurements, the quantities y, C, Γ_β are empty. However, the dimensions still need to be respected since they are at the origin of the calls to `eye(0,n)` in the parameters of the `kalman` function. The results of the predictor are shown in Figure 7.30.

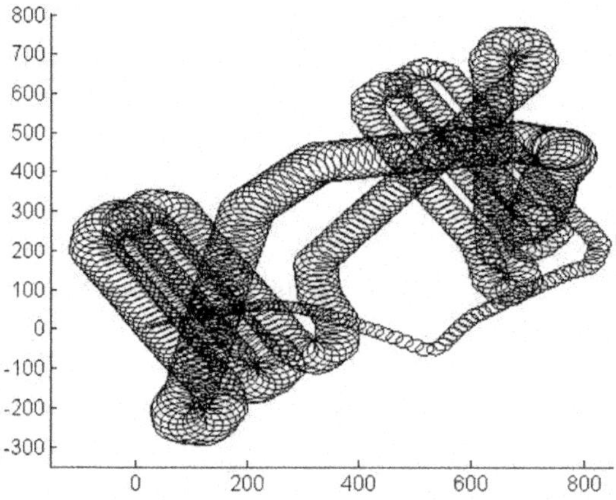

Figure 7.30. *Confidence path obtained with the predictor*

4) The observation equation is given by:

$$\underbrace{\left(y_1\right)}_{\mathbf{y}_k} = \underbrace{\left(0\,0\,1\,0\,0\,0\,0\,0\,0\,0\,0\,0 \ldots 0\,0\right)}_{\mathbf{C}(k)} \underbrace{\begin{pmatrix} \mathbf{p}_k \\ \mathbf{q}_k \end{pmatrix}}_{\mathbf{x}_k} + \beta_k$$

if no landmark is detected. It is given by:

$$\underbrace{\begin{pmatrix} y_1 \\ y_2 \\ y_3 \end{pmatrix}}_{\mathbf{y}_k} = \underbrace{\begin{pmatrix} 1\,0\,0\,0\,0\ldots\,0\,0\,{-1}\;\;0\;\;0\,0\ldots\,0\,0 \\ 0\,1\,0\,0\,0\ldots\,0\,0\;\;0\;\;{-1}\,0\,0\ldots\,0\,0 \\ 0\,0\,1\,0\,0\ldots\,0\,0\;\;0\;\;\;0\;\;0\,0\ldots\,0\,0 \end{pmatrix}}_{\mathbf{C}(k)} \underbrace{\begin{pmatrix} \mathbf{p}_k \\ \mathbf{q}_k \end{pmatrix}}_{\mathbf{x}_k} + \beta_k$$

in the case where the ith landmark $\mathbf{m}_i$ is detected. In this second case, the subvector (y_1, y_2) represents the first two components of the vector:

$$\mathbf{p} - \mathbf{m}_i = \mathbf{R}(k) \cdot \begin{pmatrix} 0 \\ -\sqrt{r_i^2(k) - a^2(k)} \\ -a(k) \end{pmatrix}$$

The y_3 component corresponds to the depth measurement given by the pressure sensor. The evolution function is the following:

```
function [y,C,Gbeta]=g(k)
y=depth(k); C=zeros(1,nx); C(1,3)=1; Gbeta=0.01;
T=[10540,10920,13740,17480,30380,36880,40240,48170,
51720,52320,
52790,56880;
1,2,1,0,1,5,4,3,3,4,5,1;
52.42,12.47,54.40,52.68,27.73,26.98,37.90,36.71,37.37,31.03,
33.51,15.05];
j=find(T(1,:)==k); % returns the column number
if (~isempty(j))
e=eulermat(phi(k),theta(k),psi(k))*[0;-sqrt(T(3,j)^2
-(alt(k))^2);-alt(k)];
y=[e(1:2);y]; C=[zeros(2,nx);C];
m=T(2,j);
C(1,1)=1; C(1,3+2*m+1)=-1; C(2,2)=1; C(2,3+2*m+2)=-1;
Gbeta=0.01*eye(3,3);
end
```

Note that the dimension of the outputs depends on the detection (or non-detection) of a landmark.

5) The evolution of the robot-landmark system can be described by:

$$\underbrace{\begin{pmatrix} \mathbf{p}_{k+1} \\ \mathbf{q}_{k+1} \end{pmatrix}}_{\mathbf{x}_{k+1}} = \underbrace{\begin{pmatrix} \mathbf{I}_3 & \mathbf{0} \\ \mathbf{0} & \mathbf{I}_{12} \end{pmatrix}}_{\mathbf{A}} \underbrace{\begin{pmatrix} \mathbf{p}_k \\ \mathbf{q}_k \end{pmatrix}}_{\mathbf{x}_k} + \underbrace{\begin{pmatrix} dt \cdot \mathbf{R}(k) \cdot \mathbf{v}_r \\ \mathbf{0}_{12\times 1} \end{pmatrix}}_{\mathbf{u}_k} + \alpha_k$$

We use lists to memorize all the intermediate results in order to be able to proceed with the smoothing (next question). We will denote by n_p the space dimension (here, equal to 3), by n_m the number of landmarks and by n_x the dimension of the state vector. The associated confidence path is drawn in Figure 7.31.

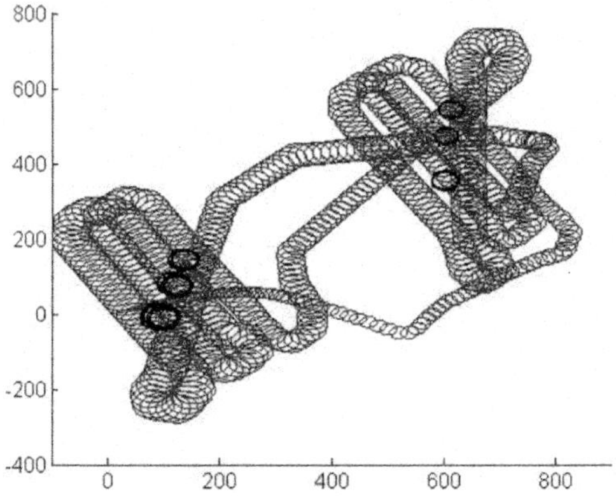

Figure 7.31. *Confidence path obtained with the Kalman filter together with the ellipses for the six landmarks*

The script corresponding to the Kalman filter is:

```
np=3; nm=6; nx=np+2*nm;
M=dlmread('slam_data.txt');
t=M(:,1); phi=M(:,2); theta=M(:,3); psi=M(:,4);
vr=M(:,5:7);
depth=M(:,8); alt=M(:,9);
dt=0.1; kmax=size(M,1);
x_forw{1}=zeros(nx,1);
G_forw{1}=diag([1,1,1,1000000*ones(1,2*nm)]);
Galpha=[0.01*eye(np,np),zeros(np,2*nm);zeros(2*nm,nx)];
```

```
for k=1:kmax,
A=eye(nx,nx);
u{k}=[dt*euler(phi(k),theta(k),psi(k))*vr(k,:)';
zeros(2*nm,1)];
[y,C,Gbeta]=slam_g(k);
[x_forw{k+1},G_forw{k+1},xup{k},Gup{k}]=kalman(x_forw{k},
G_forw{k},u{k},y,Galpha,Gbeta,A,C);
end;
```

6) We append the following instructions to the Kalman filter:

```
x_back{kmax}=xup{kmax};
G_back{kmax}=Gup{kmax};
for k=kmax-1:-1:1,
J=Gup{k}*A'/G_forw{k+1};
x_back{k}=xup{k}+J*(x_back{k+1}-x_forw{k+1});
G_back{k}=Gup{k}+J*(G_back{k+1}-G_forw{k+1})*J';
end;
```

The associated path is shown in Figure 7.32. Note that the confidence ellipses only get smaller when using the filter, especially near the end of the mission. All the programs associated with this exercise can be found in the file `slam.m`.

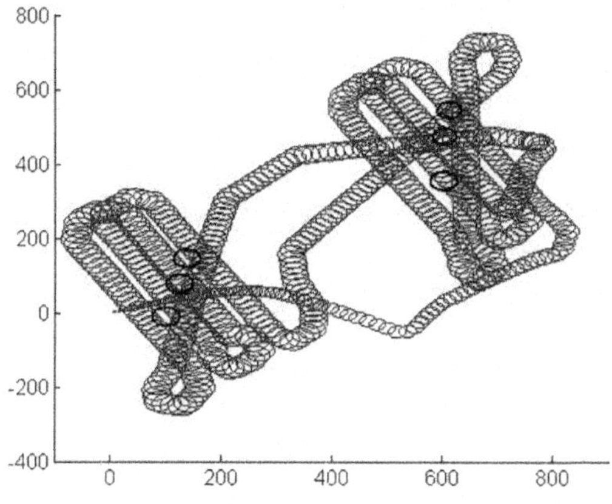

Figure 7.32. *Confidence path obtained with the Kalman smoother together with the ellipses for the six landmarks*

CORRECTION FOR EXERCISE 7.23.– (*A priori* SLAM)

The reasoning is given in the table below:

t(H)	0	1	2	3	4	5	6	7	8
landmark	0	1	2	1	3	2	1	4	0
(a)	10	110	210	310	410	510	610	710	810
(b)	10	710	610	510	410	310	210	110	10
(c)	10	110	210	310	410	310	210	110	10
(d)	10	110	210	110	310	210	110	110	10
(e)	10	110	210	110	210	210	110	110	10

Line (a): precision obtained by propagation in the direction of time; line (b): precision obtained by backpropagation in the opposite direction of time; line (c): minimum of lines (a) and (b); line (d): correspondence between the landmarks; line (e): propagation in the direct and opposite directions of time.

Bibliography

[BAZ 12] BAZEILLE S., QUIDU I., JAULIN L., "Color-based underwater object recognition using water light attenuation", *Journal of Intelligent Service Robotics*, vol. 5, no. 2, pp. 109–118, 2012.

[BEA 12] BEARD R., MCLAIN T., *Small Unmanned Aircraft, Theory and Practice*, Princeton University Press, 2012.

[BOY 06] BOYER F., POREZ M., KHALIL W., "Macro-continuous computed torque algorithm for a three-dimensional eel-like robot", *IEEE Transactions on Robotics*, vol. 22, no. 4, pp. 763–775, 2006.

[CHE 07] CHEVALLEREAU C., BESSONNET G., ABBA G. *et al.*, *Les robots marcheurs bipèdes; Modélisation, conception, synthèse de la marche, commande*, Hermes-Lavoisier, Paris, 2007.

[CRE 14] CREUZE V., *Robots marins et sous-marins; perception, modélisation, commande*, Techniques de l'ingénieur, 2014.

[DEL 93] DE LARMINAT P., *Automatique, commande des systèmes linéaires*, Hermes, Paris, France, 1993.

[DRE 11] DREVELLE V., Etude de méthodes ensemblistes robustes pour une localisation multisensorielle intègre, Application à la navigation des véhicules en milieu urbain, PhD Thesis, University of Technology of Compiègne, France, 2011.

[DUB 57] DUBINS L.E., "On curves of minimal length with a constraint on average curvature, and with prescribed initial and terminal positions and tangents", *American Journal of Mathematics*, vol. 79, no. 3, pp. 497–516, 1957.

[FAN 01] FANTONI I., LOZANO R., *Non-linear Control for Underactuated Mechanical Systems*, Springer-Verlag, 2001.

[FLI 95] FLIESS M., LIVINE J., MARTIN P. *et al.*, "Flatness and defect of non-linear systems: introductory theory and applications", *International Journal of Control*, no. 61, pp. 1327–1361, 1995.

[FLI 13] FLIESS M., JOIN C., "Model-free control", *International Journal of Control*, vol. 86, no. 12, pp. 2228–2252, 2013.

[FOS 02] FOSSEN T., Marine Control Systems: Guidance, Navigation and Control of Ships, Rigs and Underwater Vehicles, Marine Cybernetics, 2002.

[GOR 11] GORGUES T., MÉNAGE O., TERRE T. *et al.*, "An innovative approach of the surface layer sampling", *Journal des Sciences Halieutique et Aquatique*, vol. 4, pp. 105–109, 2011.

[HER 10] HERRERO P., JAULIN L., VEHI J. *et al.*, "Guaranteed set-point computation with application to the control of a sailboat", *International Journal of Control Automation and Systems*, vol. 8, no. 1, pp. 1–7, 2010.

[JAU 02] JAULIN L., KIEFFER M., WALTER E. *et al.*, "Guaranteed Robust Nonlinear Estimation with Application to Robot Localization", *IEEE Transactions on Systems, Man and Cybernetics; Part C Applications and Reviews*, vol. 32, no. 4, pp. 374–382, 2002.

[JAU 04] JAULIN L., "Modélisation et commande d'un bateau à voile", *Conférence Internationale Francophone d'Automatique (CIFA'04), CDROM*, Douz, Tunisia, 2004.

[JAU 05] JAULIN L., *Représentation d'état pour la modélisation et la commande des systèmes*, Hermes-Lavoisier, France, 2005.

[JAU 10] JAULIN L., "Commande d'un skate-car par biomimétisme", *CIFA*, Nancy, France, 2010.

[JAU 12] JAULIN L., LE BARS F., "An interval approach for stability analysis; application to sailboat robotics", *IEEE Transaction on Robotics*, vol. 27, no. 5, 2012.

[JAU 15] JAULIN L., *Automation for Robotics*, ISTE Ltd., London and John Wiley & Sons, New York, 2015.

[KAL 60] KALMAN E.R., "Contributions to the theory of optimal control", *Boletin de la Sociedad Matematica Mexicana*, vol. 5, pp. 102–119, 1960.

[KAI 80] KAILATH T, *Linear Systems*, Prentice Hall, Englewood Cliffs, 1980.

[KLE 06] KLEIN E.M.V., *Aircraft System Identification: Theory And Practice*, American Institute of Aeronautics and Astronautics, 2006.

[LAT 91] LATOMBE J., *Robot Motion Planning*, Kluwer Academic Publishers, Boston, MA, 1991.

[LAU 01] LAUMOND J., *La robotique mobile*, Hermes, France, 2001.

[LAV 06] LAVALLE S., *Planning Algorithm*, Cambridge University Press, 2006.

[MUR 89] MURATA T., "Petri nets: properties, analysis and applications", *Proceedings of the IEEE*, vol. 77, no. 4, pp. 541–580, 1989.

[PET 11] PETRES C., ROMERO RAMIREZ M., PLUMET F., " Reactive Path Planning for Autonomous Sailboat, *IEEE International Conference on Advanced Robotics*, pp. 1–6, 2011.

[WAL 14] WALTER E., *Numerical Methods and Optimization; a Consumer Guide*, Springer, London, 2014.

Index

Q, R

S

T, V

W, Y, Z